T0331708

Model Selection and Model Averaging

Given a data set, you can fit thousands of models at the push of a button, but how do you choose the best? With so many candidate models, overfitting is a real danger. Is the monkey who typed Hamlet actually a good writer?

Choosing a suitable model is central to all statistical work with data. Selecting the variables for use in a regression model is one important example. The past two decades have seen rapid advances both in our ability to fit models and in the theoretical understanding of model selection needed to harness this ability, yet this book is the first to provide a synthesis of research from this active field, and it contains much material previously difficult or impossible to find. In addition, it gives practical advice to the researcher confronted with conflicting results.

Model choice criteria are explained, discussed and compared, including Akaike's information criterion AIC, the Bayesian information criterion BIC and the focused information criterion FIC. Importantly, the uncertainties involved with model selection are addressed, with discussions of frequentist and Bayesian methods. Finally, model averaging schemes, which combine the strengths of several candidate models, are presented.

Worked examples on real data are complemented by derivations that provide deeper insight into the methodology. Exercises, both theoretical and data-based, guide the reader to familiarity with the methods. All data analyses are compatible with open-source R software, and data sets and R code are available from a companion website.

GERDA CLAESKENS is Professor in the OR & Business Statistics and Leuven Statistics Research Center at the Katholieke Universiteit Leuven, Belgium.

NILS LID HJORT is Professor of Mathematical Statistics in the Department of Mathematics at the University of Oslo, Norway.

CAMBRIDGE SERIES IN STATISTICAL AND
PROBABILISTIC MATHEMATICS

This series of high-quality upper-division textbooks and expository monographs covers all aspects of stochastic applicable mathematics. The topics range from pure and applied statistics to probability theory, operations research, optimization, and mathematical programming. The books contain clear presentations of new developments in the field and also of the state of the art in classical methods. While emphasizing rigorous treatment of theoretical methods, the books also contain applications and discussions of new techniques made possible by advances in computational practice.

Already published

Model Selection and
Model Averaging

Gerda Claeskens
K.U. Leuven

Nils Lid Hjort
University of Oslo

CAMBRIDGE
UNIVERSITY PRESS

CAMBRIDGE
UNIVERSITY PRESS

University Printing House, Cambridge CB2 8BS, United Kingdom

Cambridge University Press is part of the University of Cambridge.

It furthers the University's mission by disseminating knowledge in the pursuit of
education, learning and research at the highest international levels of excellence.

www.cambridge.org
Information on this title: www.cambridge.org/9780521852258

© G. Claeskens and N. L. Hjort 2008

First published 2008
Third printing 2010

A catalogue record for this publication is available from the British Library

ISBN 978-0-521-85225-8 Hardback

Cambridge University Press has no responsibility for the persistence or accuracy of
URLs for external or third-party internet websites referred to in this publication,
and does not guarantee that any content on such websites is, or will remain, accurate
or appropriate.

To Maarten and Hanne-Sara
– G. C.

To Jens, Audun and Stefan
– N. L. H.

Contents

Preface

Every statistician and data analyst has to make choices. The need arises especially when data have been collected and it is time to think about which model to use to describe and summarise the data. Another choice, often, is whether all measured variables are important enough to be included, for example, to make predictions. Can we make life simpler by only including a few of them, without making the prediction significantly worse?

In this book we present several methods to help make these choices easier. *Model selection* is a broad area and it reaches far beyond deciding on which variables to include in a regression model.

Two generations ago, setting up and analysing a single model was already hard work, and one rarely went to the trouble of analysing the same data via several alternative models. Thus 'model selection' was not much of an issue, apart from perhaps checking the model via goodness-of-fit tests. In the 1970s and later, proper model selection criteria were developed and actively used. With unprecedented versatility and convenience, long lists of candidate models, whether thought through in advance or not, can be fitted to a data set. But this creates problems too. With a multitude of models fitted, it is clear that methods are needed that somehow summarise model fits.

An important aspect that we should realise is that inference following model selection is, by its nature, the second step in a two-step strategy. Uncertainties involved in the first step must be taken into account when assessing distributions, confidence intervals, etc. That such themes have been largely underplayed in theoretical and practical statistics has been called 'the quiet scandal of statistics'. Realising that an analysis might have turned out differently, if preceded by data that with small modifications might have led to a different modelling route, triggers the set-up of *model averaging*. Model averaging can help to develop methods for better assessment and better construction of confidence intervals, p-values, etc. But it comprises more than that.

Each chapter ends with a brief 'Notes on the literature' section. These are not meant to contain full reviews of all existing and related literature. They rather provide some references which might then serve as a start for a fuller search. A preview of the contents of all chapters is provided in Section 1.8.

The methods used in this book are mostly based on likelihoods. To read this book it would be helpful to have at least knowledge of what a likelihood function is, and that the parameters maximising the likelihood are called maximum likelihood estimators. If properties (such as an asymptotic distribution of maximum likelihood estimators) are needed, we state the required results. We further assume that readers have had at least an applied regression course, and have some familiarity with basic matrix computations.

This book is intended for those interested in model selection and model averaging. The level of material should be accessible to master students with a background in regression modelling. Since we not only provide definitions and worked out examples, but also give some of the methodology behind model selection and model averaging, another audience for this book consists of researchers in statistically oriented fields who wish to understand better what they are doing when selecting a model. For some of the statements we provide a derivation or a proof. These can easily be skipped, but might be interesting for those wanting a deeper understanding. Some of the examples and sections are marked with a star. These contain material that might be skipped at a first reading.

This book is suitable for teaching. Exercises are provided at the end of each chapter. For many examples and methods we indicate how they can be applied using available software. For a master's level course, one could leave out most of the derivations and select the examples depending on the background of the students. Sections which can be skipped in such a course would be the large-sample analysis of Section 5.2, the average and Bayesian focussed information criteria of Sections 6.9 and 6.10, and the end of Chapter 7 (Sections 7.8, 7.9). Chapter 9 (certainly to be included) contains worked out practical examples.

All data sets used in this book, along with various computer programs (in R) for carrying out estimation and model selection via the methods we develop, are available at the following website: www.econ.kuleuven.be/gerda.claeskens/ public/modelselection.

Model selection and averaging are unusually broad areas. This is witnessed by an enormous and still expanding literature. The book is not intended as an encyclopaedia on this topic. Not all interesting methods could be covered. More could be said about models with growing numbers of parameters, finite-sample corrections, time series and other models of dependence, connections to machine learning, bagging and boosting, etc., but these topics fell by the wayside as the other chapters grew.

Acknowledgements

The authors deeply appreciate the privileges afforded to them by the following university departments by creating possibilities for meeting and working together in environments conducive to research: School of Mathematical Sciences at the Australian

National University at Canberra; Department of Mathematics at the University of Oslo; Department of Statistics at Texas A&M University; Institute of Statistics at Université Catholique de Louvain; and ORSTAT and the Leuven Statistics Research Center at the Katholieke Universiteit Leuven. N. L. H. is also grateful to the Centre of Advanced Studies at the Norwegian Academy of Science and Letters for inviting him to take part in a one-year programme on biostatistical research problems, with implications also for the present book.

More than a word of thanks is also due to the following individuals, with whom we had fruitful occasions to discuss various aspects of model selection and model averaging: Raymond Carroll, Merlise Clyde, Anthony Davison, Randy Eubank, Arnoldo Frigessi, Alan Gelfand, Axel Gandy, Ingrid Glad, Peter Hall, Jeff Hart, Alex Koning, Ian Mc-Keague, Axel Munk, Frank Samaniego, Willi Sauerbrei, Tore Schweder, Geir Storvik, and Odd Aalen.

We thank Diana Gillooly of Cambridge University Press for her advice and support.

The first author thanks her husband, Maarten Jansen, for continuing support and interest in this work, without which this book would not be here.

Gerda Claeskens and Nils Lid Hjort
Leuven and Oslo

A guide to notation

This is a list of most of the notation used in this book. The page number refers either to the first appearance or to the place where the symbol is defined.

$\phi(x, \Sigma)$	the density of a multivariate normal $N_q(0, \Sigma)$ variable	
$\chi_q^2(\lambda)$	non-central χ^2 distribution with q degrees of freedom and non-centrality parameter λ, with mean $q + \lambda$ and variance $2q + 4\lambda$	126
ω	vector of length q appearing in the asymptotic distribution of estimators under local misspecification	123
$\xrightarrow{d}, \rightarrow_d$	convergence in distribution	
$\xrightarrow{p}, \rightarrow_p$	convergence in probability	
$\sim$	'distributed according to'; so $Y_i \sim \text{Pois}(\xi_i)$ means that Y_i has a Poisson distribution with parameter ξ_i	
$\doteq_d$	$X_n \doteq_d X_n'$ indicates that their difference tends to zero in probability	

1

Model selection: data examples and introduction

This book is about making choices. If there are several possibilities for modelling data, which should we take? If multiple explanatory variables are measured, should they all be used when forming predictions, making classifications,
or attempting to summarise analysis of what influences response variables, or
will including only a few of them work equally well, or better? If so, which
ones can we best include? Model selection problems arrive in many forms and
on widely varying occasions. In this chapter we present some data examples
and discuss some of the questions they lead to. Later in the book we come back
to these data and suggest some answers. A short preview of what is to come in
later chapters is also provided.

1.1 Introduction

With the current ease of data collection which in many fields of applied science has
become cheaper and cheaper, there is a growing need for methods which point to interesting, important features of the data, and which help to build a model. The model we
wish to construct should be rich enough to explain relations in the data, but on the other
hand simple enough to understand, explain to others, and use. It is when we negotiate
this balance that model selection methods come into play. They provide formal support
to guide data users in their search for good models, or for determining which variables
to include when making predictions and classifications.

Statistical model selection is an integral part of almost any data analysis. Model
selection cannot be easily separated from the rest of the analysis, and the question 'which
model is best' is not fully well-posed until supplementing information is given about
what one plans to do or hopes to achieve given the choice of a model. The survey of data
examples that follows indicates the broad variety of applications and relevant types of
questions that arise.

Before going on to this survey we shall briefly discuss some of the key general issues
involved in model selection and model averaging.

(i) *Models are approximations:* When dealing with the issues of building or selecting a model, it needs to be realised that in most situations we will not be able to guess the 'correct' or 'true' model. This true model, which in the background generated the data we collected, might be very complex (and almost always unknown). For working with the data it might be of more practical value to work instead with a simpler, but almost-as-good model: 'All models are wrong, but some are useful', as a maxim formulated by G. E. P. Box expresses this view. Several model selection methods start from this perspective.

(ii) *The bias–variance trade-off:* The balance and interplay between variance and bias is fundamental in several branches of statistics. In the framework of model fitting and selection it takes the form of balancing simplicity (fewer parameters to estimate, leading to lower variability, but associated with modelling bias) against complexity (entering more parameters in a model, e.g. regression parameters for more covariates, means a higher degree of variability but smaller modelling bias). Statistical model selection methods must seek a proper balance between overfitting (a model with too many parameters, more than actually needed) and underfitting (a model with too few parameters, not capturing the right signal).

(iii) *Parsimony:* 'The principle of parsimony' takes many forms and has many for-mulations, in areas ranging from philosophy, physics, arts, communication, and indeed statistics. The original Ockham's razor is 'entities should not be multiplied beyond ne-cessity'. For statistical modelling a reasonable translation is that only parameters that really matter ought to be included in a selected model. One might, for example, be willing to extend a linear regression model to include an extra quadratic term if this manifestly improves prediction quality, but not otherwise.

(iv) *The context:* All modelling is rooted in an appropriate scientific context and is for a certain purpose. As Darwin once wrote, 'How odd it is that anyone should not see that all observation must be for or against some view if it is to be of any service'. One must realise that 'the context' is not always a precisely defined concept, and different researchers might discover or learn different things from the same data sets. Also, different schools of science might have different preferences for what the aims and purposes are when modelling and analysing data. Breiman (2001) discusses 'the two cultures' of statistics, broadly sorting scientific questions into respectively those of prediction and classification on one hand (where even a 'black box' model is fine as long as it works well) and those of 'deeper learning about models' on the other hand (where the discovery of a non-null parameter is important even when it might not help improve inference precision). Thus S. Karlin's statement that 'The purpose of models is not to fit the data, but to sharpen the questions' (in his R. A. Fisher memorial lecture, 1983) is important in some contexts but less relevant in others. Indeed there are differently spirited model selection methods, geared towards answering questions raised by different cultures.

(v) *The focus:* In applied statistics work it is often the case that some quantities or functions of parameters are more important than others. It is then fruitful to gear model building and model selection efforts towards criteria that favour good performance precisely for those quantities that are more important. That different aims might lead to differently selected models, for the same data and the same list of candidate models, should not be considered a paradox, as it reflects different preferences and different loss functions. In later chapters we shall in particular work with focussed information criteria that start from estimating the mean squared error (variance plus squared bias) of candidate estimators, for a given focus parameter.

(vi) *Conflicting recommendations:* As is clear from the preceding points, questions about 'which model is best' are inherently more difficult than those of the type 'for a given model, how should we carry out inference'. Sometimes different model selection strategies end up offering different advice, for the same data and the same list of candidate models. This is not a contradiction as such, but stresses the importance of learning how the most frequently used selection schemes are constructed and what their aims and properties are.

(vii) *Model averaging:* Most selection strategies work by assigning a certain score to each candidate model. In some cases there might be a clear winner, but sometimes these scores might reveal that there are several candidates that do almost as well as the winner. In such cases there may be considerable advantages in combining inference output across these best models.

1.2 Egyptian skull development

Measurements on skulls of male Egyptians have been collected from different archaeo-logical eras, with a view towards establishing biometrical differences (if any) and more generally studying evolutionary aspects. Changes over time are interpreted and discussed in a context of interbreeding and influx of immigrant populations. The data consist of four measurements for each of 30 skulls from each of five time eras, originally presented by Thomson and Randall-Maciver (1905). The five time periods are the early predy-nastic (around 4000 B.C.), late predynastic (around 3300 B.C.), 12th and 13th dynasties (around 1850 B.C.), the ptolemaic period (around 200 B.C.), and the Roman period (around 150 A.D.). For each of the 150 skulls, the following measurements are taken (all in mil-limetres): x_1 = maximal breadth of the skull (MB), x_2 = basibregmatic height (BH), x_3 = basialveolar length (BL), and x_4 = nasal height (NH); see Figure 1.1, adapted from Manly (1986, page 6). Figure 1.2 gives pairwise scatterplots of the data for the first and last time period, respectively. Similar plots are easily made for the other time periods. We notice, for example, that the level of the x_1 measurement appears to have increased while that of the x_3 measurement may have decreased somewhat over time. Statistical modelling and analysis are required to accurately validate such claims.

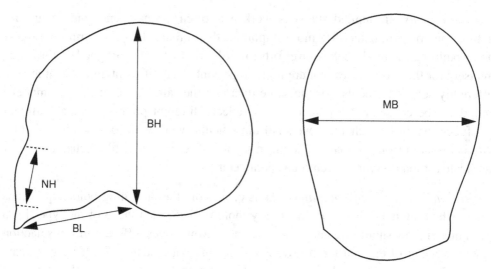

Fig. 1.1. The four skull measurements $x_1 = $ MB, $x_2 = $ BH, $x_3 = $ BL, $x_4 = $ NH; from Manly (1986, page 6).

There is a four-dimensional vector of observations $y_{t,i}$ associated with skull i and time period t, for $i = 1, \ldots, 30$ and $t = 1, \ldots, 5$, where $t = 1$ corresponds to 4000 B.C., and so on, up to $t = 5$ for 150 A.D. We use $\bar{y}_{t,\bullet}$ to denote the four-dimensional vector of averages across the 30 skulls for time period t. This yields the following summary measures:

$$\bar{y}_{1,\bullet} = (131.37, 133.60, 99.17, 50.53),$$
$$\bar{y}_{2,\bullet} = (132.37, 132.70, 99.07, 50.23),$$
$$\bar{y}_{3,\bullet} = (134.47, 133.80, 96.03, 50.57),$$
$$\bar{y}_{4,\bullet} = (135.50, 132.30, 94.53, 51.97),$$
$$\bar{y}_{5,\bullet} = (136.27, 130.33, 93.50, 51.37).$$

Standard deviations for the four measurements, computed from averaging variance estimates over the five time periods (in the order MB, BH, BL, NH), are 4.59, 4.85, 4.92, 3.19. We assume that the vectors $Y_{t,i}$ are independent and four-dimensional normally distributed, with mean vector ξ_t and variance matrix Σ_t for eras $t = 1, \ldots, 5$. However, it is not given to us how these mean vectors and variance matrices could be structured, or how they might evolve over time. Hence, although we have specified that data stem from four-dimensional normal distributions, the model for the data is not yet fully specified.

We now wish to find a statistical model that provides the clearest explanation of the main features of these data. Given the information and evolutionary context alluded to above, searching for good models would involve their ability to answer the following questions. Do the mean parameters (population averages of the four measurements)

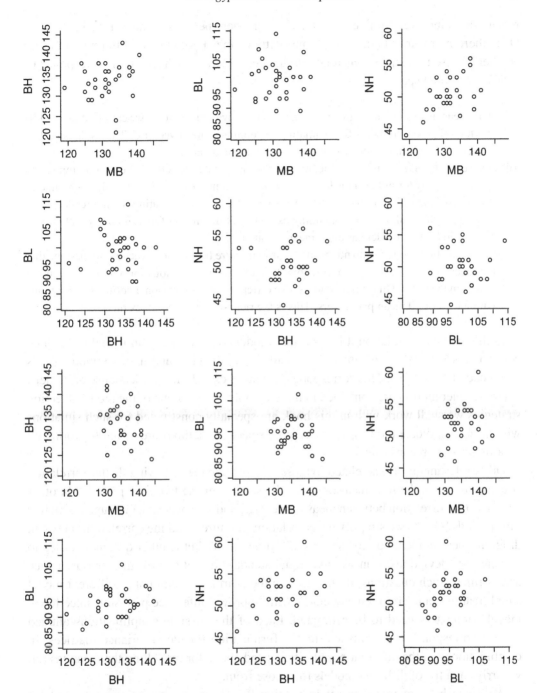

Fig. 1.2. Pairwise scatterplots for the Egyptian skull data. First two rows: early predynastic period (4000 B.C.). Last two rows: Roman period (150 A.D.).

remain the same over the five periods? If not, is there perhaps a linear trend over time? Or is there no clear structure over time, with all mean parameters different from one another? These three questions relate to the mean vector. Each situation corresponds to a different model specification:

(i) If all mean measurements remain constant over the five time periods, we can combine all 150 (5 times 30) measurements for estimating the common mean vector ξ. This is the simplest model for the mean parameters, and involves four such parameters.

(ii) If we expect a linear trend over time, we can assume that at time period t the mean components $\xi_{t,j}$ are given by formulae of the form $\xi_{t,j} = \alpha_j + \beta_j \text{time}(t)$, for $j = 1, 2, 3, 4$, where $\text{time}(t)$ is elapsed time from the first era to era t, for $t = 1, \ldots, 5$. Estimating the intercept α_j and slope β_j is then sufficient for obtaining estimates of the mean of measurement j at all five time periods. This model has eight mean parameters.

(iii) In the situation where we do not postulate any structure for the mean vectors, we assume that the mean vectors $\xi_1, \ldots, \xi_5$ are possibly different, with no obvious formula for computing one from the other. This corresponds to five different four-dimensional normal distributions, with a total of 20 mean parameters. This is the richest or most complex model.

In this particular situation it is clear that model (i) is contained in model (ii) (which corresponds to the slope parameters β_j being equal to zero), and likewise model (ii) is contained in model (iii). This corresponds to what is called a nested sequence of models, where simpler models are contained in more complex ones. Some of the model selection strategies we shall work with in this book are specially constructed for such situations with nested candidate models, whereas other selection methods are meant to work well regardless of such constraints.

Other relevant questions related to these data include the following. Is the correlation structure between the four measurements the same over the five time periods? In other words, is the correlation between measurements x_1 and x_2, and so on, the same for all five time periods? Or can we simplify the correlation structure by taking correlations between different measurements on the same skull to be equal? Yet another question relates to the standard deviations. Can we take equal standard deviations for the measurements, across time? Such questions, if answered in the affirmative, amount to different model simplifications, and are often associated with improved inference precision since fewer model parameters need to be estimated. Each of the possible simplifications alluded to here corresponds to a statistical model formulation for the covariance matrices. In combination with the different possibilities listed above for modelling the mean vector, we arrive at a list of different models to choose from.

We come back to this data set in Section 9.1. There we assign to each model a number, or a score, corresponding to a value of an information criterion. We use two such information criteria, called the AIC (Akaike's information criterion, see Chapter 2) and BIC (the Bayesian information criterion, see Chapter 3). Once each model is assigned a score, the models are ranked and the best ranked model is selected for further analysis of the data. For a multi-sample cluster analysis of the same data we refer to Bozdogan *et al.* (1994).

1.3 Who wrote 'The Quiet Don'?

The Nobel Prize in literature 1965 was awarded to Mikhail Sholokhov (1905–1984), for the epic *And Quiet Flows the Don*, or *The Quiet Don*, about Cossack life and the birth of a new Soviet society. In Russia alone his books have been published in more than a thousand editions, selling in total more than 60 million copies. But in the autumn of 1974 an article was published in Paris, The Rapids of Quiet Don: the Enigma of the Novel by the author and critic known as 'D'. He claimed that 'The Quiet Don' was not at all Sholokhov's work, but rather that it was written by Fiodor Kriukov, an author who fought against bolshevism and died in 1920. The article was given credibility and prestige by none other than Aleksandr Solzhenitsyn (a Nobel prize winner five years after Sholokhov), who in his preface to D's book strongly supported D's conclusion (Solzhenitsyn, 1974). Are we in fact faced with one of the most flagrant cases of theft in the history of literature?

An inter-Nordic research team was formed in the course of 1975, captained by Geir Kjetsaa, a professor of Russian literature at the University of Oslo, with the aim of disentangling the Don mystery. In addition to various linguistic analyses and some doses of detective work, quantitative data were also gathered, for example relating to sentence lengths, word lengths, frequencies of certain words and phrases, grammatical characteristics, etc. These data were extracted from three corpora (in the original Russian editions): (i) Sh, from published work guaranteed to be by Sholokhov; (ii) Kr, that which with equal trustworthiness came from the hand of the alternative hypothesis Kriukov; and (iii) QD, the Nobel winning text 'The Quiet Don'. Each of the corpora has about 50,000 words.

We shall here focus on the statistical distribution of the number of words used in sentences, as a possible discriminant between writing styles. Table 1.1 summarises these data, giving the number of sentences in each corpus with lengths between 1 and 5 words, between 6 and 10 words, etc. The sentence length distributions are also portrayed in Figure 1.3, along with fitted curves that are described below. The statistical challenge is to explore whether there are any sufficiently noteworthy differences between the three empirical distributions, and, if so, whether it is the upper or lower distribution of Figure 1.3 that most resembles the one in the middle.

A simple model for sentence lengths is that of the Poisson, but one sees quickly that the variance is larger than the mean (in fact, by a factor of around six). Another possibility is that of a mixed Poisson, where the parameter is not constant but varies in the space of sentences. If Y given λ is Poisson with this parameter, but λ has a Gamma (a, b) distribution, then the marginal takes the form

$$f^*(y, a, b) = \frac{b^a}{\Gamma(a)} \frac{1}{y!} \frac{\Gamma(a + y)}{(b + 1)^{a+y}} \quad \text{for } y = 0, 1, 2, \ldots,$$

Table 1.1. *The Quiet Don: number of sentences N_x in the three corpora Sh, Kr, QD of the given lengths, along with predicted numbers* pred$_x$ *under the four-parameter model (1.1), and Pearson residuals* res$_x$, *for the 13 length groups. Note: The first five columns have been compiled from tables in Kjetsaa et al. (1984).*

Words from	to	N_x Sh	Kr	QD	pred$_x$ Sh	Kr	QD	res$_x$ Sh	Kr	QD
1	5	799	714	684	803.4	717.6	690.1	−0.15	−0.13	−0.23
6	10	1408	1046	1212	1397.0	1038.9	1188.5	0.30	0.22	0.68
11	15	875	787	826	884.8	793.3	854.4	−0.33	−0.22	0.97
16	20	492	528	480	461.3	504.5	418.7	1.43	1.04	3.00
21	25	285	317	244	275.9	305.2	248.1	0.55	0.67	−0.26
26	30	144	165	121	161.5	174.8	151.1	−1.38	−0.74	−2.45
31	35	78	78	75	91.3	96.1	89.7	−1.40	−1.85	−1.55
36	40	37	44	48	50.3	51.3	52.1	−1.88	−1.02	−0.56
41	45	32	28	31	27.2	26.8	29.8	0.92	0.24	0.23
46	50	13	11	16	14.5	13.7	16.8	−0.39	−0.73	−0.19
51	55	8	8	12	7.6	6.9	9.4	0.14	0.41	0.85
56	60	8	5	3	4.0	3.5	5.2	2.03	0.83	−0.96
61	65	4	5	8	2.1	1.7	2.9	1.36	2.51	3.04
Total:		4183	3736	3760						

which is the negative binomial. Its mean is $\mu = a/b$ and its variance $a/b + a/b^2 = \mu(1 + 1/b)$, indicating the level of over-dispersion. Fitting this two-parameter model to the data was also found to be too simplistic; patterns are more variegated than those dictated by a mere negative binomial. Therefore we use the following mixture of a Poisson (a degenerate negative binomial) and another negative binomial, with a modification to leave out the possibility of having zero words in a sentence:

$$f(y, p, \xi, a, b) = p\frac{\exp(-\xi)\xi^y/y!}{1 - \exp(-\xi)} + (1 - p)\frac{f^*(y, a, b)}{1 - f^*(0, a, b)} \qquad (1.1)$$

for $y = 1, 2, 3, \ldots$ It is this four-parameter family that has been fitted to the data in Figure 1.3. The model fit is judged adequate, see Table 1.1, which in addition to the observed number N_x shows the expected or predicted number pred$_x$ of sentences of the various lengths, for length groups $x = 1, 2, 3, \ldots, 13$. Also included are Pearson residuals $(N_x - \text{pred}_x)/\text{pred}_x^{1/2}$. These residuals should essentially be on the standard normal scale if the parametric model used to produce the predicted numbers is correct; here there are no clear clashes with this hypothesis, particularly in view of the large sample sizes involved, with respectively 4183, 3736, 3760 sentences in the three corpora. The pred$_x$ numbers in the table come from minimum chi-squared fitting for each of the three

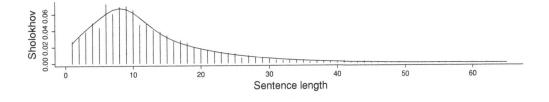

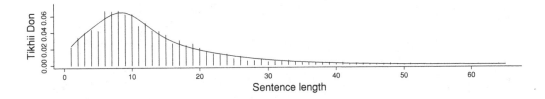

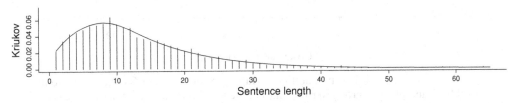

Fig. 1.3. Sentence length distributions, from 1 word to 65 words, for Sholokhov (top), Kriukov (bottom), and for 'The Quiet Don' (middle). Also shown, as continuous curves, are the distributions (1.1), fitted via maximum likelihood.

corpora, that is, finding parameter estimates to minimise

$$P_n(\theta) = \sum_x \frac{\{N_x - \text{pred}_x(\theta)\}^2}{\text{pred}_x(\theta)^2}$$

with respect to the four parameters, where $\text{pred}_x(\theta) = np_x(\theta)$ in terms of the sample size for the corpus worked with and the inferred probability $p_x(\theta)$ of writing a sentence with length landing in group x.

The statistical problem may be approached in different ways; see Hjort (2007a) for a wider discussion. Kjetsaa's group quite sensibly put up Sholokhov's authorship as the null hypothesis, and D's speculations as the alternative hypothesis, in several of their analyses. Here we shall formulate the problem in terms of selecting one of three models, inside the framework of three data sets from the four-parameter family (1.1):

M_1: Sholokhov is the rightful author, so that text corpora Sh and QD come from the same statistical distribution, while Kr represents another;

M_2: D and Solzhenitsyn were correct in denouncing Sholokhov, whose text corpus Sh is therefore not statistically compatible with Kr and QD, which are however coming from the same distribution;

M_3: Sh, Kr, QD represent three statistically disparate corpora.

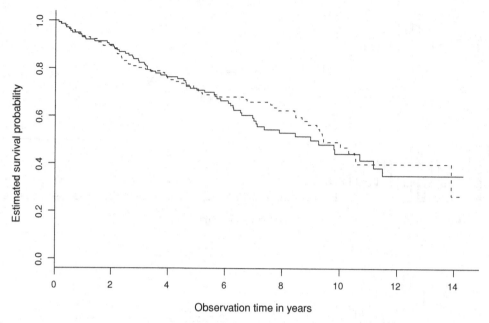

Fig. 1.4. Estimated survival probabilities (Kaplan–Meier curves) for the drug group (solid line) and placebo group (dashed line) in the study on primary biliary cirrhosis.

Selecting one of these models via statistical methodology will provide an answer to the question about who is most probably the author. (In this problem formulation we are disregarding the initial stage of model selection that is associated with using the parametric (1.1) model for the sentence distributions; the methods we shall use may be extended to encompass also this additional layer of complication, but this does not affect the conclusions we reach.) Further discussion and an analysis of this data set using a method related to the Bayesian information criterion is the topic of Section 3.3.

1.4 Survival data on primary biliary cirrhosis

PBC (primary biliary cirrhosis) is a condition which leads to progressive loss of liver function. It is commonly associated with Hepatitis C or high-volume use of alcohol, but has many other likely causes. The data set we use here for examining risk factors and treatment methods associated with PBC is the follow-up to the original PBC data set presented in appendix D of Fleming and Harrington (1991); see Murtaugh *et al.* (1994) and the data overview on page 287. This is a randomised double-blinded study where patients received either the drug D-pencillamine or placebo. Of the 280 patients for whom the information is included in this data set, 126 died before the end of the study. Figure 1.4 gives Kaplan–Meier curves, i.e. estimated survival probability curves, for the two groups. The solid line is for the drug group, the dashed line for the placebo group.

This picture already makes clear that no big difference between the two groups is to be expected.

Besides the information about age (x_1, in days, at time of registration to the study) and on whether placebo or drug is administered (x_2), other information about the patients included

- z_1, patient's gender (0 = male, 1 = female);
- z_2, presence of ascites;
- z_3, presence of hepatomegaly;
- z_4, presence of spiders;
- z_5, presence of oedema (with 0 indicating no oedema and no diuretic therapy for oedema, 1/2 for oedema present without diuretics, or oedema resolved by diuretics, and 1 for oedema despite diuretic therapy);
- z_6, serum bilirubin in mg/dl;
- z_7, serum cholesterol in mg/dl;
- z_8, albumin in gm/dl;
- z_9, alkaline phosphatase in U/l;
- z_{10}, serum glutamic-oxaloacetic transaminase (SGOT) in U/ml;
- z_{11}, platelets per cubic ml/1000;
- z_{12}, prothrombin time in seconds; and
- z_{13}, histologic stage of disease.

Here we have made a notational distinction between x_1, x_2 on the one hand and $z_1, \ldots, z_{13}$ on the other; this is because we intend to look for good survival models that always include x_1, x_2 ('protected covariates') but may or may not include any given z_j ('open covariates'). We make x_1 protected since age is known a priori to be influential for survival, while the decision to make x_2 protected too stems from the basic premise and hope that led to the large study in the first place, that one aims at seeing the effect of drug versus placebo, if any, in any selected statistical survival model.

The Cox model of proportional hazards expresses the hazard rate for individual i as

$$h_i(s) = h_0(s) \exp(x_i^t \beta + z_i^t \gamma) \quad \text{for } i = 1, \ldots, n,$$

where β has $p = 2$ component and γ is a vector of length $q = 13$. The baseline hazard function $h_0(s)$ is assumed to be continuous and positive over the range of lifetimes of interest, but is otherwise not specified. This makes the model partly parametric and partly nonparametric. When fitting the full proportional hazards regression model, we find the information on the influence of covariates given in Table 1.2. At the pointwise 5% level of significance, the significant variables are age, oedema, bilirubin, albumin, SGOT, prothrombin and stage. Using the introduced notation, these are the variables x_1, z_5, z_6, z_8, z_{10}, z_{12} and z_{13}. The variable x_2 (drug) is not significant; the corresponding p-value is equal to 0.71.

There are 15 variables measured that possibly have an effect on the lifetime of patients. The question that arises is whether a model, such as a Cox proportional hazards regression

Table 1.2. *Parameter estimates, together with their standard errors, values of the Wald Z statistic, and the corresponding p-value for the full model fit of the PBC data in a Cox regression model. (*) The shown standard errors for variables x_1, z_7 and z_9 should be multiplied by 10^{-3}.*

	Variable	coef	exp(coef)	se(coef)	z	p-Value
x_1	age	0.0001	1.000	0.029(*)	3.483	0.001
x_2	drug	0.0715	1.074	0.193	0.371	0.710
z_1	gender	−0.4746	0.622	0.270	−1.756	0.079
z_2	ascites	0.1742	1.190	0.343	0.507	0.610
z_3	hepatomegaly	0.0844	1.088	0.235	0.359	0.720
z_4	spiders	0.1669	1.182	0.217	0.769	0.440
z_5	oedema	0.7703	2.160	0.354	2.173	0.030
z_6	bilirubin	0.0849	1.089	0.023	3.648	0.000
z_7	cholesterol	0.0003	1.000	0.442(*)	0.569	0.570
z_8	albumin	−0.6089	0.544	0.288	−2.113	0.035
z_9	alkaline	0.0000	1.000	0.038(*)	0.689	0.490
z_{10}	SGOT	0.0043	1.004	0.002	2.298	0.022
z_{11}	platelets	0.0008	1.001	0.001	0.712	0.480
z_{12}	prothrombin	0.3459	1.413	0.107	3.234	0.001
z_{13}	stage	0.3587	1.431	0.162	2.211	0.027

model, needs to include all of them. Incorporating fewer variables in a model would make the clinical interpretation easier. Do we lose in statistical precision when leaving out some of the variables? Can we find a subset of the variables that explains the lifetime about equally well? Model selection methods give an answer here. An 'information criterion' assigns a value to each of the possibilities that we deem worthy of consideration. The best ranked model is then selected. This may happen to be the full model with all variables included, but does not need to be. In Chapter 9, model selection methods such as Akaike's information criterion (Chapter 2) and the Bayesian information criterion (Chapter 3) are applied to these data.

Leaving out variables will usually have the effect of introducing bias in the estimators. On the other hand, fewer variables mean fewer unknown parameters to estimate and hence a smaller variance; cf. general comments in Section 1.1. The mean squared error (mse) combines these two quantities and is defined as the sum of the squared bias and the variance. Suppose some focus parameter is studied and that different candidate models lead to different estimates of this focus parameter. We may consider the mean squared error of these candidate estimators as measures of quality of the candidate models; the lower the mse, the better. Considering the mse (or an estimator thereof) as a selection criterion, we can provide answers to questions of the following type. What are the best models for analysing respectively survival for men and survival for women, and are these necessarily the same? Is the best model for predicting the time at which at least 90%

of the patients are still alive different from the best model for estimating the cumulative hazard function? In Chapter 9, the focussed information criterion (FIC, Chapter 6) is applied to these data to provide some answers.

1.5 Low birthweight data

Establishing connections and interactions between various risk factors and the chances of giving birth to underweight children is important, as low birthweight remains a serious short- and long-term threat to the health of the child. Discovering such connections may lead to special treatments or lifestyle recommendations for different strata of mothers.

We use the low birthweight data from Hosmer and Lemeshow (1999). In this study of $n = 189$ women with newborn babies, several variables are observed which might influence low birthweight; for availability and some details regarding this data set, see the overview of data examples starting on page 287. The outcome variable Y is the indicator variable for low birthweight (low), set here as being below the threshold of 2500 grams. Recorded covariates in this study included

- the weight lwt of the mother at the last menstrual period (in pounds);
- age (in years);
- race (white, black, other);
- smoke, smoking (1/0 for yes/no);
- history of premature labour ptl (on a 0, 1, 2 scale);
- history of hypertension ht (1/0 for yes/no);
- presence of uterine irritability ui (1/0 for yes/no);
- number ftv of physician visits during the first trimester (from 0 to 6).

Aiming at building and evaluating statistical regression models for how these variables might influence the probability of low birthweight, we constructed the following covariates:

- $x_1 = 1$, an intercept constant, to be included in all models;
- x_2, the mother's weight prior to pregnancy (in kg);
- z_1, the mother's age;
- z_2, race indicator for 'black';
- z_3, race indicator for 'other' (so 'white' corresponds to $z_2 = z_3 = 0$);
- z_4, indicator for smoking;
- z_5, ptl;
- z_6, ht;
- z_7, ui;
- z_8, ftv1, indicator for ftv being one;
- z_9, ftv2p, indicator for ftv being two or more;
- $z_{10} = z_4 z_7$, smoke*ui interaction;
- $z_{11} = z_1 z_9$, age*ftv2p interaction.

The notational distinction between x_2 on the one hand and $z_1, \ldots, z_{11}$ on the other hand is as for Section 1.4, as we will study regression models where $x_1 = 1$ and x_2 are protected covariates and the z_js are open for in- or exclusion, as dictated by data and selection criteria.

In addition to z_{10} and z_{11}, also further interactions might be considered; for example, $z_{12} = z_1 z_8$ for age*ftv1 and $z_{13} = z_1 z_4$ for age*smoke. Introducing too many covariates in a regression model is, however, problematic, for several reasons; in the present case, for example, there are high correlations of respectively 0.971 and 0.963 between z_8 and z_{12} and between z_4 and z_{13}, and the more interesting first-order effects might become more difficult to spot. We learn below that smoking is positively associated with low birthweight, for example, in agreement with important epidemiological findings; if one includes each of z_4 (smoke), z_{10} (smoke*ui), z_{13} (smoke*age), however, then z_4 is no longer a significant contributor in the regression model (i.e. there is not sufficient reason to reject the null hypothesis that the associated regression coefficient is zero). Furthermore, interactions between continuous-scale and 0–1 variables have difficult interpretations, in particular in situations with high correlations, as here. For these reasons we are content here to let $x_1, x_2, z_1, \ldots, z_{11}$ be the maximal list of covariates for potential inclusion in our models.

A logistic regression model will be fitted to these data, which has the form

$$P(Y_i = 1 \mid x_i, z_i) = \frac{\exp(x_i^t \beta + z_i^t \gamma)}{1 + \exp(x_i^t \beta + z_i^t \gamma)} \quad \text{for } i = 1, \ldots, n,$$

where (x_i, z_i) is the covariate vector for mother i, with $x_i = (x_{i,1}, x_{i,2})^t$ and $z_i = (z_{i,1}, \ldots, z_{i,11})^t$ and $x_{i,1} = 1$. Also, β and γ are unknown vectors of regression coefficients. Applying the function glm(y~x, family = binomial) in the software package R gives the parameter estimates together with their standard errors, the ratio estimate/standard error, and the corresponding two-sided p-value, as exhibited in Table 1.3.

At the individual 5% level of significance, the covariates $x_2, z_4, z_6, z_7, z_9, z_{11}$ are significantly present (i.e. the corresponding regression coefficients are significantly nonzero), the others are not. This raises the question of whether we need to include all covariates for further analysis. A model that includes fewer covariates is easier to interpret and might lead to clearer recommendations for pregnant women. Including many covariates might draw attention away from the important effects.

This data set will be revisited at several places. First, we restrict attention to a subset of the covariates by only considering lwt, the weight of the mother just prior to pregnancy, age, the mother's age, and the two race indicators. Selection of covariates via AIC (see Chapter 2) is the topic of Example 2.4, while Example 3.3 applies the BIC (see Chapter 3) for this same purpose. This turns out to be an instance where there is some disagreement between the recommendations of two respectable model selection methods – which is not a contradiction as such, but which points to the importance of being aware of the

Table 1.3. *Low birthweight data: parameter estimates, together with their standard errors, the ratios estimate/standard error, and the p-values for the full model logistic regression fit.*

Parameter	Estimate	Std. error	Ratio	p-Value
x_1, intercept	−0.551	1.386	−0.397	0.691
x_2, lwt	−0.040	0.017	−2.392	0.017
z_1, age	0.035	0.043	0.826	0.409
z_2, race black	0.977	0.555	1.759	0.079
z_3, race other	0.827	0.479	1.727	0.084
z_4, smoke	1.133	0.467	2.428	0.015
z_5, ptl	0.611	0.358	1.707	0.088
z_6, ht	1.995	0.737	2.707	0.007
z_7, ui	1.650	0.652	2.529	0.011
z_8, ftv1	−0.516	0.485	−1.064	0.287
z_9, ftv2p	7.375	2.497	2.954	0.003
z_{10}, smoke*ui	−1.518	0.959	−1.583	0.114
z_{11}, age*ftv2p	−0.321	0.112	−2.882	0.004

differences in aims and behaviour for different model selectors. Different uses of the same data set may lead to different optimal models.

Another question we ask is whether there is a difference in the probability of low birthweight depending on the race of the mother, or on whether the mother smokes or not? A comparison between AIC, BIC and FIC (see Chapter 6) is made in Example 6.1, where we select a model especially useful for estimating the probability that a child has low weight at birth. Section 9.2 performs this model selection task for the full data set, including all variables mentioned at the start of this section.

1.6 Football match prediction

We have collected football (soccer) match results from five grand occasions: the 1998 World Cup held in France; the 2000 European Cup in Belgium and the Netherlands; the 2002 World Cup in Korea and Japan; the 2004 European Cup in Portugal; and finally the 2006 World Cup held in Germany. The World Cups have 64 matches among 32 national teams, while the European Cups have 31 matches among 16 teams. The results are pictured in Figure 1.5.

Along with the match results we also got hold of the official FIFA rankings for each team, one month prior to the tournaments in question. These Fédération Internationale de Football Association (FIFA) ranking scores fluctuate with time and are meant to reflect the different national teams' current form and winning chances. Table 1.4 shows the start and the end of the data matrix, with y and y' goals scored by the teams in question. The main question, of course, is how to predict the results of a football match. Different

Table 1.4. *Part of the football data, showing the number of goals scored by each team (not counting extra time or penalty shoot-outs), together with the teams' FIFA rankings and their ratio* $fifa_1/fifa_2$.

Match	y	y'	$fifa_1$	$fifa_2$	Ratio	Team 1	Team 2
1	2	1	718	480	1.49	Brazil	Scotland
2	2	2	572	597	0.96	Morocco	Norway
3	1	1	480	597	0.81	Scotland	Norway
4	3	0	718	572	1.26	Brazil	Morocco
5	3	0	572	480	1.19	Morocco	Scotland
6	2	1	597	718	0.83	Norway	Brazil
$\vdots$						$\vdots$	$\vdots$
249	0	0	741	750	0.99	England	Portugal
250	0	1	827	749	1.10	Brazil	France
251	0	0	696	728	0.96	Germany	Italy
252	0	1	750	749	1.00	Portugal	France
253	3	1	696	750	0.93	Germany	Portugal
254	1	1	728	749	0.97	Italy	France

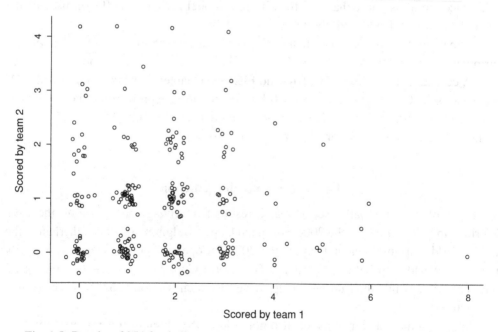

Fig. 1.5. Results of 254 football matches (jittered, to make individual match results visible), from the World Cup tournaments 1998, 2002 and 2006 and the European Championships 2000 and 2004. Thus there is one 5:2 match (Brazil–Costa Rica, 2002), one 8:0 match (Germany–Saudi Arabia, 2002), two 3:3 matches (Yugoslavia–Slovenia, 2000; Senegal–Uruguay, 2002), etc.

Table 1.5. *Top of the Adelskalenderen at the end of the 2005–2006 season, with the skaters' personal best times (in minutes:seconds).*

		500-m	1500-m	5000-m	10,000-m	Point-sum
1	C. Hedrick	35.58	1:42.78	6:09.68	12:55.11	145.563
2	S. Davis	35.17	1:42.68	6:10.49	13:05.94	145.742
3	E. Fabris	35.99	1:44.02	6:10.23	13:10.60	147.216
4	J. Uytdehaage	36.27	1:44.57	6:14.66	12:58.92	147.538
5	S. Kramer	36.93	1:46.80	6:08.78	12:51.60	147.988
6	E. Ervik	37.03	1:45.73	6:10.65	12:59.69	148.322
7	C. Verheijen	37.14	1:47.42	6:08.98	12:57.92	148.740
8	D. Parra	35.88	1:43.95	6:17.98	13:33.44	149.000
9	I. Skobrev	36.00	1:45.36	6:21.40	13:17.54	149.137
10	D. Morrison	35.34	1:42.97	6:24.13	13:45.14	149.333

assumptions lead to different models and possibly to different predictions. If we are asked to give just a single prediction for the outcome of a certain match, we have to make a choice of the model to use. This is where formal model selection methods come to help.

In Example 2.8 we construct some models for the purpose of predicting a match result and select an appropriate model. When asking this question about a specific football match, focussed prediction is one means of answering, and sometimes the best model before a given match in not identical to the model that does best on average; see Section 6.6.4.

We use this opportunity to indicate that predictions may emerge not only from a single selected statistical model, but actually also from combining different predictions, across models. With statistical model averaging techniques, developed and discussed in Chapter 7, one may use data-dictated weights over different predictions to reach a single prediction.

1.7 Speedskating

In classical long-track ice speedskating, athletes run against each other in pairs on 400-m tracks, reaching speeds of up to 60 km/h. In Table 1.5 we display the top of the *Adelskalenderen* for men, as of the end of the Olympic 2005–2006 season. This is the list of the best speedskaters ever, sorted by the so-called samalogue point-sum based on the skaters' personal bests over the four classical distances 500-m, 1500-m, 5000-m, 10,000-m. The point-sum in question is $X_1 + X_2/3 + X_3/10 + X_4/20$, where X_1, X_2, X_3, X_4 are the skated times of the skater, in seconds. The Adelskalenderen changes each time a top skater sets a new personal best. Every skater enters this list only once, with his personal best times deciding the point-sum. Five of these top ten listed skaters (so far) are Olympic gold medal winners.

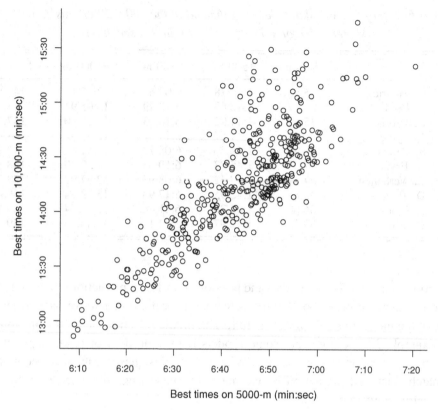

Fig. 1.6. Speedskating data. Personal best times (in minutes and seconds) for the 400 first listed skaters in the 2006 Adelskalenderen, on the 5000-m and 10,000-m distances.

One of the often-debated questions in speedskating circles, at all championships, is how well one can predict the 10,000-m time from the 5000-m time. Figure 1.6 shows, for the top 400 skaters of the Adelskalenderen, their personal best time for the 5000-m distance, versus their time for the 10,000-m distance. From the figure, there is clearly a trend visible. We wish to model the 10,000-m time Y (i.e. X_4) as a function of the 5000-m time x (i.e. X_3). The simplest possibility is ordinary linear regression of Y on x; is this perhaps too simple? The scatterplot indicates both potential quadraticity and variance heterogeneity. Linear regression leads to one prediction of a skater's 10,000-m time. Quadratic regression will give a slightly different one. Yet other values are obtained when allowing for heteroscedasticity. Here too we need to choose. The choice is between a linear or quadratic trend and with or without heteroscedasticity.

In Section 5.6 we go deeper into this issue, provide some models for these data, and then choose the better model from this list. Next to considering expected values of a skater's 10,000-m time, given his time on the 5000-m, we also pose questions such as 'can a particular skater set a new world record on the 10,000-m?'. And, which model do we best take to estimate quantiles of the 10,000-m time? The focussed information criterion (Chapter 6) is applied to provide answers, see Section 6.6.5.

Another interesting aspect of these data is the correlation structure of the 4-vector consisting of the times on all four distances per skater. This is important for relating, discussing and predicting performances on different distances. Such a comparison applies to other sporting events as well, such as decathlon and heptathlon in track and field. Several summary measures are of interest. One such is the maximal correlation between the longest distance and the other three distances. This could, for example, help to predict the result for the 10,000-m distance, when the skaters have finished the three first distances in championships. Among other quantities of interest are generalised standard deviation measures for assessing spread. When we have a particular summary measure in mind, the aim is to construct a model for the data that is best for estimating precisely this quantity. This asks for a more focussed approach than that of model selection based on overall average quality. Section 9.4 deals with this issue through application of a focussed information criterion (Chapter 6). We also go one step further by considering averages of estimators over different models, leading to model-averaged estimators (see Chapter 7).

We shall use a different set of speedskating data to illustrate the need for developing and applying model-robust procedures for estimation and model selection. In the 2004 European Championships, Eskil Ervik fell on the 1500-m, giving of course a recorded time much longer than that of his and his fellow skaters' 'normal level'. Not taking the special nature of this data point into account gives incorrect parameter estimates and a misleading ranking of candidate models. Applying robust methods that repair for the fallen skater is the topic of Example 2.14.

1.8 Preview of the following chapters

Traditional model selection methods such as Akaike's information criterion and Schwarz's Bayesian information criterion are the topics of Chapters 2 and 3. Both criteria are illustrated and discussed in a variety of examples. We provide a derivation of the criteria as well as of some related model selection methods, such as a sample size correction for AIC, and a more model-robust version, as well as an outlier-robust version. For the BIC we provide a comparison with an exact Bayesian solution, as well as the related deviance information criterion (DIC) and the minimum description length (MDL).

Chapter 4 goes deeper into some of the differences between AIC-like criteria (including Mallows's C_p and Akaike's FPE) and BIC-like criteria (including Hannan and Quinn's criterion). Classical concepts such as consistency and efficiency will be defined and explained to better understand why sometimes the criteria point to different 'best' models.

As explained in Section 1.1 models are seldom true, but some provide more useful approximations than others. We may gain understanding behind the how's and why's of model selection when considering that most (if not all) models that we use are misspecified. The challenge is then to work inside model selection frameworks where the misspecification is explicitly taken on board. In Chapter 5 we start by laying out the basic concepts for a situation with only two models; one model is simple, the other a little more complex. For example, is linear regression sufficient, or do we need an additional quadratic term? In this chapter we can find answers to, for example, how important the quadratic term should be before it is really better to include it in the model.

A different view on model selection is provided in Chapter 6. The selection criteria mentioned earlier all point to one single model, which one should then ostensibly use for all further purposes. Sometimes we can do better by narrowing our search objectives. Can we select a model specifically geared to estimating a survival probability, for example? Such questions lead to FIC, the focussed information criterion, which is constructed as an estimated mean squared error. Several worked-out examples are provided, as well as a comparison to AIC.

An alternative to model selection is to keep the estimators in different models but combine them through a weighted average to end with one estimated value. This, in a nutshell, is what model averaging is about. Chapter 7 is devoted to this topic, including both frequentist and Bayesian model averaging. One of the important side-effects of studying model-averaged estimators is that it at the same time allows one to study estimators-after-selection. Shrinkage estimators form another related topic.

Model selection is not only used in connection with inference for parameters. Chapter 8 devotes special attention to the construction of lack-of-fit and goodness-of-fit tests that use a model selection criterion as an essential ingredient.

In Chapter 9 we come back to the data examples presented above, and provide fully worked out illustrations of various methods, with information on how to compute the necessary quantities. To read that chapter it is not necessary to have worked through all details of earlier chapters.

Chapter 10 presents some further topics, including model selection in mixed models, selection criteria for use with data that contain missing observations, and for models with a growing number of parameters.

1.9 Notes on the literature

Selection among different theories or explanations of observed or imagined phenomena has concerned philosophers and scientists long before the modern era of statistics, but some of these older thoughts, paradigms and maxims still affect the way one thinks about statistical model building. The so-called Ockham's razor, mentioned in Section 1.1, is from William of Ockham's 1323 work *Summae logicae*. Einstein is often quoted as having expressed the somewhat similar rule 'Everything should be made as simple as possible, but not simpler'; the precise version is the more nuanced and guarded 'It can scarcely be denied that the supreme goal of all theory is to make the irreducible basic elements as simple and as few as possible without having to surrender the adequate representation of a single datum of experience' (Einstein, 1934).

There have been literally hundreds of journal articles published over the last 10–15 years pertaining to model selection and model averaging. Several journals have devoted special issues to these themes, such as the *Journal of Mathematical Psychology* in 2006 and *Econometric Theory* in 2008. We mention here some of the textbooks and monographs. A standard reference is Linhart and Zucchini (1986). Burnham and Anderson (2002) is a practically oriented book, with an extensive treatment of AIC and BIC. McQuarrie and Tsai (1998) concentrates on model selection in regression models and time series, with application to AIC and its 'corrected' or modified versions (like AIC_c). Also other books focussing on time series models and methods touch on issues of model selection, e.g. to select the right 'order' of memory, see for example Chatfield (2004). Miller (2002) explains subset selection in regression including forward and backward searches, branch-and-bound algorithms, the non-negative garotte, the lasso, and some Bayesian methods. The book edited by Lahiri (2001) contains four review papers on model selection, with many references.

2

Akaike's information criterion

Data can often be modelled in different ways. There might be simple approaches and more advanced ones that perhaps have more parameters. When many covariates are measured we could attempt to use them all to model their influence on a response, or only a subset of them, which would make it easier to interpret and communicate the results. For selecting a model among a list of candidates, Akaike's information criterion (AIC) is among the most popular and versatile strategies. Its essence is a penalised version of the attained maximum log-likelihood, for each model. In this chapter we shall see AIC at work in a range of applications, in addition to unravelling its basic construction and properties. Attention is also given to natural generalisations and modifications of AIC that in various situations aim at performing more accurately.

2.1 Information criteria for balancing fit with complexity

In Chapter 1 various problems were discussed where the task of selecting a suitable statistical model, from a list of candidates, was an important ingredient. By necessity there are different model selection strategies, corresponding to different aims and uses associated with the selected model. Most (but not all) selection methods are defined in terms of an appropriate *information criterion*, a mechanism that uses data to give each candidate model a certain score; this then leads to a fully ranked list of candidate models, from the ostensibly best to the worst.

The aim of the present chapter is to introduce, discuss and illustrate one of the more important of these information criteria, namely AIC (Akaike's information criterion). Its general formula is

$$\text{AIC}(M) = 2 \log\text{-likelihood}_{\max}(M) - 2 \dim(M), \tag{2.1}$$

for each candidate model M, where $\dim(M)$ is the length of its parameter vector. Thus AIC acts as a penalised log-likelihood criterion, affording a balance between good fit (high value of log-likelihood) and complexity (complex models are penalised more than simple ones). The model with the highest AIC score is then selected.

Directly comparing the values of the attained log-likelihood maxima for different models is not good enough for model comparison. Including more parameters in a model always gives rise to an increased value of the maximised log-likelihood (for an illustration, see Table 2.1). Hence, without a penalty, such as that used in (2.1), searching for the model with maximal log-likelihood would simply lead to the model with the most parameters. The penalty punishes the models for being too complex in the sense of containing many parameters. Akaike's method aims at finding models that in a sense have few parameters but nevertheless fit the data well.

The AIC strategy is admirably general in spirit, and works in principle for any situation where parametric models are compared. The method applies in particular to traditional models for i.i.d. data and regression models, in addition to time series and spatial models, and parametric hazard rate models for survival and event history analysis. Most software packages, when dealing with the most frequently used parametric regression models, have AIC values as a built-in option.

Of course there are other ways of penalising for complexity than in (2.1), and there are other ways of measuring fit of data to a model than via the maximal log-likelihood; variations will indeed be discussed later. But as we shall see in this chapter, there are precise mathematical reasons behind the AIC version. These are related to behaviour of the maximum likelihood estimators and their relation to the Kullback–Leibler distance function, as we discuss in the following section.

2.2 Maximum likelihood and the Kullback–Leibler distance

As explained above, a study of the proper comparison of candidate models, when each of these are estimated using likelihood methods, necessitates an initial discussion of maximum likelihood estimators and their behaviour, and, specifically, their relation to a certain way of measuring the statistical distance from one probability density to another, namely the Kullback–Leibler distance. This is the aim of the present section. When these matters are sorted out we proceed to a derivation of AIC as defined in (2.1).

We begin with a simple illustration of how the maximum likelihood method operates; it uses data and a given parametric model to provide an estimated model.

Example 2.1 Low birthweight data: estimation

In the data set on low birthweights (Hosmer and Lemeshow, 1999) there is a total of $n = 189$ women with newborn babies; see the introductory Section 1.5. Here we indicate how the maximum likelihood method is being used to estimate the parameters of a given model. The independent outcome variables $Y_1, \ldots, Y_n$ are binary (0–1) random variables that take the value 1 when the baby has low birthweight and 0 otherwise. Other recorded variables are $x_{2,i}$, weight; $x_{3,i}$, age of the mother; $x_{4,i}$, indicator for 'race black'; and $x_{5,i}$, indicator for 'race other'. We let $x_i = (1, x_{2,i}, x_{3,i}, x_{4,i}, x_{5,i})^t$. The most usual model for such situations is the logistic regression model, which takes the form

$$P(Y_i = 1 \mid x_i) = p_i = \frac{\exp(x_i^t \theta)}{1 + \exp(x_i^t \theta)} \quad \text{for } i = 1, \ldots, n,$$

with θ a five-dimensional parameter vector. The likelihood $\mathcal{L}_n(\theta)$ is a product of $p_i^{y_i}(1 - p_i)^{1-y_i}$ terms, leading to a log-likelihood of the form

$$\ell_n(\theta) = \sum_{i=1}^{n} \{ y_i \log p_i + (1 - y_i) \log(1 - p_i) \} = \sum_{i=1}^{n} [y_i x_i^t \theta - \log\{1 + \exp(x_i^t \theta)\}].$$

A maximum likelihood estimate for θ is found by maximising $\ell_n(\theta)$ with respect to θ. This gives $\widehat{\theta} = (1.307, -0.014, -0.026, 1.004, 0.443)^t$. ∎

In general, the models that we construct for observations $Y = (Y_1, \ldots, Y_n)$ contain a number of parameters, say $\theta = (\theta_1, \ldots, \theta_p)^t$. This translates into a joint (simultaneous) density for Y, $f_{\text{joint}}(y, \theta)$. The likelihood function is then

$$\mathcal{L}_n(\theta) = f_{\text{joint}}(y_{\text{obs}}, \theta),$$

seen as a function of θ, with $y = y_{\text{obs}}$ the observed data values. We often work with the log-likelihood function $\ell_n(\theta) = \log \mathcal{L}_n(\theta)$ instead of the likelihood itself. The maximum likelihood estimator of θ is the maximiser of $\mathcal{L}_n(\theta)$,

$$\widehat{\theta} = \widehat{\theta}_{\text{ML}} = \arg\max_{\theta} (\mathcal{L}_n) = \arg\max_{\theta} (\ell_n),$$

and is of course a function of y_{obs}. In most of the situations we shall encounter in this book, the model will be such that the maximum likelihood estimator exists and is unique, for all data sets, with probability 1. If the data Y are independent and identically distributed, the likelihood and log-likelihood functions can be written as

$$\mathcal{L}_n(\theta) = \prod_{i=1}^{n} f(y_i, \theta) \quad \text{and} \quad \ell_n(\theta) = \sum_{i=1}^{n} \log f(y_i, \theta),$$

in terms of the density $f(y, \theta)$ for an individual observation. It is important to make a distinction between the model $f(y, \theta)$ that we construct for the data, and the actual, true density $g(y)$ of the data, that is nearly always unknown. The density $g(\cdot)$ is often called the data-generating density.

There are several ways of measuring closeness of a parametric approximation $f(\cdot, \theta)$ to the true density g, but the distance intimately linked to the maximum likelihood method, as we shall see, is the Kullback–Leibler (KL) distance

$$\text{KL}(g, f(\cdot, \theta)) = \int g(y) \log \frac{g(y)}{f(y, \theta)} \, dy, \tag{2.2}$$

to be viewed as the distance from the true g to its approximation $f(\cdot, \theta)$. Applying the strong law of large numbers, one sees that for each value of the parameter vector θ,

$$n^{-1}\ell_n(\theta) \overset{\text{a.s.}}{\to} \int g(y) \log f(y, \theta) \, dy = \mathrm{E}_g \log f(Y, \theta),$$

provided only that this integral is finite; the convergence takes place 'almost surely' (a.s.), i.e. with probability 1. The maximum likelihood estimator $\widehat{\theta}$ that maximises $\ell_n(\theta)$ will therefore, under suitable and natural conditions, tend a.s. to the minimiser θ_0 of the Kullback–Leibler distance from true model to approximating model. Thus

$$\widehat{\theta} \overset{\text{a.s.}}{\to} \theta_0 = \arg\min_\theta \{\mathrm{KL}(g, f(\cdot, \theta))\}. \tag{2.3}$$

The value θ_0 is called the least false, or best approximating, parameter value. Thus the maximum likelihood estimator aims at providing the best parametric approximation to the real density g inside the parametric class $f(\cdot, \theta)$. If the parametric model is actually fully correct, then $g(y) = f(y, \theta_0)$, and the minimum Kullback–Leibler distance is zero.

Regression models involve observations (x_i, Y_i) on say n individuals or objects, where Y_i is response and x_i is a covariate vector. Maximum likelihood theory for regression models is similar to that for the i.i.d. case, but somewhat more laborious. The maximum likelihood estimators aim for least false parameter values, defined as minimisers of certain weighted Kullback–Leibler distances, as we shall see now. There is a true (but, again, typically unknown) data-generating density $g(y \mid x)$ for $Y \mid x$. The parametric model uses the density $f(y \mid x, \theta)$. Under independence, the log-likelihood function is $\ell_n(\theta) = \sum_{i=1}^n \log f(y_i \mid x_i, \theta)$. Assume furthermore that there is some underlying covariate distribution C that generates the covariate vectors $x_1, \ldots, x_n$. Then averages of the form $n^{-1} \sum_{i=1}^n a(x_i)$ tend to well-defined limits $\int a(x) \, dC(x)$, for any function a for which this integral exists, and the normalised log-likelihood function $n^{-1}\ell_n(\theta)$ tends for each θ to $\int \int g(y \mid x) \log f(y \mid x, \theta) \, dy \, dC(x)$. For given covariate vector x, consider now the Kullback–Leibler distance from the true to the approximating model, conditional on x,

$$\mathrm{KL}_x(g(\cdot \mid x), f(\cdot \mid x, \theta)) = \int g(y \mid x) \log \frac{g(y \mid x)}{f(y \mid x, \theta)} \, dy.$$

An overall (weighted) Kullback–Leibler distance is obtained by integrating KL_x over x with respect to the covariate distribution,

$$\mathrm{KL}(g, f_\theta) = \int \int g(y \mid x) \log \frac{g(y \mid x)}{f(y \mid x, \theta)} \, dy \, dC(x). \tag{2.4}$$

Under mild conditions, which must involve both the regularity of the parametric model and the behaviour of the sequence of covariate vectors, the maximum likelihood estimator $\widehat{\theta}$ based on the n first observations tends almost surely to the least false parameter value θ_0 that minimises $\mathrm{KL}(g, f_\theta)$.

Large-sample theory for the distribution of the maximum likelihood estimator is particularly well developed for the case of data assumed to follow precisely the parametric model being used; such situations certainly exist when subject-matter knowledge is well developed, but in most applied statistics contexts such an assumption would be too bold. Importantly, the large-sample likelihood theory has also been extended to the case of the density g not belonging to the assumed parametric class. We now briefly survey a couple of key results of this nature, which will be useful for later developments.

First, for the i.i.d. situation, define

$$u(y, \theta) = \frac{\partial \log f(y, \theta)}{\partial \theta} \quad \text{and} \quad I(y, \theta) = \frac{\partial^2 \log f(y, \theta)}{\partial \theta \partial \theta^t}. \tag{2.5}$$

The first expression is a p-vector function, often called the *score vector* of the model, with components $\partial \log f(y, \theta)/\partial \theta_j$ for $j = 1, \ldots, p$. The second function is a $p \times p$ matrix, sometimes called the information matrix function for the model. Its components are the mixed second-order derivatives $\partial^2 \log f(y, \theta)/\partial \theta_j \partial \theta_k$ for $j, k = 1, \ldots, p$. The score function and information matrix function are used both for numerically finding maximum likelihood estimates and for characterising their behaviour. Note that since the least false parameter minimises the Kullback–Leibler distance,

$$E_g u(Y, \theta_0) = \int g(y)u(y, \theta_0)\, dy = 0, \tag{2.6}$$

that is, the score function has zero mean at precisely the least false parameter value. We also need to define

$$J = -E_g\, I(Y, \theta_0) \quad \text{and} \quad K = \text{Var}_g\, u(Y, \theta_0). \tag{2.7}$$

These $p \times p$ matrices are identical when $g(y)$ is actually equal to $f(y, \theta_0)$ for all y. In such cases, the matrix

$$J(\theta_0) = \int f(y, \theta_0)u(y, \theta_0)u(y, \theta_0)^t\, dy = -\int f(y, \theta_0)I(y, \theta_0)\, dy \tag{2.8}$$

is called the *Fisher information matrix* of the model.

Under various and essentially rather mild regularity conditions, one may prove that

$$\widehat{\theta} = \theta_0 + J^{-1}\bar{U}_n + o_P(n^{-1/2}), \tag{2.9}$$

where $\bar{U}_n = n^{-1} \sum_{i=1}^n u(Y_i, \theta_0)$; see e.g. Hjort and Pollard (1993). This may be considered the basic asymptotic description of the maximum likelihood estimator. The size of the remainder term is concisely captured by the o_P notation; that $Z_n = o_P(n^{-1/2})$ means that $\sqrt{n}Z_n$ is $o_P(1)$ and tends to zero in probability. From the central limit theorem there is convergence in distribution $\sqrt{n}\bar{U}_n \to_d U' \sim N_p(0, K)$, which in combination with (2.9) leads to

$$\sqrt{n}(\widehat{\theta} - \theta_0) \xrightarrow{d} J^{-1}U' = N_p(0, J^{-1}KJ^{-1}). \tag{2.10}$$

Next, we deal with the regression case. To give these results, we need the $p \times 1$ score function and $p \times p$ information function of the model,

$$u(y \mid x, \theta) = \frac{\partial \log f(y \mid x, \theta)}{\partial \theta} \quad \text{and} \quad I(y \mid x, \theta) = \frac{\partial^2 \log f(y \mid x, \theta)}{\partial \theta \partial \theta^{\mathrm{t}}}.$$

Let $\theta_{0,n}$ be the least false parameter value associated with densities $g(y \mid x)$ when the covariate distribution is C_n, the empirical distribution of $x_1, \ldots, x_n$. Define the matrices

$$J_n = -n^{-1} \sum_{i=1}^{n} \int g(y \mid x_i) I(y \mid x_i, \theta_{0,n}) \, dy,$$

$$K_n = n^{-1} \sum_{i=1}^{n} \mathrm{Var}_g u(Y \mid x_i, \theta_{0,n}); \tag{2.11}$$

these are the regression model parallels of J and K of (2.7). Under natural conditions, of the Lindeberg type, there is convergence in probability of J_n and K_n to limits J and K, and $\sqrt{n} \bar{U}_n = n^{-1/2} \sum_{i=1}^{n} u(Y_i \mid x_i, \theta_{0,n})$ tends in distribution to a $U' \sim N_p(0, K)$. An important representation for the maximum likelihood estimator is $\sqrt{n}(\widehat{\theta} - \theta_{0,n}) = J_n^{-1} \sqrt{n} \bar{U}_n + o_P(1)$, which also leads to a normal limit distribution, even when the assumed model is not equal to the true model,

$$\sqrt{n}(\widehat{\theta} - \theta_{0,n}) \xrightarrow{d} J^{-1} U' \sim N_p(0, J^{-1} K J^{-1}). \tag{2.12}$$

This properly generalises (2.10). Estimators for J_n and K_n are

$$\widehat{J}_n = -n^{-1} \partial^2 \ell_n(\widehat{\theta}) / \partial \theta \partial \theta^{\mathrm{t}} = -n^{-1} \sum_{i=1}^{n} I(y_i \mid x_i, \widehat{\theta}),$$

$$\widehat{K}_n = n^{-1} \sum_{i=1}^{n} u(y_i \mid x_i, \widehat{\theta}) u(y_i \mid x_i, \widehat{\theta})^{\mathrm{t}}. \tag{2.13}$$

We note that $J_n = K_n$ when the assumed model is equal to the true model, in which case $\widehat{J}_n$ and $\widehat{K}_n$ are estimators of the same matrix, cf. (2.8). The familiar type of maximum likelihood-based inference does assume that the model is correct or nearly correct, and utilises precisely that the distribution of $\widehat{\theta}$ is approximately that of a $N_p(\theta_0, n^{-1} \widehat{J}_n^{-1})$, which follows from (2.12), leading to confidence intervals, p-values, and so on. Model-robust inference, in the sense of leading to approximately correct confidence intervals, etc., without the assumption of the parametric model being correct, uses a 'sandwich matrix' instead to approximate the variance matrix of $\widehat{\theta}$, namely $n^{-1} \widehat{J}_n^{-1} \widehat{K}_n \widehat{J}_n^{-1}$.

We now illustrate these general results for two well-known regression models.

Example 2.2 Normal linear regression

Assume $Y_i = x_i^{\mathrm{t}} \beta + \sigma \varepsilon_i$ for some p-dimensional vector β of regression coefficients, where $\varepsilon_1, \ldots, \varepsilon_n$ are i.i.d. and standard normal under traditional conditions. Then the log-likelihood function is $\sum_{i=1}^{n} \{-\frac{1}{2}(y_i - x_i^{\mathrm{t}} \beta)^2 / \sigma^2 - \log \sigma - \frac{1}{2} \log(2\pi)\}$. Assume that the ε_i are not necessarily standard normal, but that they have mean zero, standard deviation 1, skewness $\kappa_3 = E\,\varepsilon_i^3$ and kurtosis $\kappa_4 = E\,\varepsilon_i^4 - 3$. Then calculations lead to

$$J_n = \frac{1}{\sigma^2}\begin{pmatrix} \Sigma_n & 0 \\ 0 & 2 \end{pmatrix} \quad \text{and} \quad K_n = \frac{1}{\sigma^2}\begin{pmatrix} \Sigma_n & \kappa_3\bar{x}_n \\ \kappa_3\bar{x}_n^t & 2+\kappa_4 \end{pmatrix},$$

in terms of $\Sigma_n = n^{-1}\sum_{i=1}^n x_i x_i^t$. ■

Example 2.3 Poisson regression

Consider a Poisson regression model for independent count data $Y_1, \ldots, Y_n$ in terms of p-dimensional covariate vectors $x_1, \ldots, x_n$, which takes Y_i to be Poisson with parameter $\xi_i = \exp(x_i^t\beta)$. The general method outlined above leads to two matrices J_n and K_n with estimates

$$\widehat{J}_n = n^{-1}\sum_{i=1}^n \widehat{\xi}_i x_i x_i^t \quad \text{and} \quad \widehat{K}_n = n^{-1}\sum_{i=1}^n (Y_i - \widehat{\xi}_i)^2 x_i x_i^t,$$

where $\widehat{\xi}_i = \exp(x_i^t\widehat{\beta})$. When the assumed model is equal to the true model these matrices estimate the same quantity, but if there is over-dispersion, for example, then $n^{-1}\widehat{J}_n^{-1}\widehat{K}_n\widehat{J}_n^{-1}$ reflects the sampling variance of $\widehat{\beta} - \beta$ better than $n^{-1}\widehat{J}_n^{-1}$. See in this connection also Section 2.5. ■

2.3 AIC and the Kullback–Leibler distance

As we have seen, a parametric model M for data gives rise to a log-likelihood function $\ell_n(\theta) = \log\mathcal{L}_n(\theta)$. Its maximiser is the maximum likelihood estimator $\widehat{\theta}$. The value of Akaike's information criterion (Akaike, 1973) for the model is defined as in (2.1), which may also be spelled out as

$$\text{AIC}(M) = 2\ell_n(\widehat{\theta}) - 2\,\text{length}(\theta) = 2\ell_{n,\max} - 2\,\text{length}(\theta), \tag{2.14}$$

with length(θ) denoting the number of estimated parameters. To use the AIC with a collection of candidate models one computes each model's AIC value and compares these. A good model has a large value of AIC, relative to the others; cf. the general remarks made in Section 2.1. We first illustrate AIC on a data example before explaining its connection to the Kullback–Leibler distance.

Example 2.4 Low birthweight data: AIC variable selection

We continue Example 2.1. It is a priori not clear whether all variables x_i play a role in explaining low infant birthweight. Since the mother's weight is thought to be influential, we decide to include this variable x_2 in all of the possible models under investigation, as well as the intercept term ($x_1 = 1$); in other words, x_1 and x_2 are protected covariates. Let $x = (1, x_2)^t$. Subsets of $z = (x_3, x_4, x_5)^t$ are considered for potential inclusion. In

Table 2.1. *AIC values for the eight logistic regression candidate models for the low birthweight data of Example 2.4.*

Extra covariates	$\ell_n(\widehat{\theta})$	length(θ)	AIC value	Preference order	ΔAIC
none	-114.345	2	-232.691		-1.616
x_3	-113.562	3	-233.123		-2.048
x_4	-112.537	3	-231.075	(1)	0.000
x_5	-114.050	3	-234.101		-3.026
x_3, x_4	-112.087	4	-232.175	(3)	-1.100
x_3, x_5	-113.339	4	-234.677		-3.602
x_4, x_5	-111.630	4	-231.259	(2)	-0.184
x_3, x_4, x_5	-111.330	5	-232.661		-1.586

this notation the logistic regression model has the formula

$$P(\text{low birthweight} \mid x, z) = \frac{\exp(x^t \beta + z^t \gamma)}{1 + \exp(x^t \beta + z^t \gamma)},$$

with $\beta = (\beta_1, \beta_2)^t$ and $\gamma = (\gamma_1, \gamma_2, \gamma_3)^t$ the parameters to be estimated. For the estimators given in Example 2.1, and using the normal approximation for the maximum likelihood estimators $\widehat{\theta} = (\widehat{\beta}, \widehat{\gamma}) \approx_d N_p(\theta_0, n^{-1} J_n^{-1})$, we obtain the corresponding p-values $0.222, 0.028, 0.443, 0.044, 0.218$. As seen from the p-values, only γ_2 among the three γ_j is significantly different from zero at the 5% level of significance.

For this particular model it is easy to compute the maximised log-likelihood and find the required AIC values. Indeed, see Exercise 2.4,

$$\text{AIC} = 2 \sum_{i=1}^{n} \{y_i \log \widehat{p}_i + (1 - y_i) \log(1 - \widehat{p}_i)\} - 2k,$$

where $\widehat{p}_i$ is the estimated probability for $Y_i = 1$ under the model and k is the number of estimated parameters. AIC selects the model including x_4 only, see Table 2.1, with estimated low birthweight probabilities

$$\widehat{P}(\text{low birthweight} \mid x, z) = \frac{\exp(1.198 - 0.0166\, x_2 + 0.891\, x_4)}{1 + \exp(1.198 - 0.0166\, x_2 + 0.891\, x_4)}.$$

We note that AIC differences between the best ranked models are small, so we cannot claim with any degree of certainty that the AIC selected x_4 model is necessarily better than its competitors. In fact, a different recommendation will be given by the BIC method in Example 3.3.

We use this application to illustrate one more aspect of the AIC scores, namely that they are computed in a modus of comparisons across candidate models and that hence only their differences matter. For these comparisons it is often more convenient to subtract out the maximum AIC value; these are the ΔAIC scores being displayed to the right in Table 2.1. ∎

As discussed above, the AIC method has intuitive appeal in penalising the log-likelihood maxima for complexity, but it is not clear at the outset why the penalty factor should take the particular form of (2.14). We now present the theory behind the precise form of AIC, first for the i.i.d. case, then for regression models. The key is estimating the expected value of the Kullback–Leibler distance from the unknown true data-generating density $g(\cdot)$ to the parametric model.

As we saw in Section 2.2, the maximum likelihood estimator $\widehat{\theta}$ aims at the least false parameter value θ_0 that minimises the Kullback–Leibler distance (2.2). To assess how well this works, compared with other parametric models, we study the actually attained Kullback–Leibler distance

$$\mathrm{KL}(g, f(\cdot, \widehat{\theta})) = \int g(y)\{\log g(y) - \log f(y, \widehat{\theta})\}\,\mathrm{d}y = \int g \log g\,\mathrm{d}y - R_n.$$

The first term is the same across models, so we study R_n, which is a random variable, dependent upon the data via the maximum likelihood estimator $\widehat{\theta}$. Its expected value is

$$Q_n = \mathrm{E}_g R_n = \mathrm{E}_g \int g(y) \log f(y, \widehat{\theta})\,\mathrm{d}y. \tag{2.15}$$

The 'outer expectation' here is with respect to the maximum likelihood estimator, under the true density g for the Y_i. This is explicitly indicated in the notation by using the subscript g. The AIC strategy is in essence to estimate Q_n for each candidate model, and then to select the model with the highest estimated Q_n; this is equivalent to searching for the model with smallest estimated Kullback–Leibler distance.

To estimate Q_n from data, one possibility is to replace $g(y)\,\mathrm{d}y$ in R_n with the empirical distribution of the data, leading to

$$\widehat{Q}_n = n^{-1} \sum_{i=1}^{n} \log f(Y_i, \widehat{\theta}) = n^{-1}\ell_n(\widehat{\theta}),$$

i.e. the normalised log-likelihood maximum value. This estimator will tend to overshoot its target Q_n, as is made clear by the following key result. To state the result we need $V_n = \sqrt{n}(\widehat{\theta} - \theta_0)$, studied in (2.10), and involving the least false parameter θ_0; let also $\bar{Z}_n$ be the average of the i.i.d. zero mean variables $Z_i = \log f(Y_i, \theta_0) - Q_0$, writing $Q_0 = \int g(y) \log f(y, \theta_0)\,\mathrm{d}y$. The result is that

$$\widehat{Q}_n - R_n = \bar{Z}_n + n^{-1}V_n^{\mathrm{t}}JV_n + o_P(n^{-1}). \tag{2.16}$$

In view of (2.10) we have $V_n^{\mathrm{t}}JV_n \to_d W = (U')^{\mathrm{t}}J^{-1}U'$, where $U' \sim \mathrm{N}_q(0, K)$. Result (2.16) therefore leads to the approximation

$$\mathrm{E}(\widehat{Q}_n - Q_n) \approx p^*/n, \quad \text{where } p^* = \mathrm{E}\,W = \mathrm{Tr}(J^{-1}K). \tag{2.17}$$

In its turn this leads to $\widehat{Q}_n - p^*/n = n^{-1}\{\ell_n(\widehat{\theta}) - p^*\}$ as the bias-corrected version of the naive estimator $\widehat{Q}_n$.

We make some remarks before turning to the proof of (2.16).

(i) If the approximating model is correct, so that $g(y) = f(y, \theta_0)$, then $J = K$, and $p^* = p = \text{length}(\theta)$, the dimension of the model. Also, in that case, the overshooting quantity $n^{-1} V_n^t J V_n$ is close to a $n^{-1} \chi_p^2$. Taking $p^* = p$, even without any check on the adequacy of the model, leads to the AIC formula (2.14).

(ii) We may call p^* of (2.17) the generalised dimension of the model. Clearly, other approximations to or estimates of p^* than the simple $p^* = p$ are possible, and this will lead to close relatives of the AIC method; see in particular Section 2.5.

(iii) There are instances where the mean of $V_n^t J V_n$ does not tend to the mean p^* of the limit W. For the simple binomial model, for example, the mean Q_n of R_n does not even exist (since R_n is then infinite with a certain very small but positive probability); the same difficulty arises in the logistic regression model. In such cases (2.17) is formally not correct. Result (2.16) is nevertheless true and indicates that p^*/n is a sensible bias correction, with the more cautious reading 'R_n has a distribution close to that of a variable with mean equal to that of $n^{-1}\{\ell_n(\widehat{\theta}) - p^*\}$'.

Proof of (2.16). We first use a two-term Taylor expansion for R_n, using the score and information functions of the model as in (2.5), and find

$$R_n \doteq \int g(y) \left\{ \log f(y, \theta_0) + u(y, \theta_0)^t (\widehat{\theta} - \theta_0) + \tfrac{1}{2} (\widehat{\theta} - \theta_0)^t I(y, \theta_0)(\widehat{\theta} - \theta_0) \right\} dy$$

$$= Q_0 - \tfrac{1}{2} n^{-1} V_n^t J V_n.$$

Similarly, a two-term expansion for $\widehat{Q}_n$ leads to

$$\widehat{Q}_n \doteq n^{-1} \sum_{i=1}^n \left\{ \log f(Y_i, \theta_0) + u(Y_i, \theta_0)^t (\widehat{\theta} - \theta_0) + \tfrac{1}{2} (\widehat{\theta} - \theta_0)^t I(Y_i, \theta_0)(\widehat{\theta} - \theta_0) \right\}$$

$$= Q_0 + \bar{Z}_n + \bar{U}_n^t (\widehat{\theta} - \theta_0) - \tfrac{1}{2} (\widehat{\theta} - \theta_0)^t J_n (\widehat{\theta} - \theta_0),$$

where $J_n = -n^{-1} \sum_{i=1}^n I(Y_i, \theta_0) \to_p J$. This shows that $\widehat{Q}_n - R_n$ can be expressed as $\bar{Z}_n + n^{-1} \sqrt{n} \bar{U}_n^t V_n + o_P(n^{-1})$, and in conjunction with (2.10) this yields (2.16). $\square$

We next turn our attention to regression models of the general type discussed in Section 2.2. As we saw there, the distance measure involved when analysing maximum likelihood estimation in such models is the appropriately weighted Kullback–Leibler distance (2.4), involving also the distribution of x vectors in their space of covariates. For a given parametric model, with observed regression data $(x_1, y_1), \ldots, (x_n, y_n)$, the regression analogy to (2.15) is

$$Q_n = E_g R_n = E_g n^{-1} \sum_{i=1}^n \int g(y \mid x_i) \log f(y \mid x_i, \widehat{\theta}) \, dy,$$

involving the empirical distribution of the covariate vectors $x_1, \ldots, x_n$. A straightforward initial estimator of Q_n is $\widehat{Q}_n = n^{-1} \sum_{i=1}^n \log f(Y_i \mid x_i, \widehat{\theta})$, i.e. the normalised log-likelihood maximum $n^{-1}\ell_{n,\max}$. Let $\theta_{0,n}$ be the least false parameter value associated with the empirical distribution of $x_1, \ldots, x_n$, i.e. the maximiser of $n^{-1} \sum_{i=1}^{n} \int g(y \mid x_i) \log f(y \mid x_i, \theta) \, dy$. A two-term Taylor expansion leads to $R_n \doteq Q_{0,n} - \frac{1}{2} n^{-1} V_n^t J_n V_n$, where $V_n = \sqrt{n}(\widehat{\theta} - \theta_{0,n})$ and J_n is as in (2.11); also, $Q_{0,n} = n^{-1} \sum_{i=1}^{n} \int g(y \mid x_i) \log f(y \mid x_i, \theta_{0,n}) \, dy$. Similarly, a second expansion yields

$$\widehat{Q}_n = Q_{0,n} + \bar{Z}_n + \bar{U}_n^t(\widehat{\theta} - \theta_{0,n}) - \frac{1}{2}(\widehat{\theta} - \theta_{0,n})^t \widetilde{J}_n(\widehat{\theta} - \theta_{0,n})$$
$$= Q_{0,n} + \bar{Z}_n + \frac{1}{2} n^{-1} V_n^t J_n V_n + o_P(n^{-1}),$$

with $\bar{Z}_n$ being the average of the zero mean variables

$$Z_i = \log f(Y_i \mid x_i, \theta_{0,n}) - \int g(y \mid x_i) \log f(y \mid x_i, \theta_{0,n}) \, dy.$$

A clear analogy of the i.i.d. results emerges, with the help of (2.12), and with consequences parallelling those outlined above for the i.i.d. case. In particular, the AIC formula $2(\ell_{n,\max} - p)$ is valid, for the same reasons, under the same type of conditions as for i.i.d. data.

2.4 Examples and illustrations

Example 2.5 Exponential versus Weibull

For analysis of computer processes it may be important to know whether the running processes have the memory-less property or not. If they do, their failure behaviour can be described by the simple exponential model with density at failure time $= y$ equal to $\theta \exp(-\theta y)$, assuming i.i.d. data. If, on the other hand, the failure rate decreases with time (or for wear-out failures increases with time), a Weibull model may be more appropriate. Its cumulative distribution function is

$$F(y, \theta, \gamma) = 1 - \exp\{-(\theta y)^\gamma\} \quad \text{for } y > 0.$$

The density is the derivative of the cumulative distribution function, $f(y, \theta, \gamma) = \exp\{-(\theta y)^\gamma\} \theta^\gamma \gamma y^{\gamma-1}$. Note that $\gamma = 1$ corresponds to the simpler, exponential model. To select the best model, we compute

$$\text{AIC(exp)} = 2 \sum_{i=1}^{n} (\log \widetilde{\theta} - \widetilde{\theta} y_i) - 2,$$

$$\text{AIC(wei)} = 2 \sum_{i=1}^{n} \{-(\widehat{\theta} y_i)^{\widehat{\gamma}} + \widehat{\gamma} \log \widehat{\theta} + \log \widehat{\gamma} + (\widehat{\gamma} - 1) \log y_i\} - 4.$$

Here $\widetilde{\theta}$ is the maximum likelihood estimator for θ in the exponential model, while $(\widehat{\theta}, \widehat{\gamma})$ are the maximum likelihood estimators in the Weibull model. The model with the biggest value of AIC is chosen as the most appropriate one for the data at hand. See Exercise 3.1. ∎

Example 2.6 Mortality in ancient Egypt

How long is a life? A unique set of lifelengths in Roman Egypt was collected by W. Spiegelberg in 1901 and analysed by Karl Pearson (1902) in the very first volume of *Biometrika*. The data set contains the age at death for 141 Egyptian mummies in the Roman period, 82 men and 59 women, dating from around year 100 B.C. The life-lengths vary from 1 to 96, and Pearson argued that these can be considered a random sample from one of the better-living classes in that society, at a time when a fairly stable and civil government was in existence. Pearson (1902) did not try any parametric models for these data, but discussed differences between the Egyptian age distribution and that of England 2000 years later. We shall use AIC to select the most successful of a small collection of candidate parametric models for mortality rate.

For each suggested model $f(t, \theta)$ we maximise the log-likelihood $\ell_n(\theta) = \sum_{i=1}^{n} \log f(t_i, \theta)$, writing $t_1, \ldots, t_n$ for the life-lengths, and then compute AIC $= 2\ell_n(\widehat{\theta}) - 2p$, with $p = \text{length}(\theta)$. We note that finding the required maximum likelihood estimates has become drastically simpler than it used to be a decade or more ago, thanks to easily available optimisation algorithms in software packages. As demonstrated in Exercise 2.3, it does not take many lines of R code, or minutes of work, per model, to (i) program the log-likelihood function, using the `function` mechanism; (ii) find its maximiser, via the nonlinear minimisation algorithm `nlm`; and (iii) use this to find the appropriate AIC value. We have done this for five models:

- Model 1 is the exponential, with density $b \exp(-bt)$, for which we find $\widehat{b} = 0.033$.
- Model 2 is the Gamma density $\{b^a / \Gamma(a)\} t^{a-1} \exp(-bt)$, with parameter estimates $(\widehat{a}, \widehat{b}) = (1.609, 0.052)$.
- Model 3 is the log-normal, which takes a $N(\mu, \sigma^2)$ for the log-life-lengths, corresponding to a density $\phi\{(\log t - \mu)/\sigma\}/(\sigma t)$; here we find parameter estimates $(\widehat{\mu}, \widehat{\sigma}) = (3.082, 0.967)$.
- Model 4 is the Gompertz, which takes the hazard or mortality rate $h(t) = f(t)/F[t, \infty)$ to be of the form $a \exp(bt)$; this corresponds to the density $f(t) = \exp\{-H(t)\}h(t)$, with $H(t) = \int_0^t h(s)\, ds = (a/b)\{\exp(bt) - 1\}$ being the cumulative hazard rate. Parameter estimates are $(\widehat{a}, \widehat{b}) = (0.019, 0.021)$.
- Finally model 5 is the Makeham extension of the Gompertz, with hazard rate $h(t) = k + a \exp(bt)$, for k such that $k + a \exp(bt_0) > 0$, where t_0 is the minimum age under consideration (for this occasion, $t_0 = 1$ year). Estimates are $(-0.012, 0.029, 0.016)$ for (k, a, b).

We see from the AIC values for models 1–5, listed in Table 2.2, that the two parameter Gompertz model (model 4) is deemed the most successful. Figure 2.1 displays the mortality rate $\widehat{a} \exp(\widehat{b}t)$ for the Egyptian data, along with a simple nonparametric estimate. The nonparametric estimate in Figure 2.1 is of the type 'parametric start times nonparametric correction', with a bandwidth increasing with decreasing risk set; see Hjort (1992b). It indicates in this case that the Gompertz models are perhaps acceptable approximations, but that there are other fluctuations at work not quite captured by the parametric models, as for example the extra mortality at age around 25. The AIC analysis shows otherwise

Table 2.2. *Mortality in ancient Egypt: parameter estimates, the maximised log-likelihoods and the AIC scores, for the nine models. The Gompertz models are better than the others.*

Parameters	Parameter estimates				$\ell_n(\widehat{\theta})$	AIC	
model 1, b:	0.033				−623.777	−1249.553	
model 2, a, b:	1.609	0.052			−615.386	−1234.772	
model 3, μ, σ:	3.082	0.967			−629.937	−1263.874	
model 4, a, b:	0.019	0.021			−611.353	−1226.706	
model 5, k, a, b:	−0.012	0.029	0.016		−611.319	−1228.637	
model 6, a, b:	0.019	0.021			−611.353	−1226.706	
model 7, a, b_1, b_2:	0.019	0.018	0.026		−610.076	−1226.151	(3)
model 8, a_1, b, a_2:	0.016	0.024	0.022		−608.520	−1223.040	(1)
model 9, a_1, b_1, a_2, b_2:	0.016	0.024	0.022	0.020	−608.520	−1225.040	(2)

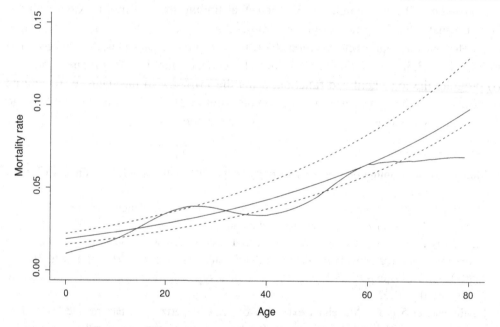

Fig. 2.1. How long is a life? For the 141 lifetimes from ancient Egypt we show the Gompertz-fitted hazard rates, for the full population (solid line), for women (dotted line, above) and for men (dotted line, below). The wiggly curve is a nonparametric hazard rate estimate.

that it does not really pay to include the extra Makeham parameter k, for example; the max log-likelihood increases merely from −611.353 to −611.319, which is not enough, as judged by the AIC value. Inclusion of one more parameter in a model is only worthwhile if the max log-likelihood is increased by at least 1.

Using the Gompertz model we attempt to separate men's and women's mortality in ancient Egypt, in spite of Pearson (1902) writing 'in dealing with [these data] I have not ventured to separate the men and women mortality, the numbers are far too insignificant'. We try out four models, corresponding to (model 6) using the same parameters (a, b) for both men and women (this is the same as model 4, of course); (model 7) using (a, b_1) and (a, b_2) for men and women (that is with the same a parameter); (model 8) using (a_1, b) and (a_2, b) for men and women (that is with the same b parameter); and (model 9) using (a_1, b_1) and (a_2, b_2) without common parameters for the two groups.

From the results listed in Table 2.2 we see that model 8 is the best one, estimating men's mortality rate as $0.016 \exp(0.024\, t)$ and women's as $0.022 \exp(0.024\, t)$; see Figure 2.1. ∎

Example 2.7 Linear regression: AIC selection of covariates

The traditional linear regression model for response data y_i in relation to covariate vectors $x_i = (x_{i,1}, \ldots, x_{i,p})^t$ for individuals $i = 1, \ldots, n$ is to take

$$Y_i = x_{i,1}\beta_1 + \cdots + x_{i,p}\beta_p + \varepsilon_i = x_i^t\beta + \varepsilon_i \quad \text{for } i = 1, \ldots, n,$$

with $\varepsilon_1, \ldots, \varepsilon_n$ independently drawn from $N(0, \sigma^2)$ and $\beta = (\beta_1, \ldots, \beta_p)^t$ a vector of regression coefficients. Typically one of the $x_{i,j}$, say the first, is equal to the constant 1, so that β_1 is the intercept parameter. The model is more compactly written in matrix form as $Y = X\beta + \varepsilon$, where $Y = (Y_1, \ldots, Y_n)^t$, $\varepsilon = (\varepsilon_1, \ldots, \varepsilon_n)^t$, and X is the $n \times p$ matrix having x_i^t as its ith row.

The log-likelihood function is

$$\ell_n(\beta, \sigma) = \sum_{i=1}^{n} \left\{ -\log \sigma - \tfrac{1}{2}(y_i - x_i^t\beta)^2/\sigma^2 - \tfrac{1}{2}\log(2\pi) \right\}.$$

Maximisation with respect to β is equivalent to

$$\text{minimisation of SSE}(\beta) = \sum_{i=1}^{n}(y_i - x_i^t\beta)^2 = \|Y - X\beta\|^2,$$

which is also the definition of the least squares estimator. The solution can be written

$$\widehat{\beta} = (X^tX)^{-1}X^tY = \Sigma_n^{-1} n^{-1} \sum_{i=1}^{n} x_i Y_i,$$

where $\Sigma_n = n^{-1}X^tX = n^{-1}\sum_{i=1}^{n} x_i x_i^t$, assuming that X has full rank p, making X^tX an invertible $p \times p$ matrix. The maximum likelihood estimator of σ is the maximiser of $\ell_n(\widehat{\beta}, \sigma)$, and is the square root of

$$\widehat{\sigma}^2 = n^{-1}\text{SSE}(\widehat{\beta}) = n^{-1}\sum_{i=1}^{n} \text{res}_i^2 = n^{-1}\|\text{res}\|^2,$$

involving the residuals $\text{res}_i = Y_i - x_i^t \widehat{\beta}$. Plugging in $\widehat{\sigma}$ in $\ell_n(\widehat{\beta}, \sigma)$ gives $\ell_{n,\max} = -n \log \widehat{\sigma} - \frac{1}{2}n - \frac{1}{2} \log(2\pi)$ and

$$\text{AIC} = -2n \log \widehat{\sigma} - 2(p + 1) - n - n \log(2\pi). \tag{2.18}$$

Thus the best subset of covariates to use, according to the AIC method, is determined by minimising $n \log \widehat{\sigma} + p$, across all candidate models. In Section 2.6 we obtain a different estimator of the expected Kullback–Leibler distance between the estimated normal regression model and the unknown true model, which leads to a stricter complexity penalty. ∎

Example 2.8 Predicting football match results

To what extent can one predict results of football matches via statistical modelling? We use the data on scores of 254 football matches; see Section 1.6 for more details regarding these data.

Denote by y and y' the number of goals scored by the teams in question. A natural type of model for outcomes (y, y') is to take these independent and Poisson distributed, with parameters (λ, λ'), with different possible specialisations for how λ and λ' should depend on the FIFA ranking scores of the two teams, say fifa and fifa$'$. A simple possibility is

$$\lambda = \lambda_0(\text{fifa}/\text{fifa}')^\beta \quad \text{and} \quad \lambda' = \lambda_0(\text{fifa}'/\text{fifa})^\beta,$$

where λ_0 and β are unknown parameters that have to be estimated. The Norwegian Computing Centre (see vm.nr.no), which produces predictions before and during these championships, uses models similar in spirit to the model above. This is a log-linear Poisson regression model in $x = \log(\text{fifa}/\text{fifa}')$, with $\lambda = \exp(\alpha + \beta x)$.

We shall in fact discuss four different candidate models here. The most general is model M_3, which takes

$$\lambda(x) = \begin{cases} \exp\{a + c(x - x_0)\} & \text{for } x \leq x_0, \\ \exp\{a + b(x - x_0)\} & \text{for } x \geq x_0, \end{cases} \tag{2.19}$$

where x_0 is a threshold value on the x scale of logarithmic ratios of FIFA ranking scores. In our illustration we are using the fixed value $x_0 = -0.21$. This value was found via separate profile likelihood analysis of other data, and affects matches where the ratio of the weaker FIFA score to the stronger FIFA score is less than $\exp(x_0) = 0.811$. Model M_3 has three free parameters. Model M_2 is the hockey-stick model where $c = 0$, and gives a constant rate for $x \leq x_0$. Model M_1 takes $b = c$, corresponding to the traditional log-linear Poisson rate model with $\exp\{a + b(x - x_0)\}$ across all x values. Finally M_0 is the simplest one, with $b = c = 0$, leaving us with a constant $\lambda = \exp(a)$ for all matches. The point of the truncated M_2 is that the log-linear model M_1 may lead to too small goal scoring rates for weaker teams meeting stronger teams.

Models M_0 and M_1 are easily handled using Poisson regression routines, like the glm($y \sim x$, family $=$ poisson) algorithm in R, as they correspond directly to

Table 2.3. *AIC and BIC scores for the four football match models of*
Example 2.8. The two criteria agree that model M_2 is best.

Model	$\widehat{a}$	$\widehat{b}$	$\widehat{c}$	AIC	BIC
M_0	0.211	0.000	0.000	−1487.442	−1491.672
M_1	−0.174	1.690	1.690	−1453.062	−1461.523
M_2	−0.208	1.811	0.000	−1451.223	−1459.684
M_3	−0.235	1.893	−1.846	−1452.488	−1465.180

constant and log-linear modelling in x. To show how also model M_3 can be dealt with, write the indicator function $I(x) = I\{x \leq x_0\}$. Then

$$\log \lambda(x) = a + cI(x)(x - x_0) + b\{1 - I(x)\}(x - x_0)$$
$$= a + b(x - x_0) + (c - b)I(x)(x - x_0),$$

which means that this is a log-linear Poisson model in the two covariates $x - x_0$ and $I(x)(x - x_0)$. Model M_2 can be handled similarly.

Table 2.3 gives the result of the AIC analysis, indicating in particular that model M_2 is judged the best one. (We have also included the BIC scores, see Chapter 3; these agree with AIC that model M_2 is best.) The reason why the hockey-stick model M_2 is better than, for example, the more traditional model M_1 is that even when teams with weak FIFA score tend to lose against teams with stronger FIFA score, they still manage, sufficiently often, to score say one goal. This 'underdog effect' is also seen in Figure 2.2, which along with the fitted intensity for models M_1 and M_2 displays a nonparametric estimate of the $\lambda(x)$. Such an estimator is constructed locally and without any global parametric model specification. For this reason it is often used to make a comparison with parametric estimators, such as the ones obtained from models M_1 and M_2. A good parametric estimator will roughly follow the same trend as the nonparametric estimator. The latter one is defined as follows. The estimator $\widehat{\lambda}(x) = \exp(\widehat{a}_x)$, where $(\widehat{a}_x, \widehat{b}_x)$ are the parameter values maximising the kernel-smoothed log-likelihood function

$$\sum_{i=1}^{n} K_h(x_i - x)\{y_i(a + bx_i) - \exp(a + bx_i) - \log(y_i!)\},$$

where $K_h(u) = h^{-1}K(h^{-1}u)$ is a scaled version of a kernel function K. In this illustration we took K equal to the standard normal density function and selected h via a cross-validation argument. For general material on such local log-likelihood smoothing of parametric families, see Fan and Gijbels (1996), Hastie and Tibshirani (1990), and Hjort and Jones (1996).

We return to the football prediction problem in Example 3.4 (using BIC) and in Section 6.6.4 (using FIC). Interestingly, while both AIC and BIC agree that model M_2

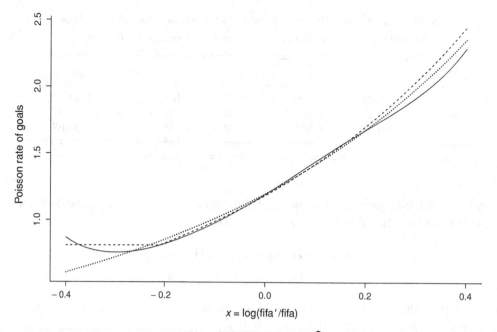

Fig. 2.2. The figure shows the fitted Poisson intensity rate $\widehat{\lambda}(x)$ of goals scored per match, as a function of $x = \log(\text{fifa/fifa}')$, where fifa and fifa$'$ are the FIFA ranking scores for the team and its opponent, for two different parametric models. These are the log-linear model $\lambda = \exp\{a + b(x - x_0)\}$ (dotted line) and the hockey-stick model (dashed line) where $\lambda(x)$ is $\exp(a)$ for $x \leq x_0$ and $\exp\{a + b(x - x_0)\}$ for $x \geq x_0$, and $x_0 = -0.21$. Also shown is a nonparametric kernel-smoothed log-likelihood-based estimate (solid line).

is best, we find using the FIC methods of Chapter 6 that model M_1 may sometimes be best for estimating the probability of the event 'team 1 defeats team 2'. ∎

Example 2.9 Density estimation via AIC

Suppose independent data $X_1, \ldots, X_n$ come from an unknown density f. There is a multitude of nonparametric methods for estimating f, chief among them methods using kernel smoothing. Parametric methods in combination with a model selection method are an easy-to-use alternative. We use AIC to select the right degree of complexity in the description of the density, starting out from

$$f_m(x) = f_0(x) \exp\left\{ \sum_{j=1}^{m} a_j \psi_j(x) \right\} \Big/ c_m(a),$$

where f_0 is some specified density and the normalising constant is defined as $c_m(a) = \int f_0 \exp(\sum_{j=1}^{m} a_j \psi_j) \, dx$. The basis functions $\psi_1, \psi_2, \ldots$ are orthogonal with respect to f_0, in the sense that $\int f_0 \psi_j \psi_k \, dx = \delta_{j,k} = I\{j = k\}$. We may for example take $\psi_j(x) = \sqrt{2} \cos(j \pi F_0(x))$, where F_0 is the cumulative distribution with f_0 as density. Within this family f_m the maximum likelihood estimators $\widehat{a} = (\widehat{a}_1, \ldots, \widehat{a}_m)$ are those that maximise

$$\ell_n(a) = n\left\{ \sum_{j=1}^{m} a_j \bar{\psi}_j - \log c_m(a) \right\},$$

where $\bar{\psi}_j = n^{-1} \sum_{i=1}^{n} \psi_j(X_i)$ and where we disregard terms not depending on a. This function is concave and has a unique maximiser as long as $n > m$. AIC selects its optimal order $\widehat{m}$ to maximise $\mathrm{AIC}(m) = 2\,\ell_n(\widehat{a}) - 2m$, perhaps among all $m \leq m_{\max}$ for a reasonable upper bound of complexity. The end result is a semiparametric density estimator $\widehat{f}(x) = f_{\widehat{m}}(x)$, which may well do better than full-fledged nonparametric estimators in cases where a low-order sum captures the main aspects of an underlying density curve. See also Example 3.5 for an extension of the present method, Exercise 7.6 for further analysis, and Chapter 8 for more on order selection in combination with hypothesis testing. ∎

Example 2.10 Autoregressive order selection

When a variable is observed over time, the correlation between observations needs to be carefully modelled. For a stationary time series, that is a time series for which the statistical properties such as mean, autocorrelation and variance do not depend on time, an autoregressive (AR) model is often suitable; see e.g. Brockwell and Davis (1991). In such a model, the observation at time t is written in the form

$$X_t = a_1 X_{t-1} + a_2 X_{t-2} + \cdots + a_k X_{t-k} + \varepsilon_t,$$

where the error terms ε_t are independent and identically distributed as $\mathrm{N}(0, \sigma^2)$ and the variables X_t have been centred around their mean. We make the assumption that the coefficients a_j are such that the complex polynomial $1 - a_1 z - \cdots - a_k z^k$ is different from zero for $|z| \leq 1$; this ensures that the time series is stationary. The value k is called the order of the autoregressive model, and the model is denoted by $\mathrm{AR}(k)$. If the value of k is small, only observations in the nearby past influence the current value X_t. If k is large, long-term effects in the past will still influence the present observation X_t. Knowing the order of the autoregressive structure is especially important for making predictions about the future, that is, predicting values of $X_{t+1}, X_{t+2}, \ldots$ when we observe the series up to and including time t; AIC can be used to select an appropriate order k. For a number of candidate orders $k = 1, 2, \ldots$ we construct $\mathrm{AIC}(k)$ by taking twice the value of the maximised log-likelihood for that $\mathrm{AR}(k)$ model, penalised with twice the number of estimated parameters, which is equal to $k + 1$ (adding one for the estimated standard deviation σ). Leaving out constants not depending on k, AIC takes a similar formula as for linear regresssion models, see (2.18). Specifically, for selecting the order k in $\mathrm{AR}(k)$ time series models, AIC boils down to computing

$$\mathrm{AIC}(k) = -2n \log \widehat{\sigma}_k - 2(k + 1),$$

where $\widehat{\sigma}_k$ is the maximum likelihood standard deviation estimator in the model with order k. The value of the autoregressive order k which corresponds to the largest $\mathrm{AIC}(k)$

identifies the best model. Being 'best' here, according to the AIC method, should be interpreted as the model that has the smallest estimated expected Kullback–Leiber distance from the true data-generating model. Order selection for autoregressive moving average (ARMA) models is treated extensively in Choi (1992). ∎

Example 2.11 The exponential decay of beer froth*

Three German beers are investigated for their frothiness: Erdinger Weißbier, Augustinerbräu München, and Budweiser Budvar. For a given beer make, the amount of froth $V_j^0(t_i)$ is measured, in centimetres, after time t_i seconds has passed since filling. The experiment is repeated $j = 1, \ldots, m$ times with the same make (where m was 7, 4, 4 for the three brands). Observation time points $t_0 = 0, t_1, \ldots, t_n$ spanned 6 minutes (with spacing first 15 seconds, later 30 and 60 seconds). Since focus here is on the decay, let $V_j(t_i) = V_j^0(t_i)/V_j^0(t_0)$; these ratios start at 1 and decay towards zero. Leike (2002) was interested in the exponential decay hypothesis, which he formulated as

$$\mu(t) = \mathrm{E}\, V_j(t) = \exp(-t/\tau) \quad \text{for } t \geq 0, \; j = 1, \ldots, m.$$

His main claims were that (i) data supported the exponential decay hypothesis; (ii) precise estimates of the decay parameter τ can be obtained by a minimum χ^2 type procedure; and (iii) that different brands of beers have decay parameters that differ significantly from each other. Leike's 2002 paper was published in the *European Journal of Physics*, and landed the author the Ig Nobel Prize for Physics that year.

Here we shall compare three models for the data, and reach somewhat sharper conclusions than Leike's. Model M_1 is the one indirectly used in Leike (2002), that observations are independent with

$$V_j(t_i) \sim \mathrm{N}(\mu_i(\tau), \sigma_i^2) \quad \text{where } \mu_i(\tau) = \exp(-t_i/\tau)$$

for time points t_i and repetitions $j = 1, \ldots, m$. This is an example of a nonlinear normal regression model. The log-likelihood function is

$$\ell_n = \sum_{i=1}^{n} \sum_{j=1}^{m} \left[-\frac{1}{2} \frac{\{V_j(t_i) - \mu_i(\tau)\}^2}{\sigma_i^2} - \log \sigma_i - \frac{1}{2} \log(2\pi) \right]$$

$$= m \sum_{i=1}^{n} \left[-\frac{1}{2} \frac{\widetilde{\sigma}_i^2 + \{\bar{V}(t_i) - \mu_i(\tau)\}^2}{\sigma_i^2} - \log \sigma_i - \frac{1}{2} \log(2\pi) \right],$$

where $\widetilde{\sigma}_i^2 = m^{-1} \sum_{j=1}^{m} \{V_j(t_i) - \bar{V}(t_i)\}^2$. To find the maximum likelihood estimates $(\widehat{\tau}, \widehat{\sigma}_1, \ldots, \widehat{\sigma}_n)$ we first maximise for fixed τ, and find

$$\widehat{\sigma}_i(\tau)^2 = \widetilde{\sigma}_i^2 + \{\bar{V}(t_i) - \mu_i(\tau)\}^2 \quad \text{for } i = 1, \ldots, n,$$

Table 2.4. *Decay parameter estimates and AIC scores for three beer froth decay models, for the three German beer brands. Model 1 has 15 parameters per beer whereas models 2 and 3 have two parameters per beer. Model 3 is judged the best.*

	Model 1		Model 2		Model 3	
	τ	AIC	τ	AIC	τ	AIC
beer 1	275.75	235.42	277.35	254.18	298.81	389.10
beer 2	126.50	78.11	136.72	92.62	112.33	66.59
beer 3	167.62	130.21	166.07	150.80	129.98	128.60
Total:		443.75		497.59		584.28

leading to a log-likelihood profile of the form $-\frac{1}{2}m$, $-m\sum_{i=1}^{n}\log\widehat{\sigma}_i(\tau) - \frac{1}{2}mn\log(2\pi)$. This is then maximised over τ, yielding maximum likelihood estimates along with

$$\ell_{\max,1} = -\tfrac{1}{2}mn - m\sum_{i=1}^{n}\log\widehat{\sigma}_i - \tfrac{1}{2}mn\log(2\pi)$$

and $\mathrm{AIC}(M_1) = 2(\ell_{\max,1} - n - 1)$. This produces decay parameter estimates shown in Table 2.4, for the three brands. Leike used a quite related estimation method, and found similar values. (More specifically, his method corresponds to the minimum χ^2 type method that is optimal provided the 14 σ_i parameters were known, with inserted estimates for these.)

Model M_1 employs different standard deviation parameters for each time point, and has accordingly a rather high number of parameters. Model M_2 is the simplification where the σ_i are set equal across time. The log-likelihood function is then

$$\ell_n = m\sum_{i=1}^{n}\left[-\tfrac{1}{2}\frac{\widetilde{\sigma}_i^2 + \{\bar{V}(t_i) - \mu_i(\tau)\}^2}{\sigma^2} - \log\sigma - \tfrac{1}{2}\log(2\pi)\right].$$

Maximising for fixed τ gives the equation

$$\widehat{\sigma}(\tau)^2 = n^{-1}\sum_{i=1}^{n}[\widetilde{\sigma}_i^2 + \{\bar{V}(t_i) - \mu_i(\tau)\}^2]$$

and a corresponding log-profile function in τ. Maximising with respect to τ gives

$$\ell_{\max,2} = -\tfrac{1}{2}mn - mn\log\widehat{\sigma}(\widehat{\tau}) - \tfrac{1}{2}mn\log(2\pi),$$

and $\mathrm{AIC}(M_2) = 2(\ell_{\max,2} - 2)$.

Models M_1 and M_2 both use an assumption of independence from one time observation point to the next. This is unreasonable in view of the decay character of the data, even when the time difference between observations is 15 seconds and more. A more direct probability model M_3, that takes the physical nature of the decay process into account, uses $V_j(t) = \exp\{-Z_j(t)\}$, where Z_j is taken to have independent, non-negative

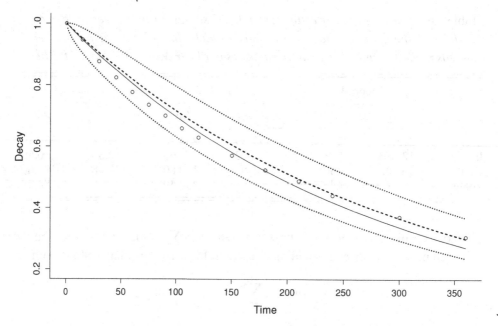

Fig. 2.3. Exponential decay of beer froth: the small circles indicate $\bar{V}(t)$, the average observed decay for seven mugs of Erdinger Weißbier, over 6 minutes. The solid line and broken line indicate respectively Leike's estimate of $\exp(-t/\tau)$ and our estimate of $\exp\{-at\log(1 + 1/b)\}$. The lower and upper curves form a pointwise 90% prediction band for the decay of the next mug of beer.

increments. Thus $Z_j(t) = -\log V_j(t)$ is a Lévy process, for which probability theory guarantees that

$$\mathrm{E}\,\exp\{-\theta Z(t)\} = \exp\{-K(t, \theta)\} \quad \text{for } t \geq 0 \text{ and } \theta \geq 0,$$

where $K(t, \theta)$ can be represented as $\int_0^\infty \{1 - \exp(-\theta s)\}\,\mathrm{d}N_t(s)$. See, for example, Hjort (1990) for such Lévy theory. Exponential decay corresponds to the Lévy measures N_t being time-homogeneous, i.e. of the form $\mathrm{d}N_t(s) = t\,\mathrm{d}N(s)$ for a single Lévy measure N. Then

$$\mathrm{E}\,\exp\{-\theta Z(t)\} = \exp\{-tK(\theta)\} \quad \text{for } t \geq 0,$$

which also implies $\tau = 1/K(1)$. This is the Khintchine formula for Lévy processes. The simplest such Lévy model is that of a Gamma process. So we take the increments $\Delta Z_j(t_i) = Z_j(t_i) - Z_j(t_{i-1})$ to be independent Gamma variables, with parameters $(a(t_i - t_{i-1}), b)$. Note that

$$\mathrm{E}\,V(t) = \mathrm{E}\,\exp\{-Z(t)\} = \{b/(b+1)\}^{at} = \exp\{-at\log(1 + 1/b)\},$$

in agreement with the exponential decay hypothesis. Maximum likelihood estimates have been found numerically, leading also to estimates of τ and to AIC scores, as presented in Table 2.4.

The conclusion is that the Lévy approach to beer froth watching is more successful than those of nonlinear normal regression (whether homo- or heteroscedastic); see the AIC values in the table. Figure 2.3 shows Leike's original decay curve estimate, along with that associated with model 3. Also given is a pointwise 90% confidence band for the froth decay of the next mug of Erdinger Weißbier. For more details and for comparisons with yet other models for such decay data, including Brownian bridge goodness-of-fit methods for assessing the exponential decay hypothesis, see Hjort (2007c). ∎

2.5 Takeuchi's model-robust information criterion

The key property underlying the AIC method, as identified in (2.16)–(2.17), is that the bias of the estimator $\widehat{Q}_n$ can be approximated by the generalised dimension p^*/n, more precisely $\mathrm{E}(\widehat{Q}_n - Q_n) = p^*/n + o(1/n)$. Different approximations to the bias of $\widehat{Q}_n$ are obtained by using different estimators $\widehat{p}^*$ of p^*, leading to the bias correction $n^{-1}(\ell_{n,\max} - \widehat{p}^*)$ for estimating Q_n.

Using AIC in its most familiar form (2.14) amounts to simply setting the p^* of (2.17) equal to the dimension of the model $p = \mathrm{length}(\theta)$. In case the model used is equal to the true model that generated the data, it indeed holds that both dimensions are equal, thus $p^* = p$, but this is not true in general. A more model-robust version can be used in case one does not want to make the assumption that the model used is the true model. Therefore we estimate p^* by plugging in estimates of the matrices J and K. Takeuchi (1976) proposed such an estimator and the corresponding criterion,

$$\mathrm{TIC} = 2\,\ell_n(\widehat{\theta}) - 2\widehat{p}^* \quad \text{with} \quad \widehat{p}^* = \mathrm{Tr}(\widehat{J}^{-1}\widehat{K}), \tag{2.20}$$

with estimators $\widehat{J}$ and $\widehat{K}$ as in (2.13). One should consider (2.20) as an attempt at making an AIC-type selection criterion that is robust against deviations from the assumption that the model used is correct. Note that this model-robustness issue is different from that of achieving robustness against outliers; the TIC as well as AIC rest on the use of maximum likelihood estimators and as such are prone to being overly influenced by outlying data values, in many models. Selection criteria made to be robust in this sense are developed in Section 2.10.

We now give an example in which the candidate model is incorrect. Suppose that the Y_i come from some g and that the $\mathrm{N}(\mu, \sigma^2)$ model is used as a candidate model. Then it can be shown that

$$J = \frac{1}{\sigma_0^2}\begin{pmatrix} 1 & 0 \\ 0 & 2 \end{pmatrix} \quad \text{and} \quad K = \frac{1}{\sigma_0^2}\begin{pmatrix} 1 & \kappa_3 \\ \kappa_3 & 2+\kappa_4 \end{pmatrix},$$

in terms of the skewness $\kappa_3 = \mathrm{E}U^3$ and kurtosis $\kappa_4 = \mathrm{E}U^4 - 3$ of $U = (Y - \mu_0)/\sigma_0$. This leads to $p^* = 2 + \frac{1}{2}\kappa_4$ and $\widehat{p}^* = 2 + \frac{1}{2}\widehat{\kappa}_4$, with $\widehat{\kappa}_4$ an estimator for the kurtosis of the residual distribution.

More generally we can consider the linear regression model $Y_i = x_i^t \beta + \sigma \varepsilon_i$, involving a k-dimensional parameter $\beta = (\beta_1, \ldots, \beta_k)^t$, where the ε_i are i.i.d. with mean zero and unit variance, but not necessarily normal. Then we find the matrices J_n and K_n as in Example 2.2. This gives the formula $p^* = k + 1 + \frac{1}{2}\kappa_4$ and $\widehat{p}^* = k + 1 + \frac{1}{2}\widehat{\kappa}_4$ for the generalised dimension of the model.

Sometimes there would be serious sampling variability for the trace of $\widehat{J}^{-1}\widehat{K}$, particularly if the dimension $p \times p$ of the two matrices is not small. In such cases one looks for less variable estimates for p^* of (2.17). Consider for illustration the problem of selecting regressors in Poisson count models, in the framework of Example 2.3. The TIC method may be used, with $\widehat{J}_n$ and $\widehat{K}_n$ given in that example. With certain over-dispersion models one might infer that $K_n = (1 + d)J_n$, for a dispersion parameter $1 + d$, and methods are available for estimating d and the resulting $p^* = \text{Tr}(J_n^{-1} K_n) = p(1 + d)$ more precisely than with the (2.20) formula. See also Section 2.7.

It is also worth noting that both the AIC and TIC methods may be generalised to various other types of parametric model, with appropriate modifications of the form of $\widehat{J}$ and $\widehat{K}$ and hence $\widehat{p}^*$ above; see Example 3.10 for such an illustration, in a context of parametric hazard regression models.

2.6 Corrected AIC for linear regression and autoregressive time series

It is important to realise that AIC typically will select more and more complex models as the sample size increases. This is because the maximal log-likelihood will increase linearly with n while the penalty term for complexity is proportional to the number of parameters. We now examine the linear regression model in more detail. In particular we shall see how some exact calculations lead to sample-size modifications of the direct AIC.

In Example 2.7 we considered the general linear regression model $Y = X\beta + \varepsilon$ and found that the direct AIC could be expressed as

$$\text{AIC} = -2n \log \widehat{\sigma} - 2(p + 1) - n - n \log(2\pi), \tag{2.21}$$

with $\widehat{\sigma}^2 = \|\text{res}\|^2/n$ from residuals $\text{res} = Y - X\widehat{\beta}$; in particular, the AIC advice is to choose the candidate model that minimises $n \log \widehat{\sigma} + p$ across candidate models. The aim of AIC is to estimate the expected Kullback–Leibler distance from the true data-generating mechanism $g(y \mid x)$ to the estimated model $f(y \mid x, \widehat{\theta})$, see Section 2.3, where in this situation $\theta = (\beta, \sigma)$. Assume here that $g(y \mid x)$ has mean $\xi(x)$ and constant standard deviation σ_0. If the assumed model is equal to the true model, then $\xi_i = \xi(x_i) = x_i^t \beta$.

We are not required to always use the maximum likelihood estimators. Here it is natural to modify $\widehat{\sigma}^2$ above, for example, since for the case that the assumed model is equal to the true model $\text{SSE} = \|\text{res}\|^2 \sim \sigma^2 \chi_{n-p}^2$, which means that $\|\text{res}\|^2$ should be divided by $n - p$ rather than n to make it an unbiased estimator. This is commonly done when computing estimates, but is not typical practice when working with AIC.

Write in general

$$\widehat{\sigma}^2 = \frac{\|\text{res}\|^2}{n-a} = \frac{1}{n-a} \sum_{i=1}^{n} (Y_i - x_i^t \widehat{\beta})^2, \tag{2.22}$$

with the cases $a = 0$ and $a = p$ corresponding to maximum likelihood and unbiased estimation respectively. In the spirit of the derivation of AIC, we examine how much

$$\widehat{Q}_n = n^{-1} \sum_{i=1}^{n} \log f(Y_i \mid x_i, \widehat{\beta}, \widehat{\sigma})$$

$$= n^{-1} \sum_{i=1}^{n} \left\{ -\log \widehat{\sigma} - \tfrac{1}{2}(Y_i - x_i^t \widehat{\beta})^2 / \widehat{\sigma}^2 - \tfrac{1}{2} \log(2\pi) \right\}$$

$$= -\log \widehat{\sigma} - \tfrac{1}{2} \frac{n-a}{n} - \tfrac{1}{2} \log(2\pi)$$

can be expected to overestimate

$$R_n = n^{-1} \sum_{i=1}^{n} \int g(y \mid x_i) \log f(y \mid x_i, \widehat{\beta}, \widehat{\sigma}) \, dy$$

$$= -\log \widehat{\sigma} - \tfrac{1}{2} n^{-1} \sum_{i=1}^{n} \frac{(\xi_i - x_i^t \widehat{\beta})^2 + \sigma_0^2}{\widehat{\sigma}^2} - \tfrac{1}{2} \log(2\pi).$$

We find

$$\mathrm{E}_g(\widehat{Q}_n - R_n) = -\tfrac{1}{2} \frac{n-a}{n} + \tfrac{1}{2} \mathrm{E}_g \left[\frac{\sigma_0^2}{\widehat{\sigma}^2} \left\{ n^{-1} \sum_{i=1}^{n} (x_i^t \widehat{\beta} - \xi_i)^2 / \sigma_0^2 + 1 \right\} \right].$$

Under model circumstances, where $\xi_i = x_i^t \beta$ and $\sigma_0 = \sigma$, it is well known that $\widehat{\sigma}^2 / \sigma^2$ is distributed as a $\chi_{n-p}^2 / (n-a)$ and is independent of $\widehat{\beta}$. For the fitted values,

$$X\widehat{\beta} = X(X^t X)^{-1} X^t Y = X(X^t X)^{-1} X^t (X\beta + \varepsilon) = X\beta + H\varepsilon$$

using the 'hat matrix' $H = X(X^t X)^{-1} X^t$. This shows that

$$n^{-1} \sum_{i=1}^{n} (x_i^t \widehat{\beta} - x_i^t \beta)^2 = n^{-1} \|X\widehat{\beta} - X\beta\|^2 = n^{-1} \varepsilon^t H \varepsilon$$

has mean equal to $n^{-1} \mathrm{E} \operatorname{Tr}(H\varepsilon\varepsilon^t) = \sigma^2 \operatorname{Tr}(H)/n = (p/n)\sigma^2$. Thus

$$\mathrm{E}_g(\widehat{Q}_n - R_n) = -\tfrac{1}{2} \frac{n-a}{n} + \frac{1}{2} \frac{n-a}{n-p-2} \frac{p+n}{n}$$

$$= \tfrac{1}{2} \frac{n-a}{n} \frac{2p+2}{n-p-2} = \frac{p+1}{n} \frac{n-a}{n-p-2},$$

using that $\mathrm{E}(1/\chi_{n-p}^2) = 1/(n-p-2)$ for $n > p+2$.

This leads to a couple of strategies for modifying the direct version (2.21) to obtain a more precise penalty. The first is to keep the maximum likelihood estimator $\widehat{\sigma}$, that is, using $a = 0$ above, but to penalise the maximised log-likelihood with a more precisely calibrated factor; the result is

$$\text{AIC}_c = 2\ell_n(\widehat{\beta}, \widehat{\sigma}) - 2(p+1)\frac{n}{n-p-2}. \tag{2.23}$$

This is equivalent to what has been suggested by Sugiura (1978) and Hurvich and Tsai (1989). Note that this penalises complexity (the number of parameters $p+1$) more strongly than with the standard version of AIC, with less chance of overfitting the model. The second modification is actually simpler. It consists of using $a = p + 2$ in (2.22) and keeping the usual penalty at $2(p+1)$:

$$\text{AIC}_c^* = 2\ell_n(\widehat{\beta}, \widehat{\sigma}^*) - 2(p+1), \tag{2.24}$$

where $(\widehat{\sigma}^*)^2 = \|\text{res}\|^2/(n-p-2)$. This is like the ordinary AIC but with a corrected σ estimate. In particular, this corrected AIC procedure amounts to picking the model with smallest $n \log \widehat{\sigma}^* + p$. Note that the question of which AIC correction method works the best may be made precise in different ways, and there is no 'clear winner'; see Exercise 2.6.

Of the two modifications AIC_c and AIC_c^* only the first has an immediate, but ad hoc, generalisation to general parametric regression models. The suggestion is to use the penalty term obtained for normal linear regression models also for general likelihood models, leading to

$$\text{AIC}_c = 2\ell_n(\widehat{\theta}) - 2 \, \text{length}(\theta)\frac{n}{n - \text{length}(\theta) - 1}. \tag{2.25}$$

Hurvich and Tsai (1989) show that this form is appropriate when searching for model order in normal autoregressive models. For autoregressive models $\text{AR}(k)$ (see Example 2.10) the formula of AIC_c is again the same as that for linear regression models, namely (leaving out constants not depending on the number of parameters)

$$\text{AIC}_c = -n \log(\widehat{\sigma}_k^2) - \frac{n(n+k)}{n-k-2}.$$

Outside linear regression and autoregressive models the (2.25) formula should be used with care since there is no proof of this statement for general likelihood models. The more versatile bootstrapping method, described at the end of the next section, can be used instead.

2.7 AIC, corrected AIC and bootstrap-AIC for generalised linear models*

While in a traditional linear model the mean of the response $\text{E}(Y \mid x) = x^t\beta$ is a linear function, in a *generalised linear model* there is a monotone and smooth *link function* $g(\cdot)$

such that

$$g(\mathrm{E}(Y_i \mid x_i)) = g(\xi_i) = \eta_i = x_i^{\mathrm{t}}\beta = \sum_{j=1}^{p} x_{i,j}\beta_j$$

for $i = 1, \ldots, n$. The Y_i are independent, and the likelihood contribution for individual i has the form

$$f(y_i, \theta_i, \phi) = \exp\left\{\frac{y_i\theta_i - b(\theta_i)}{a(\phi)} + c(y_i, \phi)\right\}.$$

The functions $a(\cdot)$, $b(\cdot)$ and $c(\cdot, \cdot)$ are known. The $b(\cdot)$ plays an important role since its derivatives yield the mean and variance function. The parameter ϕ is a scale parameter, and θ_i is the main parameter of interest, as it can be written as a function of the mean $\mathrm{E}(Y_i \mid x)$. Since $\mathrm{E}(\partial \log f(Y_i; \theta_i, \phi)/\partial\theta_i) = 0$, a formula valid for general models, it follows that

$$\xi_i = \mathrm{E}(Y_i \mid x_i) = b'(\theta_i) = \partial b(\theta_i)/\partial\theta_i.$$

Also, $\mathrm{Var}(Y_i \mid x_i) = a(\phi)b''(\theta_i)$. Many familiar density functions can be written in this form. The class of generalised linear models includes as important examples the normal, Poisson, binomial, and Gamma distribution. For more information on generalised linear models, we refer to McCullagh and Nelder (1989) and Dobson (2002).

The log-likelihood function $\ell_n(\beta, \phi)$ is a sum of $\{y_i\theta_i - b(\theta_i)\}/a(\phi) + c(y_i, \phi)$ contributions. Maximum likelihood estimators $(\widehat{\beta}, \widehat{\phi})$ determine fitted values $\widehat{\theta}_i$ and $\widehat{\xi}_i$. Let as before $g(y \mid x)$ denote the true density of Y given x. We now work with the Kullback–Leibler related quantity

$$R_n = n^{-1} \sum_{i=1}^{n} \int g(y \mid x_i) \log f(y_i \mid x_i, \widehat{\theta}) \, \mathrm{d}y.$$

The definitions above give

$$R_n = n^{-1} \sum_{i=1}^{n} \int g(y \mid x_i)[\{y\widehat{\theta}_i - b(\widehat{\theta}_i)\}/\widehat{a} + c(y, \widehat{\phi})] \, \mathrm{d}y$$

$$= n^{-1} \sum_{i=1}^{n} \left[\{\xi_i^0\widehat{\theta}_i - b(\widehat{\theta}_i)\}/\widehat{a} + \int c(y, \widehat{\phi})g(y \mid x_i) \, \mathrm{d}y\right],$$

where ξ_i^0 denotes the real mean of Y_i given x_i, and with $\widehat{a} = a(\widehat{\phi})$

$$\widehat{Q}_n = n^{-1}\ell_n(\widehat{\beta}, \widehat{\phi}) = n^{-1} \sum_{i=1}^{n} [\{Y_i\widehat{\theta}_i - b(\widehat{\theta}_i)\}/\widehat{a} + c(Y_i, \widehat{\phi})].$$

From this follows, upon simplification, that

$$\mathrm{E}_g(\widehat{Q}_n - R_n) = n^{-1} \sum_{i=1}^{n} \mathrm{E}_g\left\{\frac{(Y_i - \xi_i^0)\widehat{\theta}_i}{\widehat{a}}\right\}.$$

Let us define the quantity

$$\alpha_n = n \, E_g(\widehat{Q}_n - R_n) = E_g \, A_n, \qquad (2.26)$$

with $A_n = \sum_{i=1}^n (Y_i - \xi_i^0)\widehat{\theta}_i / \widehat{a}$. Thus α_n/n is the expected bias for the normalised max-imised log-likelihood, as an estimator of the estimand $Q_n = E_g R_n$, and can be estimated from data. AIC takes $\alpha_n = p$. We derive approximations to α_n for the class of generalised linear models. For simplicity of presentation we limit discussion to generalised linear models with canonical link function, which means that $\theta_i = \eta_i = x_i^t \beta$. Let

$$\Sigma_n = n^{-1} \sum_{i=1}^n v_i x_i x_i^t \quad \text{and} \quad \Sigma_n^* = n^{-1} \sum_{i=1}^n v_i^* x_i x_i^t,$$

where $v_i = b''(\theta_i)$ and $v_i^* = a(\phi)^{-1} \text{Var}(Y_i \mid x_i)$. Then $\widehat{\beta} - \beta_0$ may up to an $O_P(n^{-1/2})$ error be represented as $\Sigma_n^{-1} n^{-1} \sum_{i=1}^n \{Y_i - b'(\theta_i)\} x_i$, as a consequence of the property exhibited in connection with (2.12). When the assumed model is equal to the true model, $v_i^* = v_i$. We find, also outside the model conditions, that

$$\alpha_n \doteq E_g(1/\widehat{a}) \sum_{i=1}^n E_g \big[(Y_i - \xi_i^0) x_i^t \widehat{\beta}\big]$$

$$\doteq E_g(1/\widehat{a}) \sum_{i=1}^n E_g \big[(Y_i - \xi_i^0) x_i^t \Sigma_n^{-1} \{Y_i - b'(\theta_i)\} x_i\big]$$

$$\doteq E_g(a/\widehat{a}) n^{-1} \sum_{i=1}^n v_i^* x_i^t \Sigma_n^{-1} x_i = E_g(a/\widehat{a}) \text{Tr}\big(\Sigma_n^{-1} \Sigma_n^*\big).$$

We have used the parameter orthogonality property for generalised linear models, which implies approximate independence between the linear part estimator $\widehat{\beta}$ and the scale esti-mator $\widehat{a} = a(\widehat{\phi})$. For the linear normal model, for example, there is exact independence, and $a/\widehat{a} = \sigma^2/\widehat{\sigma}^2$ has exact mean $m/(m-2)$ if the unbiased estimator $\widehat{\sigma}^2$ is used with degrees of freedom $m = n - p$.

These approximations suggest some corrections to the ordinary AIC version, which uses $\alpha_n = p$. Some models for dealing with overdispersion use ideas that in the present terminology amount to $\Sigma_n^* = (1 + d)\Sigma_n$, with d an overdispersion parameter. For Poisson-type regression there is overdispersion when the observed variance is too big in comparison with the variance one would expect for a Poisson variable. In such a case the dispersion is $1 + d = \text{Var}_g Y / E_g Y > 1$. Including such an overdispersion parameter corresponds to modelling means with $\exp(x_i^t \beta)$ but variances with $(1 + d) \exp(x_i^t \beta)$. An estimate of this d leads to $\alpha_n \doteq (1 + d)p$ and to a corrected

$$\text{AIC}_c = 2\ell_n(\widehat{\theta}) - 2p(1 + d)$$

for use in model selection when there is overdispersion.

Table 2.5. *AIC and bootstrap AIC values for the eight candidate models of Example 2.4 on low birthweights. The model rankings agree on the top three models but not on the rest.*

Model	ℓ_{max}	length(β)	$\widehat{\alpha}_n$	AIC	Ranking		AIC_{boot}
$\emptyset$	−114.345	2	1.989	−232.691	5	4	−232.669
x_3	−113.562	3	3.021	−233.123	6	6	−233.165
x_4	−112.537	3	3.134	−231.075	1	1	−231.344
x_5	−114.050	3	3.033	−234.101	7	7	−234.167
x_3, x_4	−112.087	4	4.078	−232.175	3	3	−232.331
x_3, x_5	−113.339	4	4.096	−234.677	8	8	−234.834
x_4, x_5	−111.630	4	4.124	−231.259	2	2	−231.507
x_3, x_4, x_5	−111.330	5	5.187	−232.661	4	5	−233.034

Using the bootstrap is another alternative for estimating α_n. Simulate $Y_1^*, \ldots, Y_n^*$ from an estimated bigger model, say one that uses all available covariates, with estimated means $\widehat{\xi}_1, \ldots, \widehat{\xi}_n$, keeping $x_1, \ldots, x_n$ fixed. For this simulated data set, produce estimates $\widehat{\beta}^*$ and $\widehat{a}^* = a(\widehat{\phi}^*)$, along with $\widehat{\theta}_i^*$. Then form $A_n^* = \sum_{i=1}^n (Y_i^* - \widehat{\xi}_i) x_i^t \widehat{\beta}^* / \widehat{a}^*$. The α_n estimate is formed by averaging over a high number of simulated A_n^* values. This procedure can be used for each candidate model, leading to the bootstrap corrected

$$\text{AIC}_{boot} = 2\ell_n(\widehat{\theta}) - 2\widehat{\alpha}_n. \tag{2.27}$$

Again, the model with highest such value would then be selected.

Example 2.12 Low birthweight data: bootstrap AIC*
As an illustration, let us return to the low birthweight data of Example 2.4. For the logistic regression model, A_n of (2.26) is $\sum_{i=1}^n (Y_i - p_i^0) x_i^t \widehat{\beta}$, where p_i^0 is the mean of Y_i under the true (but unknown) model. For each candidate model, we simulate a large number of $A_n^* = \sum_{i=1}^n (Y_i^* - \widehat{p}_i) x_i^t \widehat{\beta}^*$, where Y_i^* are 0–1 variables with probabilities set at $\widehat{p}_i$ values found from the biggest model, and where $\widehat{\beta}^*$ is the maximum likelihood estimator based on the simulated data set, in the given candidate model. For the situation of Example 2.4, with 25,000 simulations for each of the eight candidate models, the results are presented in Table 2.5. The $\widehat{\alpha}_n$ numbers, which may be seen as the exact penalties (modulo simulation noise) from the real perspective of AIC, agree well with the AIC default values in this particular illustration. Also, the AIC and AIC_{boot} values are in essential agreement. ∎

2.8 Behaviour of AIC for moderately misspecified models*

Consider a local neighbourhood framework where data stem from a density

$$f_n(y) = f(y, \theta_0, \gamma_0 + \delta/\sqrt{n}), \tag{2.28}$$

with a p-dimensional θ and a one-dimensional γ. The null value γ_0 corresponds to the narrow model, so (2.28) describes a one-parameter extension. This framework will be extended and discussed further in Sections 5.2, 5.7 and 6.3, also for situations with $q \geq 2$ extra parameters. The AIC method for selecting one of the two models compares

$$\text{AIC}_{\text{narr}} = 2\,\ell_n(\widetilde{\theta}, \gamma_0) - 2p \quad \text{with} \quad \text{AIC}_{\text{wide}} = 2\,\ell_n(\widehat{\theta}, \widehat{\gamma}) - 2(p+1),$$

where maximum likelihood estimators are used in each model. To understand these random AIC scores better, introduce first

$$\begin{pmatrix} U(y) \\ V(y) \end{pmatrix} = \begin{pmatrix} \partial \log f(y, \theta_0, \gamma_0)/\partial\theta \\ \partial \log f(y, \theta_0, \gamma_0)/\partial\gamma \end{pmatrix},$$

along with $\bar{U}_n = n^{-1}\sum_{i=1}^n U(Y_i)$ and $\bar{V}_n = n^{-1}\sum_{i=1}^n V(Y_i)$. The $(p+1) \times (p+1)$-size information matrix of the model is

$$J_{\text{wide}} = \text{Var}_0\begin{pmatrix} U(Y) \\ V(Y) \end{pmatrix} = \begin{pmatrix} J_{00} & J_{01} \\ J_{10} & J_{11} \end{pmatrix}, \quad \text{with } J_{\text{wide}}^{-1} = \begin{pmatrix} J^{00} & J^{01} \\ J^{10} & J^{11} \end{pmatrix}.$$

Here the $p \times p$-size J_{00} is simply the information matrix of the narrow model, evaluated at θ_0, and the scalar J_{11} is the variance of $V(Y_i)$, also computed under the narrow model. We define $\kappa^2 = J^{11}$.

Coming back to AIC for the two models, one finds via result (2.9) and Taylor expansions that

$$\text{AIC}_{\text{narr}} \doteq_d 2\sum_{i=1}^n \log f(Y_i, \theta_0, \gamma_0) + n\bar{U}_n^t J_{00}^{-1}\bar{U}_n - 2p,$$

$$\text{AIC}_{\text{wide}} \doteq_d 2\sum_{i=1}^n \log f(Y_i, \theta_0, \gamma_0) + n\begin{pmatrix} \bar{U}_n \\ \bar{V}_n \end{pmatrix}^t J_{\text{wide}}^{-1}\begin{pmatrix} \bar{U}_n \\ \bar{V}_n \end{pmatrix} - 2(p+1).$$

Further algebraic calculations give, with $D \sim \text{N}(\delta, \kappa^2)$,

$$\begin{aligned}
\text{AIC}_{\text{wide}} - \text{AIC}_{\text{narr}} &\doteq_d n\begin{pmatrix} \bar{U}_n \\ \bar{V}_n \end{pmatrix}^t J_{\text{wide}}^{-1}\begin{pmatrix} \bar{U}_n \\ \bar{V}_n \end{pmatrix} - n\bar{U}_n^t J_{00}^{-1}\bar{U}_n - 2 \\
&= n\{\bar{U}_n^t(J^{00} - J_{00}^{-1})\bar{U}_n + 2\bar{U}_n^t J^{01}\bar{V}_n + \bar{V}_n^2 J^{11}\} - 2 \\
&= n(\bar{V}_n - J_{10}J_{00}^{-1}\bar{U}_n)^2\kappa^2 - 2 \\
&\xrightarrow{d} D^2/\kappa^2 - 2 \sim \chi_1^2(\delta^2/\kappa^2) - 2.
\end{aligned}$$

The probability that AIC prefers the narrow model over the wide model is therefore approximately $\text{P}(\chi_1^2(\delta^2/\kappa^2) \leq 2)$. In particular, if the narrow model is perfect, the probability is 0.843, and if $\delta = \kappa$, the probability is 0.653.

It is also instructive to see that $\text{AIC}_{\text{wide}} - \text{AIC}_{\text{narr}}$ is asymptotically equivalent to $D_n^2/\kappa^2 - 2$, where $D_n = \sqrt{n}(\widehat{\gamma} - \gamma_0)$. This gives a connection between the Akaike criterion and a certain pre-test strategy. Pre-testing is a form of variable selection which consists of checking the coefficients in the wide model and keeping only those which are significant. In this example with only two models, we check whether the extra parameter γ is significant and decide to keep the wide model when it indeed is significant. There are various ways to test for significance (t-statistics, z-statistics with different significance level, etc.). According to AIC we would use the wide model when $D_n^2/\widehat{\kappa}^2 > 2$. Generalisations of these results to AIC behaviour in situations with more candidate models than only two, as here, are derived and discussed in Section 5.7.

2.9 Cross-validation

An approach to model selection that at least at the outset is of a somewhat different spirit from that of AIC and TIC is that of *cross-validation*. Cross-validation has a long history in applied and theoretical statistics, but was first formalised as a general methodology in Stone (1974) and Geisser (1975). The idea is to split the data into two parts: the majority of data are used for model fitting and development of prediction algorithms, which are then used for estimation or prediction of the left-out observations. In the case of leave-one-out cross-validation, perhaps the most common form, only one observation is left out at a time. Candidate models are fit with all but this one observation, and are then used to predict the case which was left out.

We first explain a relation between leave-one-out cross-validation, the Kullback–Leibler distance and Takeuchi's information criterion. For concreteness let us first use the i.i.d. framework. Since the Kullback–Leibler related quantity R_n worked with in (2.15) may be expressed as $\int g(y) \log f(y, \widehat{\theta}) \, dy = \mathrm{E}_g \log f(Y_{\mathrm{new}}, \widehat{\theta})$, where Y_{new} is a new datum independent of $Y_1, \ldots, Y_n$, an estimator of its expected value is

$$\mathrm{xv}_n = n^{-1} \sum_{i=1}^{n} \log f(Y_i, \widehat{\theta}_{(i)}). \tag{2.29}$$

Here $\widehat{\theta}_{(i)}$ is the maximum likelihood estimator computed based on the data set where Y_i is omitted; this should well emulate the log $f(Y_{\mathrm{new}}, \widehat{\theta})$ situation, modulo the small difference between estimators based on sample sizes n versus $n - 1$. The cross-validation model selection method, in this context, is to compute the (2.29) value for each candidate model, and select the one with highest value.

There is a connection between cross-validation and the model-robust AIC. Specifically,

$$\mathrm{xv}_n \doteq n^{-1}\{\ell_n(\widehat{\theta}) - \mathrm{Tr}(\widehat{J}^{-1}\widehat{K})\}, \tag{2.30}$$

which means that $2n \, \mathrm{xv}_n$ is close to Takeuchi's information criterion (2.20). To prove (2.30) we use properties of influence functions.

Let in general terms $T = T(G)$ be a function of some probability distribution G. The influence function of T, at position y, is defined as the limit

$$\mathrm{infl}(G, y) = \lim_{\varepsilon \to 0}\{T((1 - \varepsilon)G + \varepsilon\delta(y)) - T(G)\}/\varepsilon, \tag{2.31}$$

when it exists. Here $\delta(y)$ denotes a unit point-mass at position y, which means that $(1 - \varepsilon)G + \delta(y)$ is the distribution of a variable that with probability $1 - \varepsilon$ is drawn from G and with probability ε is equal to y. These influence functions are useful for several purposes in probability theory and statistics, for example in the study of robustness, see e.g. Huber (1981), Hampel *et al.* (1986), Basu *et al.* (1998) and Jones *et al.* (2001).

Consider some parameter functional $\theta = T(G)$. Data points $y_1, \ldots, y_n$ give rise to the empirical distribution function G_n and an estimate $\widehat{\theta} = T(G_n)$. The primary use of influence functions in our context is the fact that under general and weak conditions,

$$T(G_n) = T(G) + n^{-1} \sum_{i=1}^{n} \text{infl}(G, Y_i) + o_P(n^{-1/2}). \tag{2.32}$$

This implies the fundamental large-sample result

$$\sqrt{n}\{T(G_n) - T(G)\} \xrightarrow{d} N_p(0, \Omega), \tag{2.33}$$

where Ω is the variance matrix of $\text{infl}(G, Y)$ when $Y \sim G$. For $T(G)$ we now take the Kullback–Leibler minimiser $\theta_0 = \theta_0(g)$ that minimises $\text{KL}(g, f(\cdot, \theta))$ of (2.2), with g the density of G. Then $T(G_n)$ is the maximum likelihood estimator $\widehat{\theta}$. The influence function for maximum likelihood estimation can be shown to be

$$\text{infl}(G, y) = J^{-1}u(y, \theta_0), \tag{2.34}$$

for precisely $\theta_0 = T(G)$, with J as defined in (2.7) and $u(y, \theta)$ the score function; see Exercise 2.11.

We record one more useful consequence of influence functions. For a given data point y_i, consider the leave-one-out estimator $\widehat{\theta}_{(i)}$, constructed by leaving out y_i. Note that $G_n = (1 - 1/n)G_{n,(i)} + n^{-1}\delta(y_i)$, with $G_{n,(i)}$ denoting the empirical distribution of the $n - 1$ data points without y_i. From (2.31), therefore, with $\varepsilon = 1/n$, the approximation $T(G_n) \doteq T(G_{n,(i)}) + n^{-1}\,\text{infl}(G_{n,(i)}, y_i)$ emerges, that is

$$\widehat{\theta}_{(i)} \doteq \widehat{\theta} - n^{-1}\,\text{infl}(G_{n,(i)}, y_i) \doteq \widehat{\theta} - n^{-1}\,\text{infl}(G_n, y_i). \tag{2.35}$$

From (2.35) and (2.34), $\widehat{\theta}_{(i)} \doteq \widehat{\theta} - n^{-1}\widehat{J}^{-1}u(y_i, \widehat{\theta})$. Using Taylor expansion, $\log f(y_i, \widehat{\theta}_{(i)})$ is close to $\log f(y_i, \widehat{\theta}) + u(y_i, \widehat{\theta})^{\text{t}}(\widehat{\theta}_{(i)} - \widehat{\theta})$. Combining these observations, we reach

$$\text{xv}_n \doteq n^{-1} \sum_{i=1}^{n} \{\log f(y_i, \widehat{\theta}) + u(y_i, \widehat{\theta})^{\text{t}}(\widehat{\theta}_{(i)} - \widehat{\theta})\}$$

$$\doteq n^{-1}\ell_n(\widehat{\theta}) - n^{-1} \sum_{i=1}^{n} u(y_i, \widehat{\theta})^{\text{t}}n^{-1}\widehat{J}^{-1}u(y_i, \widehat{\theta}),$$

which leads to (2.30) and ends the proof. The approximation holds in the sense that the difference between the two sides goes to zero in probability.

The case of regression models can be handled similarly, inside the framework of Section 2.2. We again have

$$\text{xv}_n = n^{-1} \sum_{i=1}^{n} \log f(y_i \mid x_i, \widehat{\theta}_{(i)}) \doteq n^{-1} \ell_{n,\max} + n^{-1} \sum_{i=1}^{n} u(y_i \mid x_i, \widehat{\theta})^{t}(\widehat{\theta}_{(i)} - \widehat{\theta}),$$

where we need a parallel result to that above for the difference $\widehat{\theta}_{(i)} - \widehat{\theta}$. Such an approximation emerges from similar arguments, where the work tool is that of the influence function $\text{infl}((G, C), (x, y))$ defined in a manner similar to (2.31), but now viewed as associated with the functional $T = T(G, C)$ that operates on the combination of true densities $g(y \mid x)$ and the covariate distribution C. For the maximum likelihood functional, which is also the minimiser of the weighted Kullback–Leibler divergence (2.4), one may prove that the influence function takes the form $J^{-1}u(y \mid x, \theta_0)$, with J as in (2.12). With these ingredients one may copy the earlier arguments to reach the approximation $\widehat{\theta}_{(i)} - \widehat{\theta} \approx -n^{-1}\widehat{J}_n^{-1}u(y_i \mid x_i, \widehat{\theta})$. In conclusion,

$$\text{xv}_n \doteq n^{-1}\ell_{n,\max} - n^{-2} \sum_{i=1}^{n} u(y_i \mid x_i, \widehat{\theta})^{t}\widehat{J}_n^{-1}u(y_i \mid x_i, \widehat{\theta})$$

$$= n^{-1}\{\ell_{n,\max} - \text{Tr}(\widehat{J}_n^{-1}\widehat{K}_n)\},$$

essentially proving that cross-validation in regression models is first-order large-sample equivalent to the model-robust AIC method discussed in Section 2.5.

We have so far discussed cross-validation as a tool in connection with assessing the expected size of $\int \int g(y \mid x) \log f(y \mid x, \widehat{\theta}) \, dy \, dC(x)$, associated with maximum likelihood estimation and the weighted Kullback–Leibler distance (2.4). Sometimes a more immediate problem is to assess the quality of the more direct prediction task that estimates a new y_{new} with say $\widehat{y}_{\text{new}} = \widehat{\xi}(x_{\text{new}})$, where $\widehat{\xi}(x) = \xi(x, \widehat{\theta})$ estimates $\xi(x, \theta) = E_\theta(Y \mid x)$. This leads to the problem of estimating the mean of $(y_{\text{new}} - \widehat{y}_{\text{new}})^2$ and related quantities. Cross-validation is also a tool for such problems.

Consider in general terms $E_g h(Y_{\text{new}} \mid x, \widehat{\theta})$, for a suitable h function, for example of the type $\{Y_{\text{new}} - \xi(x, \widehat{\theta})\}^2$. We may write this as

$$\pi_n = \int \int g(y \mid x) h(y \mid x, \widehat{\theta}) \, dy \, dC_n(x) = n^{-1} \sum_{i=1}^{n} E_{g(y \mid x_i)} h(Y_{\text{new},i} \mid x_i, \widehat{\theta}),$$

where $Y_{\text{new},i}$ is a new observation drawn from the distribution associated with covariate vector x_i. The direct but naive estimator is $\bar{\pi}_n = n^{-1} \sum_{i=1}^{n} h(y_i \mid x_i, \widehat{\theta})$, but we would prefer the nearly unbiased cross-validated estimator

$$\widehat{\pi}_n = n^{-1} \sum_{i=1}^{n} h(y_i \mid x_i, \widehat{\theta}_{(i)}).$$

Assume that h is smooth in θ and that the estimation method used has influence function of the type $J(G, C)^{-1}a(y \mid x, \theta_0)$, for a suitable matrix $J(G, C)$ defined in terms of both the distribution G of $Y \mid x$ and of the covariate distribution C of x. Such influence representations hold for various minimum distance estimators, including the maximum likelihood method. Then Taylor expansion, as above, leads to

$$\widehat{\pi}_n \doteq \bar{\pi}_n - n^{-1}\mathrm{Tr}(\widetilde{J}^{-1}\widetilde{L}), \qquad (2.36)$$

where $\widetilde{J} = J(G_n, C_n)$ is the empirical version of $J(G, C)$ and

$$\widetilde{L} = n^{-1}\sum_{i=1}^{n} a(y_i \mid x_i, \widehat{\theta})h'(y_i \mid x_i, \widehat{\theta})^{\mathrm{t}},$$

with $h'(y \mid x, \theta) = \partial h(y \mid x, \theta)/\partial\theta$. The cross-validation result (2.30) corresponds to the special case where the h function is $\log f(y \mid x, \theta)$ and where the estimation method is the maximum likelihood; in this situation, both $a(y \mid x, \theta)$ and $h'(y \mid x, \theta)$ are equal to $u(y \mid x, \theta)$, and $\widetilde{L}$ is identical to $\widehat{K}$. For an illustration of various cross-validation methods, see Example 5.10.

Example 2.13 Two views on selecting Poisson models

Assume as in Example 2.3 that count data Y_i are associated with covariate vectors x_i, and that a Poisson model is being considered with parameters $\xi_i = \exp(x_i^{\mathrm{t}}\beta)$. To assess the prediction quality of such a candidate model, at least two methods might be put forward, yielding two different model selectors. The first is that implied by concentrating on maximum likelihood and the Kullback–Leibler divergence (2.4), and where the above results lead to

$$\mathrm{xv}_n = n^{-1}\sum_{i=1}^{n} \log f(y_i \mid x_i, \widehat{\beta}_{(i)}) \doteq n^{-1}\ell_{n,\max} - n^{-1}\mathrm{Tr}(\widehat{J}_n^{-1}\widehat{K}_n).$$

As we know, using the right-hand side scores for candidate models corresponds to the model-robust AIC, cf. TIC in Section 2.5. The second option is that of comparing predicted $\widehat{y}_i = \exp(x_i^{\mathrm{t}}\widehat{\beta})$ with observed y_i directly. Let therefore $h(y \mid x, \beta) = \{y - \exp(x^{\mathrm{t}}\beta)\}^2$. Here $h'(y \mid x, \beta) = -2\{y - \exp(x^{\mathrm{t}}\beta)\}\exp(x^{\mathrm{t}}\beta)x$, and the techniques above yield

$$\widehat{\pi}_n = n^{-1}\sum_{i=1}^{n}\left\{y_i - \exp(x_i^{\mathrm{t}}\widehat{\beta}_{(i)})\right\}^2 \doteq n^{-1}\sum_{i=1}^{n}(\widehat{y}_i - y_i)^2 + 2n^{-1}\mathrm{Tr}(\widehat{J}_n^{-1}\widehat{M}_n),$$

where $\widehat{M}_n = n^{-1}\sum_{i=1}^{n}\widehat{\xi}_i(y_i - \widehat{\xi}_i)^2 x_i x_i^{\mathrm{t}}$. This example serves as a reminder that 'model selection' should not be an automated task, and that what makes a 'good model' must depend on the context. ∎

The first-order asymptotic equivalence of cross-validation and AIC has first been explained in Stone (1977). There is a large literature on cross-validation. For more information we refer to, for example, Stone (1978), Efron (1983), Picard and Cook (1984), Efron and Tibshirani (1993) and Hjorth (1994).

Allen (1974) proposed a method of model selection in linear regression models which is based on the sum of squared differences of the leave-one-out predictions $\widehat{y}_i$ with their true observed values y_i. This gives the PRESS statistic, an abbreviation for prediction sum of squares. In a linear regression model $Y_i = x_i^t \beta + \varepsilon_i$ with a vector β of regression coefficients, leave out observation i, then fit the model, obtain the estimated coefficients $\widehat{\beta}_{(i)}$ and form the leave-one-out prediction $\widehat{y}_{i,xv} = x_i^t \widehat{\beta}_{(i)}$. This gives the statistic PRESS $= \sum_{i=1}^n (y_i - \widehat{y}_{i,xv})^2$. This score can then be computed for different subsets of regression variables. The model with the smallest value of PRESS is selected. Alternatively one may compute the more robust $\sum_{i=1}^n |y_i - \widehat{y}_{i,xv}|$. There are ways of simplifying the algebra here, making the required computations easy to perform. Let $s_1, \ldots, s_n$ be the diagonal entries of the matrix $I_n - H$, where $H = X(X^t X)^{-1} X^t$ is the hat matrix. Then the difference $y_i - \widehat{y}_{i,xv}$ is identical to $(y_i - x_i^t \widehat{\beta})/s_i$, where $\widehat{\beta}$ is computed using all data in the model; see Exercise 2.13. Thus PRESS $= \sum_{i=1}^n (y_i - x_i^t \widehat{\beta})^2 / s_i^2$. We note that PRESS only searches for models among those that have linear mean functions and constant variance, and that sometimes alternatives with heterogeneous variability are more important; see e.g. Section 5.6.

2.10 Outlier-robust methods

We start with an example illustrating the effect that outlying observations can have on model selection.

Example 2.14 How to repair for Ervik's 1500-m fall?

In the 2004 European Championships for speedskating, held in Heerenveen, the Netherlands, the Norwegian participant Eskil Ervik unfortunately fell in the third distance, the 1500-m. This cost Norway a lost spot at the World Championships later that season. In the result list, his 2:09.20 time, although an officially registered result, is accordingly a clear outlier, cf. Figure 2.4. Including it in a statistical analysis of the results might give misleading results. We shall work with the times in seconds computed for the two distances, and will try to relate the 1500-m time via a linear regression model to the 5000-m time. We perform our model fitting using results from the 28 skaters who completed the two distances. Including Ervik's result seriously disturbs the maximum likelihood estimators here. We fit polynomial regression models, linear, quadratic, cubic and quartic, $Y = \beta_0 + \beta_1 x + \cdots + \beta_4 x^4 + \varepsilon$, where x and Y are 1500-m time and 5000-m time, respectively, and assuming the errors to come from a $N(0, \sigma^2)$ distribution with unknown variance.

Table 2.6 contains the results of applying AIC to these data. When all observations are included, AIC ranks the linear model last, and prefers a quadratic model. When leaving out the outlying observation, the linear model is deemed best by AIC. For model selection purposes, the 'best model' should be considered the one which is best for non-outlying data. The estimated linear trend is very different in these two cases. Without the

Table 2.6. *AIC for selecting a polynomial regression model for the speedskating data of the 2004 championship: First with all 28 skaters included, next leaving out the result of Ervik who fell in the 1500-m.*

Model:	linear	quadratic	cubic	quartic
AIC (all skaters):	−227.131	−213.989	−215.913	−217.911
preference order:	(4)	(1)	(2)	(3)
AIC (without Ervik):	−206.203	−207.522	−209.502	−210.593
preference order:	(1)	(2)	(3)	(4)

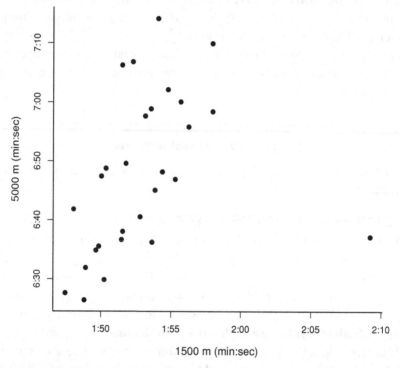

Fig. 2.4. Ervik's fall in the 2004 European Championship: 1500-m versus 5000-m results, with Ervik's outlying 1500-m time very visible to the right.

outlying observation the estimated slope is 3.215, while with Ervik included, the value is only 1.056, pushing the line downwards. Since the estimators are so much influenced by this single outlying observation, also the model selection methods using such estimators suffer. In this example it is quite obvious that Ervik's time is an outlier (not because of his recorded time per se, but because he fell), and that this observation can be removed before the analysis. In general, however, it might not always be so clear which observations are outliers; hence outlier-robust model selection methods are called for. ∎

2.10.1 AIC from weighted penalised likelihood methods

Let us consider a linear regression model of the form $Y = X\beta + \varepsilon$, for a p-vector of regression coefficients β, where the errors ε have common variance σ^2, which is unknown. We wish to select components of β. For normal regression models, where $\varepsilon_i \sim \mathrm{N}(0, \sigma^2)$, maximum likelihood methods coincide with least squares estimators. In general, both maximum likelihood estimators and least squares estimators are known to be non-robust against outlying observations in many models. This means that a few observations far from the remaining part of the observations may seriously influence estimators constructed from the data. Consequently a model selected using methods based on maximum likelihood might select a model that does not fit well, or one that may be too complex for the majority of the data. Agostinelli (2002) studies weighted versions of likelihood estimators to construct weighted model selection criteria. These methods are available in the R package `wle`.

While maximum likelihood estimators for $\theta = (\beta, \sigma)$ are found via solving the set of score equations $\sum_{i=1}^{n} u(Y_i \mid x_i, \beta, \sigma) = 0$, where $u(y \mid x, \theta) = \partial \log f(y \mid x, \theta)/\partial\theta$, weighted maximum likelihood estimators solve a set of equations of the form

$$\sum_{i=1}^{n} w(y_i - x_i^{\mathrm{t}}\beta, \sigma)u(y_i \mid x_i, \beta, \sigma) = 0. \tag{2.37}$$

We now summarise details of the construction of the weights, see Agostinelli (2002). With $\widehat{e}_i$ denoting smoothed Pearson residuals (see below), the weight functions $w(\cdot)$ are defined as $w(y_i - x_i^{\mathrm{t}}\beta, \sigma) = \min\{1, \max(0, r(\widehat{e}_i) + 1)/(\widehat{e}_i + 1)\}$. The option $r(a) = a$ brings us back to classical maximum likelihood estimators, since then all weights equal one. Some robust choices for the residual adjustment function r are discussed in Agostinelli (2002). Instead of the usual residuals $e_i = y_i - x_i^{\mathrm{t}}\beta$, smoothed Pearson residuals are used to construct the weights. These are defined as $\widehat{e}_i = \widehat{f}_e(e_i)/\widehat{m}(e_i) - 1$. More precisely, for a kernel function K, a given univariate density function like the standard normal, the kernel density estimator of the residuals $e_i = y_i - x_i^{\mathrm{t}}\beta$ is defined as $\widehat{f}_e(t) = n^{-1}\sum_{i=1}^{n} h^{-1}K(h^{-1}(e_i - t))$, where h is a bandwidth parameter (see for example Wand and Jones, 1995). And for a normal regression model, the kernel smoothed model density equals $\widehat{m}(t) = \int h^{-1}K(h^{-1}(t - s))\phi(s, \sigma^2)\,\mathrm{d}s$; here $\phi(s, \sigma^2)$ is the $\mathrm{N}(0, \sigma^2)$ density. When K is the standard normal kernel, $\widehat{m}(t) = \phi(t, \sigma^2 + h^2)$.

Suggestions have been made for adjusting model selection methods based on the likelihood function with weights constructed as above, to lead to robust selection criteria. For example, Agostinelli (2002) proposes to modify the AIC $= 2\,\ell_n(\widehat{\theta}) - 2(p + 1)$ of (2.14) to

$$\mathrm{wleAIC} = 2\sum_{i=1}^{n} w(y_i - x_i^{\mathrm{t}}\widehat{\beta}, \widehat{\sigma}) \log f(y_i \mid x_i, \widehat{\beta}, \widehat{\sigma}) - 2(p + 1).$$

The parameters are estimated by solving (2.37). We return to Example 2.14 and apply the wleAIC using the R function `wle.aic` with its built-in default choice for selection of the bandwidth h. This leads to the following wleAIC values for the four models (linear to quartic): $-199.033, -208.756, -218.612, -228.340$. These are quite sensible values in view of Table 2.6, and in particular indicate preference for the simplest linear model. The wleAIC analysis agrees also with the robust model selection method that we discuss in the following subsection.

2.10.2 Robust model selection from weighted Kullback–Leibler distances

The methods discussed in the previous subsection involve robust weighting of the log-likelihood contributions, but maximising the resulting expression is not necessarily a fruitful idea for models more general than the linear regression one. We shall in fact see below that when log-likelihood terms are weighted, then, in general, a certain additional parameter-dependent factor needs to be taken into account, in order for the procedures to have a minimum distance interpretation.

Let g be the true data-generating density and f_θ a candidate model. For any non-negative weight function $w(y)$,

$$d_w(g, f_\theta) = \int w\{g \log(g/f_\theta) - (g - f_\theta)\} \, dy \qquad (2.38)$$

defines a divergence, or distance from the true g to the model density f_θ; see Exercise 2.14. When w is constant, this is equivalent to the Kullback–Leibler distance. Minimising the d_w distance is the same as maximising $H(\theta) = \int w(g \log f_\theta - f_\theta) \, dy$. This invites the maximum weighted likelihood estimator $\widehat{\theta}$ that maximises

$$H_n(\theta) = n^{-1} \sum_{i=1}^{n} w(y_i) \log f(y_i, \theta) - \int w f_\theta \, dy. \qquad (2.39)$$

This is an estimation method worked with by several authors, from different perspectives, including Hjort (1994b), Hjort and Jones (1996), Loader (1996) and Eguchi and Copas (1998).

In principle any non-negative weight function may be used, for example to bring extra estimation efforts into certain regions of the sample space or to downweight more extreme values. Robustness, in the classical first-order sense of leading to bounded influence functions, is achieved for any weight function $w(y)$ with the property that $w(y)u(y, \theta)$ is bounded at the least false value θ_0 that minimises d_w; see below. We note that simply maximising the weighted likelihood function $\sum_{i=1}^{n} w(y_i) \log f(y_i, \theta)$ will not work without the correction term $-\int w f_\theta \, dy$, as it may easily lead to inconsistent estimators.

The argmax $\widehat{\theta}$ of H_n may be seen as a minimum distance estimator associated with the generalised Kullback–Leibler distance d_w. As such it is also an M-estimator, obtained by solving $H_n^{(1)}(\theta) = 0$, where $H_n^{(1)}(\theta) = n^{-1} \sum_{i=1}^{n} w(y_i)u(y_i, \theta) - \xi(\theta)$ is the derivative of H_n; here $\xi(\theta) = \int w f_\theta u_\theta \, dy$ is the derivative of $\int w f_\theta \, dy$ with respect to θ. Here and below we let the '(j)' indicate the jth partial derivative with respect to θ, for $j = 1, 2$. The estimator is consistent for the least false value θ_0 that minimises the d_w distance from truth to model density. Note that $H_n^{(1)}(\theta)$ converges in probability to $\int w(g - f_\theta)u_\theta \, dy$, so the least false θ_0 may also be characterised as the solution to $\int w g u_\theta \, dy = \int w f_\theta u_\theta \, dy$. Next consider the second derivative matrix function

$$H_n^{(2)}(\theta) = n^{-1} \sum_{i=1}^{n} w(y_i) I(y_i, \theta) - \int w f_\theta u_\theta u_\theta^t \, dy - \int w f_\theta I_\theta \, dy,$$

and define $J_n = -H_n^{(2)}(\theta_0)$. Here $J_n \to_p J$ as n grows, where $J = \int w f_\theta u_\theta u_\theta^t \, dy + \int w(f_\theta - g)I_\theta \, dy$, evaluated at $\theta = \theta_0$. Furthermore, $\sqrt{n} H_n^{(1)}(\theta_0) \to_d U' \sim N_p(0, K)$, by the central limit theorem, with

$$K = \text{Var}_g\{w(Y)u(Y, \theta_0) - \xi(\theta_0)\} = \int g(y)w(y)^2 u(y, \theta_0)u(y, \theta_0)^t \, dy - \xi\xi^t,$$

and where $\xi = \int g(y)w(y)u(y, \theta_0) \, dy = \int w(y)f(y, \theta_0)u(y, \theta_0) \, dy$. With these definitions,

$$\sqrt{n}(\widehat{\theta} - \theta_0) = J^{-1}\sqrt{n} H_n^{(1)}(\theta_0) + o_P(1) \overset{d}{\to} J^{-1}U' \sim N_p(0, J^{-1}KJ^{-1}). \qquad (2.40)$$

The influence function for the associated $\widehat{\theta} = T(G_n)$ functional, where $T(G)$ is the argmax of $\int w(g \log f_\theta - f_\theta) \, dy$, is $\text{infl}(G, y) = J^{-1}\{w(y)u(y, \theta_0) - \xi(\theta_0)\}$, see Exercise 2.14.

To judge the prediction quality of a suggested parametric model f_θ, via the weighted Kullback–Leibler distance d_w of (2.38), we need to assess the expected distance $d_w(g(\cdot), f(\cdot, \widehat{\theta}))$ in comparison with other competing models. This quantity is a constant away from $Q_n = E_g R_n$, where $R_n = \int w\{g \log f(\cdot, \widehat{\theta}) - f(\cdot, \widehat{\theta})\} \, dy$, with the outer expectation in Q_n referring to the distribution of the maximum weighted likelihood estimator $\widehat{\theta}$. We have $Q_n = E_g\{w(Y_{\text{new}}) \log f(Y_{\text{new}}, \widehat{\theta}) - \int w(y)f(y, \widehat{\theta}) \, dy\}$, with Y_{new} denoting a new observation generated from the same distribution as the previous ones. A natural estimator is therefore the cross-validated leave-one-out statistic

$$\widehat{Q}_{n,\text{xv}} = n^{-1} \sum_{i=1}^{n} w(Y_i) \log f(Y_i, \widehat{\theta}_{(i)}) - \int w(y)f(y, \widehat{\theta}) \, dy, \qquad (2.41)$$

where $\widehat{\theta}_{(i)}$ is the maximum weighted likelihood estimator found from the reduced data set that omits the ith data point.

We show now that the (2.41) estimator is close to a penalised version of the directly available statistic $\widehat{Q}_n = H_n(\widehat{\theta}) = H_{n,\text{max}}$ associated with (2.39). Arguments similar to those used to exhibit a connection from cross-validation to AIC lead under the present circumstances first to

$$\widehat{\theta}_{(i)} - \widehat{\theta} \doteq -n^{-1}\mathrm{infl}(G_n, y_i) \doteq -n^{-1}\widetilde{J}_n^{-1}\{w(y_i)u(y_i, \widehat{\theta}) - \xi(\widehat{\theta})\},$$

where $\widetilde{J}_n = -H_n^{(2)}(\widehat{\theta})$, and then to

$$\widehat{Q}_{n,\mathrm{xv}} \doteq H_n(\widehat{\theta}) + n^{-1}\sum_{i=1}^{n} w(y_i)u(y_i, \widehat{\theta})^{\mathrm{t}}(\widehat{\theta}_{(i)} - \widehat{\theta}) \doteq H_n(\widehat{\theta}) - n^{-1}\mathrm{Tr}(\widetilde{J}_n^{-1}\widetilde{K}_n),$$

in terms of $\widetilde{K}_n = n^{-1}\sum_{i=1}^{n} w(y_i)^2 u(y_i, \widehat{\theta})u(y_i, \widehat{\theta})^{\mathrm{t}} - \xi(\widehat{\theta})\xi(\widehat{\theta})^{\mathrm{t}}$. This is a generalisation of results derived and discussed in Section 2.9.

The approximation to the leave-one-out statistic (2.41) above can be made to resemble AIC, by multiplying with $2n$, so we define the w-weighted AIC score as

$$\begin{aligned}
\mathrm{wAIC} &= 2n H_{n,\mathrm{max}} - 2\widetilde{p}^* \\
&= 2\left\{\sum_{i=1}^{n} w(y_i)\log f(y_i, \widehat{\theta}) - n\int w(y)f(y, \widehat{\theta})\,\mathrm{d}y\right\} - 2\widetilde{p}^*,
\end{aligned} \tag{2.42}$$

where now $\widetilde{p}^* = \mathrm{Tr}(\widetilde{J}_n^{-1}\widetilde{K}_n)$. This appropriately generalises the model-robust AIC, which corresponds to $w = 1$. The wAIC selection method may in principle be put to work for any non-negative weight function w, but its primary use is in connection with weight functions that downscale the more extreme parts of the sample space, to avoid the sometimes severely non-robust aspects of ordinary maximum likelihood and AIC strategies.

There is another path leading to wAIC of (2.42), more akin to the derivation of the AIC formula given in Section 2.3. The task is to quantify the degree to which the simple direct estimator $\widehat{Q}_n = H_n(\widehat{\theta}) = H_{n,\mathrm{max}}$ needs to be penalised, in order to achieve approximate unbiasedness for $Q_n = \mathrm{E}_g R_n$. Translating arguments that led to (2.16) to the present more general case of weighted likelihood functions, one arrives after some algebra at

$$\widehat{Q}_n - R_n = \bar{Z}_n + H_n^{(1)}(\theta_0)^{\mathrm{t}}(\widehat{\theta} - \theta_0) + o_P(n^{-1}) = \bar{Z}_n + n^{-1}V_n^{\mathrm{t}}JV_n + o_P(n^{-1}),$$

with $V_n = \sqrt{n}(\widehat{\theta} - \theta_0)$, where $\bar{Z}_n$ is the average of the zero mean variables $Z_i = w(Y_i)\log f(Y_i, \theta_0) - \int g(y)w(y)\log f(y, \theta_0)\,\mathrm{d}y$, and where (2.40) is used. This result, in conjunction with $W_n = V_n^{\mathrm{t}}JV_n \to_d W = V^{\mathrm{t}}JV$, where $V = J^{-1}U'$ and $U' \sim \mathrm{N}_p(0, K)$. Estimating the mean of W_n with $\mathrm{Tr}(\widetilde{J}_n^{-1}\widetilde{K}_n)$ leads therefore, again, to wAIC of (2.42), as the natural generalisation of AIC.

Just as we were able in Section 2.9 to generalise easier results for the i.i.d. model to the class of regression models, we learn here, with the appropriate efforts, that two paths of arguments both lead to the robust model selection criterion

$$\mathrm{wAIC} = 2n H_n(\widehat{\theta}) - 2\widetilde{p}^* = 2n H_{n,\mathrm{max}} - 2\widetilde{p}^*. \tag{2.43}$$

Here the criterion function to maximise is

$$H_n(\theta) = n^{-1} \sum_{i=1}^{n} w(x_i, y_i) \log f(y_i \mid x_i, \theta) - n^{-1} \sum_{i=1}^{n} \int w(x_i, y) f(y \mid x_i, \theta) \, dy,$$

and

$$\widetilde{p}^* = \mathrm{Tr}(\widetilde{J}_n^{-1} \widetilde{K}_n), \quad \text{with } \widetilde{J}_n = -H_n^{(2)}(\widehat{\theta}) \text{ and } \widetilde{K}_n = n^{-1} \sum_{i=1}^{n} v_i v_i^t,$$

where we write $v_i = w(x_i, y_i) u(y_i \mid x_i, \widehat{\theta}) - \xi(\widehat{\theta} \mid x_i)$ and $\xi(\theta \mid x_i) = \int w(x_i, y) f(y \mid x_i, \theta) u(y \mid x_i, \theta) \, dy$. The traditional AIC is again the special case of a constant weight function.

Example 2.15 (2.14 continued) How to repair for Ervik's 1500-m fall?

To put the robust weighted AIC method into action, we need to specify a weight function $w(x, y)$ for computing the maxima and penalised maxima of the $H_n(\beta, \sigma)$ criterion functions. For this particular application we have chosen

$$w(x, y) = w_1 \Big(\frac{x - \mathrm{med}(x)}{\mathrm{sd}(x)} \Big) w_2 \Big(\frac{z - \mathrm{med}(z)}{\mathrm{sd}(z)} \Big),$$

involving the median and standard deviations of the x_i and the z_i, where $z_i = y_i - (5/1.5) x_i$, for appropriate w_1 and w_2 functions that are equal to 1 inside standard areas but scale down towards zero for values further away. Specifically, $w_1(u)$ is 1 to the left of a threshold value λ and $(\lambda/u)^2$ to the right of λ; while $w_2(u)$ is 1 inside $[-\lambda, \lambda]$ and $(\lambda/|u|)^2$ outside. In other applications we might have let also $w_1(u)$ be symmetric around zero, but in this particular context we did not wish to lower the value of $w_1(u)$ from its standard value 1 to the left, since unusually strong performances (like national records) should enter the analysis without any down-sizing in importance, even when they might look like statistical outliers. The choice of the $1/u^2$ factor secures that the estimation methods used have bounded influence functions for all y and for all x outlying to the right, with respect to both mean and standard deviation parameters in the linear-normal models. The threshold value λ may be selected in different ways; there are, for example, instances in the robustness literature where such a value is chosen to achieve say 95% efficiency of the associated estimation method, if the parametric model is correct. For this application we have used $\lambda = 2$.

We computed robust wAIC scores via the (2.43) formula, for each of the candidate models, operating on the full data set with the $n = 28$ skaters that include the fallen Ervik. This required using numerical integration and minimisation algorithms. Values of H_n maxima and $\widetilde{p}^*$ penalties are given in Table 2.7, and lead to the satisfactory conclusion that the four models are ranked in exactly the same order as for AIC used on the reduced $n = 27$ data set that has omitted the fallen skater. In this case the present wAIC and the wleAIC analysis reported on above agree fully on the preference order for the four

Table 2.7. *AIC and robust weighted AIC for selecting a polynomial regression model for predicting the 5000-m time from the 1500-m time, with preference order, with data from the 2004 European Championships. The four models have 3, 4, 5, 6 parameters, respectively.*

Model:	linear	quadratic	cubic	quartic
AIC (all skaters):	−227.130 (4)	−213.989 (1)	−215.913 (2)	−217.911 (3)
AIC (w/o Ervik):	−206.203 (1)	−207.522 (2)	−209.502 (3)	−210.593 (4)
$n H_{n,\max}$:	−100.213	−99.782	−99.781	−99.780
$\widetilde{p}^*$	2.312	2.883	3.087	3.624
wAIC (all skaters):	−205.049 (1)	−205.329 (2)	−205.747 (3)	−206.811 (4)

models, and also agree well on the robustly estimated selected model: the wleAIC method yields estimates 3.202, 9.779 for slope b and spread parameter σ in the linear model, while the wAIC method finds estimates 3.193, 9.895 for the same parameters. ■

2.10.3 Robust AIC using M-estimators

Solutions to the direct likelihood equations lead to non-robust estimators for many models. We here abandon the likelihood as a starting point and instead start from M-estimating equations. As in Section 2.10.1 we restrict attention to linear regression models.

An M-estimator $\widehat{\beta}$ for the regression coefficients β is the solution to an equation of the form

$$\sum_{i=1}^{n} \Psi(x_i, y_i, \beta, \sigma) = \sum_{i=1}^{n} \eta(x_i, (y_i - x_i^t \beta)/\sigma) x_i = 0, \qquad (2.44)$$

see Hampel *et al.* (1986, Section 6.3). An example of such a function η is Huber's ψ function, in which case $\eta(x_i, \varepsilon_i/\sigma) = \max\{-1.345, \min(\varepsilon_i/\sigma, 1.345)\}$, using a certain default value associated with a 0.95 efficiency level under normal conditions. With this function we define weights for each observation, $\widehat{w}_i = \eta(x_i, (y_i - x_i^t \widehat{\beta})/\sigma)/\{(y_i - x_i^t \widehat{\beta})/\sigma\}$. Ronchetti (1985) defines a robust version of AIC, though not via direct weighting. The idea is the following. AIC is based on a likelihood function, and maximum likelihood estimators are obtained (in most regular cases) by solving the set of score equations $\sum_{i=1}^{n} u(Y_i \mid x_i, \beta, \sigma) = 0$. M-estimators solve equations (2.44). This set of equations can (under some conditions) be seen as the derivative of a function τ, which then might take the role of the likelihood function in AIC, to obtain Ronchetti's

$$\text{AICR} = 2 \sum_{i=1}^{n} \tau\big((y_i - x_i^t \widehat{\beta})/\widehat{\sigma}\big) - 2 \operatorname{Tr}(\widehat{J}_n^{-1} \widehat{K}_n),$$

where here $\widehat{J}_n$ and $\widehat{K}_n$ are estimators of $J = \mathrm{E}(-\partial\Psi/\partial\theta)$ (with $\theta = (\beta, \sigma)$) and $K = \mathrm{E}(\Psi\Psi^t)$. Note that the type of penalty above is also used for the model-robust TIC, see (2.20), though there used with the likelihood instead of τ. For the set of equations (2.44) with Huber's ψ function, the corresponding τ function is equal to (see Ronchetti, 1985)

$$\tau\big((y_i - x_i^t\widehat{\beta})/\widehat{\sigma}\big) = \begin{cases} \frac{1}{2}(y_i - x_i^t\widehat{\beta})^2/\widehat{\sigma}^2 & \text{if } |y_i - x_i^t\widehat{\beta}| < 1.345\widehat{\sigma}, \\ 1.345|y_i - x_i^t\widehat{\beta}|/\widehat{\sigma} - 1.345^2/2 & \text{otherwise.} \end{cases}$$

For a further overview of these methods, see Ronchetti (1997).

2.10.4 The generalised information criterion

Yet another approach for robustification of AIC by changing its penalty is obtained via the generalised information criterion (GIC) of Konishi and Kitagawa (1996). This criterion is based on the use of influence functions and can not only be applied for robust models, but also works for likelihood-based methods as well as for some Bayesian procedures.

We use notation for influence functions as in Section 2.9. Denote with G the true distribution of the data and G_n the empirical distribution. The data are modelled via a density function $f(y, \theta)$ with corresponding distribution F_θ. The GIC deals with functional estimators of the type $\widetilde{\theta} = T(G_n)$, for suitable estimator functionals $T = T(G)$. Maximum likelihood is one example, where the T is the minimiser of the Kullback–Leibler distance from the density of G to the parametric family. We assume that $T(F_\theta) = \theta$, i.e. T is consistent at the model itself, and let $\theta_{0,T} = T(G)$, the least false parameter value as implicitly defined by the $T(G_n)$ estimation procedure. To explain the GIC, we return to the derivation of AIC in Section 2.3. The arguments that led to $\mathrm{E}_g(\widehat{Q}_n - Q_n) = p^*/n + o(1/n)$ of (2.17), for the maximum likelihood estimator, can be used to work through the behaviour of $\widetilde{Q}_n = n^{-1}\ell_n(\widetilde{\theta})$ (which is now not the normalised log-likelihood maximum, since we employ a different estimator). Konishi and Kitagawa (1996) found that $\mathrm{E}_g(\widetilde{Q}_n - Q_n) = p^*_{\mathrm{infl}}/n + o(1/n)$, where

$$p^*_{\mathrm{infl}} = \mathrm{Tr}\left\{\int \mathrm{infl}(G, y)u(y, \theta_{0,T})^t \, \mathrm{d}G(y)\right\},$$

with infl the influence function of $T(G)$, see (2.31). The GIC method estimates p^*_{infl} to arrive at

$$\mathrm{GIC} = 2\ell_n(\widetilde{\theta}) - 2\sum_{i=1}^{n} \mathrm{Tr}\{\mathrm{infl}(G_n, Y_i)^t u(Y_i, \widetilde{\theta})\}.$$

For M-estimators $\widetilde{\theta} = T(G_n)$ which solve a set of equations $\sum_{i=1}^{n} \Psi(Y_i, \widetilde{\theta}) = 0$, the influence function can be shown to be

$$\mathrm{infl}(G, y) = J_T^{-1}\Psi(y, T(G)) = \left\{-\int \Psi^*(y, \theta_{0,T}) \, \mathrm{d}G(y)\right\}^{-1}\Psi(y, \theta_{0,T}),$$

where $\Psi^*(y, \theta) = \partial \Psi(y, \theta) / \partial \theta$. This makes p_{infl}^* take the form $\text{Tr}(J_T^{-1} K_T)$, with $K_T = \int \Psi(y, \theta_{0,T}) u(y, \theta_{0,T})^t \, dy$. Clearly the maximum likelihood method is a special case, corresponding to using $\Psi(y, \theta) = u(y, \theta)$, in which case p_{infl}^* coincides with p^* of (2.17) and GIC coincides with TIC of Section 2.5.

The above brief discussion was kept to the framework of i.i.d. models for observations $y_1, \ldots, y_n$, but the methods and results can be extended to regression models without serious obstacles. There are various M-estimation schemes that make the resulting $\widetilde{\theta}$ more robust than the maximum likelihood estimator. Some such are specifically constructed for robustification of linear regression type models, see e.g. Hampel *et al.* (1986), while others are more general in spirit and may work for any parametric models, like the minimum distance method of Section 2.10.2; see also Basu *et al.* (1998) and Jones *et al.* (2001).

Remark 2.1 One or two levels of robustness

Looking below the surface of the simple AIC formula (2.14), we have learned that AIC is inextricably linked to the Kullback–Leibler divergence in two separate ways; it aims at estimating the expected Kullback–Leibler distance from the true data generator to the parametric family, *and* it uses maximum likelihood estimators. The GIC sticks to the non-robust Kullback–Leibler as the basic discrepancy measure for measuring prediction quality, but inserts robust estimates for the parameters. Reciprocally, one might also consider the strategy of using a robustified distance measure, like that of (2.38), but inserting maximum likelihood estimators when comparing their minima. It may however appear more natural to simultaneously robustify both steps of the process, which the generalised Kullback–Leibler methods of Section 2.10.2 manage to do in a unified manner. ∎

2.11 Notes on the literature

Maximum likelihood theory belongs to the core of theoretical and applied statistics; see e.g. Lehmann (1983) and Bickel and Doksum (2001) for comprehensive treatments. The standard theory associated with likelihoods relies on the assumption that the true data-generating mechanism is inside the parametric model in question. The necessary generalisations to results that do not require the parametric model to hold appeared some 50 years later than the standard theory, but are now considered standard, along with terms like least false parameter values and the sandwich matrix; see e.g. White (1982, 1994); Hjort (1986b, 1992a).

Applications of AIC have a long history. A large traditional application area is time series analysis, where one has studied how best to select the order of autoregressive and autoregressive moving average models. See for example Shibata (1976), Hurvich and Tsai (1989) or general books on time series analysis such as Brockwell and Davis (2002),

Chatfield (2004). The book by McQuarrie and Tsai (1998) deals with model selection in both time series and regression models. For multiple regression models, see also Shibata (1981) or Nishii (1984). Burnham and Anderson (2002) deal extensively with AIC, also in the model averaging context. The form of the penalty in TIC has also been observed by Stone (1977).

Criteria related to AIC have sometimes been renamed to reflect the particular application area. The network information criterion (NIC) is such an example (Murata *et al.*, 1994), where AIC is adjusted for application with model selection in neural networks. The question of selecting the optimal number of parameters here corresponds to whether or not more neurons should be added to the network. Uchida and Yoshida (2001) construct information criteria for stochastic processes based on estimated Kullback–Leibler information for mixing processes with a continuous time parameter. As examples they include diffusion processes with jumps, mixing point processes and nonlinear time series models.

The GIC of Konishi and Kitagawa (1996) specified to M-estimators belongs to the domain of robust model selection criteria. Robust estimation is treated in depth in the books by Huber (1981) and Hampel *et al.* (1986). See also Carroll and Ruppert (1988) for an overview. Choi *et al.* (2000) use empirical tilting for the likelihood function to make maximum likelihood methods more robust. An idea of reweighted down-scaled likelihoods is explored in Windham (1995). Different classes of minimum distance estimators that contain Windham's method as a special case are developed in Basu *et al.* (1998) and Jones *et al.* (2001). Ronchetti and Staudte (1994) construct a robust version of Mallow's C_p criterion, see Chapter 4. Several of the robust model selection methods are surveyed in Ronchetti (1997). We refer to that paper for more references. Ronchetti *et al.* (1997) develop a robust version of cross-validation for model selection. The sample is split in two parts, one part is used for parameter estimation and outliers are here dealt with by using optimal bounded influence estimators. A robust criterion for prediction error deals with outliers in the validation set. Müller and Welsh (2005) use a stratified bootstrap to estimate a robust conditional expected prediction loss function, and combine this with a robust penalised criterion to arrive at a consistent model selection method. Qian and Künsch (1998) perform robust model selection via stochastic complexity. For a stochastic complexity criterion for robust linear regression, see Qian (1999).

Extensions of AIC that deal with missing data are proposed by Shimodaira (1994), Cavanaugh and Shumway (1998), Hens *et al.* (2006) and Claeskens and Consentino (2008); see also Section 10.4.

Hurvich *et al.* (1998) developed a version of the AIC_c for use in nonparametric regression to select the smoothing parameter. This is further extended for use in semiparametric and additive models by Simonoff and Tsai (1999). Hart and Yi (1998) develop one-sided cross-validation to find the smoothing parameter. Their approach only uses data at one side of x_i when predicting the value for Y_i.

Exercises

2.1 *The Kullback–Leibler distance:*

   (a) Show that $\mathrm{KL}(g, f) = \int g \log(g/f)\,\mathrm{d}y$ is always non-negative, and is equal to zero only if $g = f$ almost everywhere. (One may e.g. use the Jensen inequality which states that for a convex function h, $\mathrm{E}\,h(X) \geq h(\mathrm{E}X)$, and strict inequality holds when h is strictly convex and X is nondegenerate.)

   (b) Find the KL distance from a $\mathrm{N}(\xi_1, \sigma_1^2)$ to a $\mathrm{N}(\xi_2, \sigma_2^2)$. Generalise to the case of d-dimensional normal distributions.

   (c) Find the KL distance from a $\mathrm{Bin}(n, p)$ model to a $\mathrm{Bin}(n, p')$ model.

2.2 *Subset selection in linear regression:* Show that in a linear regression model $Y_i = \beta_0 + \beta_1 x_{i,1} + \cdots + \beta_p x_{i,p} + \varepsilon_i$, with independent normally distributed errors $\mathrm{N}(0, \sigma^2)$, AIC is given by

$$\mathrm{AIC}(p) = -n \log(\mathrm{SSE}_p/n) - n\{1 + \log(2\pi)\} - 2(p + 2),$$

where SSE_p is the residual sum of squares. Except for the minus sign, this is the result of an application of the function $\mathtt{AIC()}$ in R. The R function $\mathtt{stepAIC()}$, on the other hand, uses as its AIC the value $\mathrm{AIC}_{\mathrm{step}}(p) = n \log(\mathrm{SSE}_p/n) + 2(p + 1)$. Verify that the maximiser of $\mathrm{AIC}_{\mathrm{step}}(p)$ is identical to the maximiser of $\mathrm{AIC}(p)$ over p.

2.3 *ML and AIC computations in R:* This exercise is meant to make users familiar with ways of finding maximum likelihood estimates and AIC scores for given models via algorithms in R.

   Consider the Gompertz model first, the following is all that is required to compute estimates and AIC score. One first defines the log-likelihood function

```
logL = function(para, x) {
a = para[1]
b = para[2]
return(sum(log(a) + b * x - (a/b) * (exp(b * x) - 1)))}
```

   where x is the data set, and then proceeds to its maximisation, using the nonlinear minimisation algorithm $\mathtt{nlm}$, perhaps involving some trial and error for finding an effective starting point:

```
minuslogL = function(para, x){-logL(para, x)}
nlm(minuslogL, c(0.03, 0.03), x)
```

   This then leads to $\mathtt{parahat} = \mathtt{c(0.0188, 0.0207)}$, and afterwards to $\mathtt{maxlogL} = \mathtt{logL(parahat, x)}$ and $\mathtt{aic} = \mathtt{2 * maxlogL - 4}$. Carry out analysis, as above, for the other models discussed in Example 2.6.

2.4 *Model-robust AIC for logistic regression:* Obtain TIC for the logistic regression model

$$P(Y = 1 \mid x, z) = \frac{\exp(x^t\beta + z^t\gamma)}{1 + \exp(x^t\beta + z^t\gamma)},$$

where $x = (1, x_2)^t$ and $z = (x_3, x_4, x_5)^t$. Verify that the approximation of $\mathrm{Tr}(J^{-1}K)$ by k, the number of free parameters in the model, leads to the AIC expression for logistic regression as in Example 2.4. Apply TIC to the low birthweight data, including in the model the same variables as used in Example 2.4.

2.5 *The AIC$_c$ for linear regression:* Verify that for the linear regression models $Y_i = \beta_0 + \beta_1 x_{i,1} + \cdots + \beta_p x_{i,p} + \varepsilon_i$, with independent normally distributed errors $N(0, \sigma^2)$, the corrected AIC is

$$\mathrm{AIC}_c = \mathrm{AIC} - 2(p+2)(p+3)/(n-p-3).$$

2.6 *Two AIC correction methods for linear regression:* Simulate data from a linear regression model with say $p = 5$ and $n = 15$, and implement in addition to the AIC score both correction methods AIC$_c$ of (2.23) and AIC$_c^*$ of (2.24). Compare values of these three scores and their ability to pick the 'right' model in a little simulation study. Try to formalise what it should mean that one of the two correction methods works better than the other.

2.7 *Stackloss:* Read the stackloss data in R by typing data(stackloss). Information on the variables in this data set is obtained by typing ?stackloss. We learn that there are three regression variables 'Air.Flow', 'Water.Temp' and 'Acid.Conc.'. The response variable is 'stack.loss'.

  (a) Consider first a model without interactions, only main effects. With three regression variables, there are $2^3 = 8$ possible models to fit. Fit separately all eight models (no regression variables included; only one included; only two included; and all three included). For each model obtain AIC (using function AIC()). Make a table with these values and write down the model which is selected.

  (b) Consider the model with main effects and pairwise interactions. Leave all main effects in the model, and search for the best model including interactions. This may be done using the R function stepAIC in library(MASS).

  (c) Now also allow for the main effects to be left out of the final selected model and search for the best model, possibly including interactions. The function stepAIC may again be used for this purpose.

2.8 *Switzerland in 1888:* Perform model selection for the R data set called swiss, which gives a standardised fertility measure and socio-economic indicators for each of the 47 French-speaking provinces of Switzerland, at around 1888. The response variable is called 'fertility', and there are five regression variables. Type ?swiss in R for more information about the data set.

2.9 *The Ladies Adelskalenderen:* Consider the data from the Adelskalenderen for ladies' speed-skating (available from the book's website). Times on four distances 500 m, 1500 m, 3000 m and 5000-m are given. See Section 1.7 for more information, there concerning the data set for men. Construct a scatterplot of the 1500-m time versus the 500-m time, and of the 5000-m time versus the 3000-m time. Try to find a good model to estimate the 1500-m time from the 500-m time, and a second model to estimate the 5000-m time from the 3000-m time. Suggest several possible models and use AIC and AIC$_c$ to perform the model selection.

2.10 *Hofstedt's highway data:* This data set is available via data(highway) in the R library alr3 (see also Weisberg, 2005, Section 7.2). There are 39 observations on several highway-related measurements. The response value of interest is Rate, which is the accident rate per million vehicle miles in the year 1973. Eleven other variables $(x_1, \ldots, x_{11})$ are possibly needed in a linear regression model with Rate as response.

(a) Fit the full model $Y_i = \beta_0 + \beta_1 x_{i,1} + \cdots + \beta_{11} x_{i,11} + \varepsilon_i$ using robust regression with M-estimators for the regression coefficients, using Huber's ψ function. The function `rlm` in the R library `MASS` may be used for this purpose. Obtain for each observation the value of the weight $\widehat{w}_i$ using the residuals of the fitted full model. Weights below one indicate downweighted observations. Identify these cases.

(b) Use AICR to perform variable selection.

2.11 *Influence function for maximum likelihood:* For a regular parametric family $f(y, \theta)$, consider the maximum likelihood functional $T = T(G)$ that for a given density g (e.g. the imagined truth) finds the least false value $\theta_0 = \theta_0(g)$ that minimises the Kullback–Leibler distance $\int g \log(g/f_\theta) \, dy$. Two continuous derivatives are assumed to exist, as is the matrix $J = -\int g(y) \mathrm{infl}(y, \theta_0) \, dy$, taken to be positive definite.

(a) Show that $T(G)$ also may be expressed as the solution θ_0 to $\int g(y) u(y, \theta_0) \, dy = 0$.

(b) Let y be fixed and consider the distribution $G_\varepsilon = (1 - \varepsilon)G + \varepsilon \delta(y)$, where $\delta(y)$ denotes unit point mass at position y. Write $\theta_\varepsilon = \theta_0 + z$ for the least false parameter when f_θ is compared to G_ε. Show that this is the solution to

$$(1 - \varepsilon) \int g(y') u(y', \theta_\varepsilon) \, dy' + \varepsilon u(y, \theta_\varepsilon) = 0.$$

By Taylor expansion, show that z must be equal to $J^{-1} u(y, \theta_0) \varepsilon$ plus smaller terms. Show that this implies $\mathrm{infl}(G, y) = J^{-1} u(y, \theta_0)$, as claimed in (2.34).

2.12 *GIC for M-estimation:* Start from Huber's ψ function for M-estimation: for a specified value $b > 0$, $\psi(y) = y$ if $|y| \le b$, $\psi(y) = -b$ if $y < -b$ and $\psi(y) = b$ if $y > b$, and show that the influence function $\mathrm{infl}(G, y)$ at the standard normal distribution function $G = \Phi$ equals $\psi(y)/\{2\Phi(b) - 1\}$. Next, specify GIC for M-estimation in this setting.

2.13 *Leave-one-out for linear regression:* Consider the linear regression model where y_i has mean $x_i^t \beta$ for $i = 1, \ldots, n$, i.e. the vector y has mean $X\beta$, with β a parameter vector of length p and with X of dimension $n \times p$, assumed of full rank p. The direct predictor of y_i is $\widehat{y}_i = x_i^t \widehat{\beta}$, while its cross-validated predictor is $\widehat{y}_{(i)} = x_i^t \widehat{\beta}_{(i)}$, where $\widehat{\beta}$ and $\widehat{\beta}_{(i)}$ are the ordinary least squares estimates of β based on respectively the full data set of size n and the reduced data set of size $n - 1$ that omits (x_i, y_i).

(a) Let $A = X^t X = \sum_{i=1}^{n} x_i x_i^t = A_i + x_i x_i^t$, with $A_i = \sum_{j \ne i} x_j x_j^t$. Show that

$$A^{-1} = A_i^{-1} - \frac{A_i^{-1} x_i x_i^t A_i^{-1}}{1 + x_i^t A_i^{-1} x_i},$$

assuming that also A_i has full rank p.

(b) Let $s_1, \ldots, s_n$ be the diagonal elements of $I_n - H = I_n - X(X^t X)^{-1} X$. Show that $s_i = 1 - x_i^t A^{-1} x_i = 1/(1 + x_i^t A_i^{-1} x_i)$.

(c) Use $\widehat{\beta} = (A_i + x_i x_i^t)^{-1}(w + x_i y_i)$ and $x_i^t \widehat{\beta}_{(i)} = x_i^t A_i^{-1} w$, with $w = \sum_{j \ne i} x_j y_j$, to show that $y_i - \widehat{y}_i = (y_i - \widehat{y}_{(i)})/(1 + x_i^t A_i^{-1} x_i)$. Combine these findings to conclude that

$y_i - \widehat{y}_{(i)} = (y_i - \widehat{y}_i)/s_i$ for $i = 1, \ldots, n$. This makes it easy to carry out leave-one-out cross-validation for linear regression models.

2.14 *The robustified Kullback–Leibler divergence:* For a given non-negative weight function w defined on the sample space, consider $d_w(g, f_\theta)$ of (2.38).

(a) Show that $d_w(g, f_\theta)$ is always non-negative, and that it is equal to zero only when $g(y) = f(y, \theta_0)$ almost everywhere, for some θ_0. Show also that the case of a constant w is equivalent to the ordinary Kullback–Leibler divergence.

(b) Let $T(G) = \theta_0$ be the minimiser of $d_w(g, f_\theta)$, for a proposed parametric family $\{f_\theta : \theta \in \Theta\}$. With G_n the empirical distribution of data $y_1, \ldots, y_n$, show that $T(G_n)$ is the estimator that maximises $H_n(\theta)$ of (2.39). Show that the influence function of T may be expressed as $J^{-1}a(y, \theta_0)$, where $a(y, \theta) = w(y)u(y, \theta) - \xi(\theta)$ and J is the limit of $J_n = -H_n^{(2)}(\theta_0)$; also, $\xi(\theta) = \int w f_\theta u_\theta \, dy$.

(c) It is sometimes desirable to let the data influence which weight function w to use in (2.38). Suppose a parametric $w(y, \alpha)$ is used for weight function, with $\hat{\alpha}$ some estimator with the property that $\sqrt{n}(\hat{\alpha} - \alpha)$ has a normal limit. Extend the theory of Section 2.10.2 to cover also these cases of an estimated weight function, and, in particular, derive an appropriate generalisation of the robust model selection criterion wAIC of (2.42).

3

The Bayesian information criterion

One approach to model selection is to pick the candidate model with the highest probability given the data. This chapter shows how this idea can be formalised inside a Bayesian framework, involving prior probabilities on candidate models along with prior densities on all parameter vectors in the models. It is found to be related to the criterion $\mathrm{BIC} = 2\ell_n(\widehat{\theta}) - (\log n)\,\mathrm{length}(\theta)$, with n the sample size and $\mathrm{length}(\theta)$ the number of parameters for the model studied. This is like AIC but with a more severe penalty for complexity. More accurate versions and modifications are also discussed and compared. Various applications and examples illustrate the BIC at work.

3.1 Examples and illustrations of the BIC

The Bayesian information criterion (BIC) of Schwarz (1978) and Akaike (1977, 1978) takes the form of a penalised log-likelihood function where the penalty is equal to the logarithm of the sample size times the number of estimated parameters in the model. In detail,

$$\mathrm{BIC}(M) = 2\,\text{log-likelihood}_{\max}(M) - (\log n)\dim(M), \qquad (3.1)$$

for each candidate model M, again with $\dim(M)$ being the number of parameters estimated in the model, and with n the sample size of the data. The model with the largest BIC value is chosen as the best model. The BIC of (3.1) is clearly constructed in a manner quite similar to the AIC of (2.1), with a stronger penalty for complexity (as long as $n \geq 8$).

There are various advantages and disadvantages when comparing these methods. We shall return to such comparisons in Chapter 4, but may already point out that the BIC successfully addresses one of the shortcomings of AIC, namely that the latter will not succeed in detecting 'the true model' with probability tending to 1 when the sample size increases. At this moment, however, we start out showing the BIC at work in a list of examples.

Table 3.1. *Mortality in ancient Egypt: maximised log-likelihoods, and the BIC scores, for each of the nine models.*

Parameters	$\ell_n(\widehat{\theta})$	BIC	Order
model 1, b:	−623.777	−1252.503	(7)
model 2, a, b:	−615.386	−1240.670	(6)
model 3, μ, σ:	−629.937	−1269.772	(8)
model 4, a, b:	−611.353	−1232.604	(2)
model 5, k, a, b:	−611.319	−1237.484	(5)
model 6, a, b:	−611.353	−1232.604	(2)
model 7, a, b_1, b_2:	−610.076	−1234.998	(3)
model 8, a_1, b, a_2:	−608.520	−1231.886	(1)
model 9, a_1, b_1, a_2, b_2:	−608.520	−1236.835	(4)

Example 3.1 Exponential versus Weibull

Consider again Example 2.5, where independent failure time data $Y_1, \ldots, Y_n$ are modelled either via an exponential distribution or via the Weibull model. To select the best model according to the BIC, we compute

$$\mathrm{BIC(exp)} = 2 \sum_{i=1}^{n} (\log \widetilde{\theta} - \widetilde{\theta} y_i) - \log n,$$

$$\mathrm{BIC(wei)} = 2 \sum_{i=1}^{n} \{-(\widehat{\theta} y_i)^{\widehat{\gamma}} + \widehat{\gamma} \log \widehat{\theta} + \log \widehat{\gamma} + (\widehat{\gamma} - 1) \log y_i\} - 2 \log n.$$

Here $\widetilde{\theta}$ is the maximum likelihood estimator for θ in the first model, while $(\widehat{\theta}, \widehat{\gamma})$ are the maximum likelihood estimators in the Weibull model. The best model has the largest BIC value. For an illustration, check Exercise 3.1. ∎

Example 3.2 Mortality in ancient Egypt

BIC values for each of the models considered in Example 2.6 are readily obtained using Table 2.2; the results are presented in Table 3.1. The maximised log-likelihood values

$$\ell_n(\widehat{\theta}) = \sum_{i=1}^{n} \log f(t_i, \widehat{\theta})$$

are found in the column labelled '$\ell_n(\widehat{\theta})$'. We compute $\mathrm{BIC} = 2 \log \ell_n(\widehat{\theta}) - p \log n$, with $p = \mathrm{length}(\theta)$ and $n = 141$, or $\log n \approx 4.949$. The BIC penalty is stricter than the AIC penalty, which equals $2p$. Model 1 has only one parameter, resulting in $\mathrm{BIC}_1 = 2(-623.777) - \log 141 = -1252.503$. Models 2, 3 and 4 (the latter is also equal to model 6) have two parameters. Amongst these four models, the two-parameter Gompertz model is again the best one since it has the largest BIC score. Models 5, 7, 8 each have three parameters, with BIC values given in the table. Only model 8, which includes

Table 3.2. *BIC values for the low birthweight data. The smallest model only contains an intercept, the fullest model adds covariates x_2, x_3, x_4 and x_5.*

Covariates	BIC value	Order		Covariates	BIC value
–	−239.914	(2)		x_3, x_4	−246.471
x_2	−239.174	(1)		x_3, x_5	−246.296
x_3	−242.395	(4)		x_4, x_5	−245.387
x_4	−243.502			x_2, x_3, x_5	−247.644
x_5	−243.382			x_2, x_4, x_5	−244.226
x_2, x_3	−242.849	(5)		x_3, x_4, x_5	−249.094
x_2, x_4	−240.800	(3)		x_2, x_3, x_4	245.142
x_2, x_5	−243.826			full	−248.869

separate hazard rate proportionality parameters a_1 and a_2 for men and women, improves on the Gompertz model. This is found the best model in the list of candidate models according to selection by the BIC. This BIC value is not further improved by model 9, which has $\text{BIC}_9 = -1236.835$.

For this application, the conclusion about a best model coincides with that obtained via AIC model choice. Since the penalty of the BIC for these data is bigger than that of AIC (4.9489 as compared to 2), bigger models receive a heavier 'punishment'. This is clearly observed by considering model 9, which has rank number (2) for AIC, while receiving the lower rank equal to (4) for the BIC. When the sample n gets bigger, the heavier the penalty used in the BIC. Especially for large sample sizes we expect to find a difference in the ranks when comparing the selection by AIC and BIC. Chapter 4 provides a theoretical comparison of these criteria. ∎

Example 3.3 Low birthweight data: BIC variable selection

We consider the same variables as in Example 2.4. That is, a constant intercept $x_1 = 1$; x_2, weight of mother just prior to pregnancy; x_3, age of mother; x_4, indicator for race 'black'; x_5, indicator for race 'other'; and $x_4 = x_5 = 0$ indicates race 'white'. For the logistic regression model one finds the BIC formula

$$\text{BIC} = 2 \sum_{i=1}^{n} \{y_i \log \widehat{p}_i + (1 - y_i) \log(1 - \widehat{p}_i)\} - \text{length}(\beta) \log n,$$

where $\widehat{p}_i$ is the estimated probability for $Y_i = 1$ under the model and length(β) is the number of estimated regression coefficients. The sample size is $n = 189$, with $\log 189 \approx 5.2417$. The values of $-\text{BIC}$ can easily be obtained from this formula, or in R via the function $\text{AIC}(\text{fitted.object}, k = \log(\text{sample.size}))$. The default argument for the penalty k is the value 2, corresponding to AIC. The model with the biggest value of the BIC is preferred. In Table 3.2 we examine the 2^4 models that always include an intercept term, that is, $x_1 = 1$ is always included.

According to Table 3.2, the best BIC model is the one containing only variable x_2 in addition to an intercept. The estimated intercept coefficient for this model equals 0.998, with estimated slope parameter -0.014 for x_2, leading to the following fitted model:

$$\widehat{P}(\text{low birthweight} \mid x_2) = \frac{\exp(0.998 - 0.014\,x_2)}{1 + \exp(0.998 - 0.014\,x_2)}.$$

Second best is the intercept-only model, followed closely by the model containing both x_2 and x_4 as extra parameters. If we insist on variable x_2 being part of the model, the best model does not change, though second best is now the model labelled (3) above. This is the model selected as the best one by AIC. The model which was second best in the AIC model choice is the model containing x_2, x_4 and x_5.

As we saw in Example 2.1, both x_2 and x_4 are individually significant at the 5% level of significance. For this particular illustration, AIC model choice is more in line with the individual testing approach. We note here the tendency of the BIC to choose models with fewer variables than those chosen by AIC. ∎

Example 3.4 Predicting football match results

In Example 2.8 we were interested in finding a good model for predicting the results of football matches. The table given there provides both AIC and BIC scores, both pointing to the same best model; the hockey-stick model M_2 is better than the other models. On the second place BIC prefers model M_1 above M_3 (the order was reversed for AIC), while both AIC and BIC agree on the last place for model M_0. ∎

Example 3.5 Density estimation via the BIC

By analogy with the density estimators based on AIC (see Example 2.9), we may construct a class of estimators that use the BIC to select the degree of complexity. We illustrate this with a natural extension of the framework of Example 2.9, defining a sequence of approximators to the density of the data by

$$f_m(x) = \phi\left(\frac{x - \xi}{\sigma}\right) \frac{1}{\sigma} \frac{1}{c_m(a)} \exp\left\{\sum_{j=1}^{m} a_j \psi_j\left(\frac{x - \xi}{\sigma}\right)\right\} \quad \text{for } m = 0, 1, 2, 3, \ldots$$

This creates a nested sequence of models starting with the usual normal density $\sigma^{-1}\phi(\sigma^{-1}(x - \xi))$. As previously, $c_m(a) = \int \phi \exp(\sum_{j=1}^{m} a_j \psi_j)\,dx$ is the normalising constant, in terms of basis functions $\psi_1, \psi_2, \ldots$ that are now taken to be orthogonal with respect to the standard normal density ϕ. The BIC selects the optimal order $\widehat{m}$ to maximise $\text{BIC}(m) = 2\ell_n(\widehat{\xi}, \widehat{\sigma}, \widehat{a}) - (m + 2)\log n$, in terms of the attained log-likelihood maximum in the $(m + 2)$-parameter family f_m. As sample size n increases, the BIC preferred order $\widehat{m}$ will increase (but slowly). The method amounts to a nonparametric density estimator with a parametric start, similar in spirit to the kernel method of Hjort and Glad (1995). For the different problem of using the BIC for testing the null

Table 3.3. *Bernstein's blood group data from 1924, for 502 Japanese living in Korea, along with probabilities and predicted numbers under the one-locus and two-loci theories.*

Observed	Probability	One-locus	fits1	Two-loci	fits2
212	θ_A	$p(2 - p - 2q)$	229.4	$a(1 - b)$	180
103	θ_B	$q(2 - 2p - q)$	97.2	$(1 - a)b$	71
39	θ_{AB}	$2pq$	45.5	ab	71
148	θ_O	$(1 - p - q)^2$	152.7	$(1 - a)(1 - b)$	180

hypothesis that f_0 is the true density of the data, we refer to Section 8.6; see also Ledwina (1994). ∎

Example 3.6 Blood groups A, B, AB, O

The blood classification system invented by the Nobel Prize winner Karl Landsteiner relates to certain blood substances or antigenes 'a' and 'b'. There are four blood groups A, B, AB, O, where A corresponds to 'a' being present, B corresponds to 'b' being present, AB to both 'a' and 'b' being present, while O (for 'ohne') is the case of neither 'a' nor 'b' present.

Landsteiner and others were early aware that the blood group of human beings followed the Mendelian laws of heredity, but it was not clear precisely which mechanisms were at work. Until around 1924 there were two competing theories. The first theory held that there were three alleles for a single locus that determined the blood group. The second theory hypothesised two alleles at each of two loci. Letting $\theta_A, \theta_B, \theta_{AB}, \theta_O$ be the frequencies of categories A, B, AB, O in a large and genetically stable population, the two theories amount to two different parametric forms for these. Leaving details aside, Table 3.3 gives these, using parameters (p, q) for the one-locus theory and parameters (a, b) for the two-loci theory; (a, b) can vary freely in $(0, 1)^2$ while the probabilities p and q are restricted by $p + q < 1$. Also shown in Table 3.3 is an important set of data that actually settled the discussion, as we shall see. These are the numbers N_A, N_B, N_{AB}, N_O in the four blood categories, among 502 Japanese living in Korea, collected by Bernstein (1924).

The two log-likelihood functions in question take the form

$$\ell_{n,1}(p, q) = N_A\{\log p + \log(2 - p - 2q)\} + N_B\{\log q + \log(2 - 2p - q)\} + N_{AB}(\log 2 + \log p + \log q) + 2N_O \log(1 - p - q)$$

and

$$\ell_{n,2}(a, b) = (N_A + N_{AB}) \log a + (N_B + N_O) \log(1 - a) + (N_B + N_{AB}) \log b + (N_A + N_O) \log(1 - b).$$

One finds $(\widehat{p}, \widehat{q}) = (0.294, 0.154)$ and $(\widehat{a}, \widehat{b}) = (0.500, 0.283)$, which leads to $\ell_{n,1}(\widehat{p}, \widehat{q}) = -627.104$ and $\ell_{n,2}(\widehat{a}, \widehat{b}) = -646.972$, and to

	1-locus	2-loci
AIC:	-1258.21	-1297.95
BIC:	-1266.65	-1306.38

So the one-locus theory is vastly superior, the criteria agree, as is also quite clear from the predicted counts, see Table 3.3. See also Exercise 3.9. We also note that goodness-of-fit analysis of the chi squared type will show that the one-locus model is fully acceptable whereas the two-loci one is being soundly rejected. ∎

Example 3.7 Beta regression for MHAQ and HAQ data

When response data Y_i are confined to a bounded interval, like $[0, 1]$, then ordinary linear regression methods are not really suitable and may work poorly. Consider instead the following Beta regression set-up, where $Y_i \sim \text{Beta}(ka_i, k(1 - a_i))$ and $a_i = H(x_i^t \beta)$ for $i = 1, \ldots, n$, writing $H(u) = \exp(u)/\{1 + \exp(u)\}$ for the inverse logistic transform. Then the Y_i have means $a_i = H(x_i^t \beta)$ and nonconstant variances $a_i(1 - a_i)/(k + 1)$. Here k can be seen as a concentration parameter, with tighter focus around a_i if k is large than if k is small. The log-likelihood function becomes

$$\sum_{i=1}^{n} \{\log \Gamma(k) - \log \Gamma(ka_i) - \log \Gamma(k\bar{a}_i) + (ka_i - 1) \log y_i + (k\bar{a}_i - 1) \log \bar{y}_i\},$$

where we put $\bar{a}_i = 1 - a_i$ and $\bar{y}_i = 1 - y_i$. Inference can be carried out using general likelihood techniques summarised in Chapter 2, and in particular confidence intervals (model-dependent as well as model-robust) can be computed for each of the parameters.

Figure 3.1 relates to an application of such Beta regression methodology, in the study of so-called HAQ and MHAQ data; these are from certain standardised health assessment questionnaires, in original (HAQ) and modified (MHAQ) form. In the study at hand, the interest was partly geared towards predicting the rather elaborate $y = \text{HAQ}$ score (which has range $[0, 3]$) from the easier to use $x = \text{MHAQ}$ score (which has range $[1, 4]$). The study involved 1018 patients from the Division for Women and Children at the Oslo University Hospital at Ullevål. For this application we treated the 219 most healthy patients separately, those having $x = 1$, since they fall outside the normal range for which one wishes to build regression models. We carried out Beta regression analysis as above for the remaining 799 data pairs (x_i, y_i). Models considered included $Y_i \mid x_i \sim 3 \text{Beta}(ka_i, k(1 - a_i))$, where $a_i = H(\beta_0 + \beta_1 x + \cdots + \beta_q x^q)$, of order up to $q = 4$. AIC and BIC analysis was carried out, and preferred respectively the fourth-order and the third-order model. Importantly, the model selection criteria showed that the Beta regressions were vastly superior to both ordinary polynomial regressions and to non-negative Gamma-type regressions. The figure displays the median prediction line, say

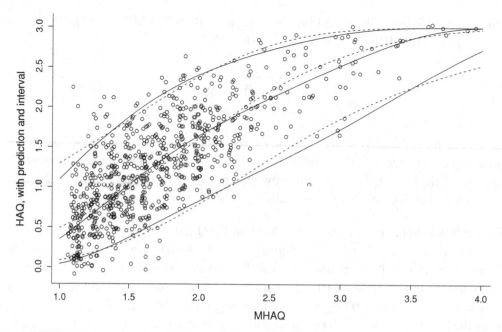

Fig. 3.1. Data on $(x, y) = $ (MHAQ, HAQ) are shown for 799 patients at the Division
for Women and Children at the Ulleväl University Hospital in Oslo; the data have been
modestly jittered for display purposes. Also shown are the median and lower and upper
5% curves for predicting HAQ from MHAQ, for the third-order (full lines) and first-order
(dashed lines) Beta regression models.

$\widehat{\xi}_{0.50}(x) = 3G^{-1}(\frac{1}{2}, \widehat{ka}(x), \widehat{k}\{1 - \widehat{a}(x)\})$, where $G^{-1}(\cdot, a, b)$ is the inverse cumulative distribution function for the Beta(a, b) distribution, along with lower and upper 5% lines. ∎

Example 3.8 A Markov reading of *The Raven*

'Once upon a midnight dreary, while I pondered weak and weary, . . . ' is the opening
line of Edgar Allan Poe's celebrated 1845 poem *The Raven*. In order to learn about the
narrative's poetic rhythm, for example in a context of comparisons with other poems by
Poe, or with those of other authors, we have studied statistical aspects of *word lengths*.
Thus the poem translates into a sequence 4, 4, 1, 8, 6, 5, 1, 8, 4, 3, 5, . . . of lengths of in
total $n = 1085$ words. (For our conversion of words to lengths we have followed some
simple conventions, like reading 'o'er' as 'over', 'bosom's' as 'bosoms', and 'foot-falls'
as 'footfalls'.) For this particular illustration we have chosen to break these lengths into
categories '1', short words (one to three letters); '2', middle length words (four of five
letters); and '3', long words (six or more letters). Thus *The Raven* is now read as a
sequence 2, 2, 1, 3, 3, 2, 1, 3, 2, 1, 2, . . . The proportions of short, middle, long words
are 0.348, 0.355, 0.297. In some linguistic studies one is interested in the ways words
or aspects of words depend on the immediately preceding words. The approach we shall
report on here is to model and analyse the sequence $X_1, \ldots, X_n$ of short, middle, long
words as a Markov chain.

Table 3.4. *A Markov reading of The Raven: maximal log-likelihood values,*
along with AIC and BIC scores and their rankings, for the Markov chain models
of order 0, 1, 2, 3, 4.

Markov order	Dimension	$\ell_{\max}$	AIC			BIC
0	2	−1188.63	−2381.27	(5)	(3)	−2391.25
1	6	−1161.08	−2334.16	(3)	(1)	−2364.09
2	18	−1121.62	−2279.24	(1)	(2)	−2369.02
3	54	−1093.47	−2294.94	(2)	(4)	−2564.21
4	162	−1028.19	−2380.38	(4)	(5)	−3188.05

An ordinary Markov chain has memory length 1, which means that the distribution of X_k given all previous $X_1, \ldots, X_{k-1}$ depends only on X_{k-1}. The distributional aspects of the chain are then governed by one-step transition probabilities
$p_{a,b} = P(X_k = b \mid X_{k-1} = a)$, for $a, b = 1, 2, 3$. There are $3 \cdot 2 = 6$ free parameters
in this model. The Markov model with memory length 2 has X_k given all previous
$X_1, \ldots, X_{k-1}$ depending on (X_{k-2}, X_{k-1}); in this case one needs transition probabilities
of the type $p_{a,b,c} = P(X_k = c \mid (X_{k-2}, X_{k-1}) = (a, b))$. This model has $3^2 \cdot 2 = 18$ free
parameters, corresponding to $3^2 = 9$ unknown probability distributions on $\{1, 2, 3\}$. In
our analyses we also include the Markov models with memory length 3 and 4, with
respectively $3^3 \cdot 2 = 54$ and $3^4 \cdot 2 = 162$ free parameters; and finally also the trivial
Markov model of order 0, which corresponds to an assumption of full independence
among the X_ks, and only two free parameters.

Maximum likelihood analysis is relatively easy to carry out in these models, using
the simple lemma that if $z_1, \ldots, z_k$ are given positive numbers, then the maximum of
$\sum_{j=1}^{k} z_j \log p_j$, where $p_1, \ldots, p_k$ are probabilities with sum 1, takes place for $\widehat{p}_j =$
$z_j/(z_1 + \cdots + z_k)$ for $j = 1, \ldots, k$. For the second-order Markov model, for example,
the likelihood may be written $\prod_{a,b,c} p_{a,b,c}^{N_{a,b,c}}$, with $N_{a,b,c}$ denoting the number of observed
consecutive triples (X_{k-2}, X_{k-1}, X_k) that equal (a, b, c). This leads to a log-likelihood

$$\ell_n^{(2)} = \sum_{a,b,c} N_{a,b,c} \log p_{a,b,c} = \sum_{a,b} N_{a,b,\bullet} \sum_c \frac{N_{a,b,c}}{N_{a,b,\bullet}} \log p_{a,b,c},$$

with maximal value $\sum_{a,b} N_{a,b,\bullet} \sum_c \widehat{p}_{a,b,c} \log \widehat{p}_{a,b,c}$, where $\widehat{p}_{a,b,c} = N_{a,b,c}/N_{a,b,\bullet}$. Here
$N_{a,b,\bullet} = \sum_c N_{a,b,c}$. Similar analysis for the other Markov models leads to Table 3.4.
We see that the BIC prefers the one-step memory length model, which has estimated
transition probabilities as given in the left part of Table 3.5.

We learn from this that transitions from middle to short, from long to middle,
and from short to long, are all more frequent than what the simpler memory-less
view explains, namely that the words are merely short, middle, long with probabilities $(0.348, 0.355, 0.297)$ (which is also the equilibrium distribution of the one-step

Table 3.5. *A Markov reading of The Raven. Transition probabilities for the model selected by the BIC (left) and AIC (right).*

	1	2	3
1	0.241	0.376	0.384
2	0.473	0.286	0.242
3	0.327	0.411	0.262

	to 1	to 2	to 3
from 1,1	0.121	0.451	0.429
from 1,2	0.401	0.289	0.310
from 1,3	0.276	0.462	0.262
from 2,1	0.143	0.429	0.429
from 2,2	0.436	0.291	0.273
from 2,3	0.376	0.323	0.301
from 3,1	0.514	0.219	0.267
from 3,2	0.583	0.273	0.144
from 3,3	0.361	0.422	0.217

Markov chain). On the other hand, AIC prefers the richer two-step memory length model, which corresponds to the estimated transition probabilities, from (a, b) to (b, c), also given in Table 3.5.

These investigations indicate that a first-order Markov chain gives a satisfactory description of the poem's basic rhythm, but that the richer structure captured by the second-order Markov model is significantly present. In line with general properties for AIC and BIC, discussed in Chapter 4, this might be interpreted as stating that if different poems (say by Poe and contemporaries) are to be rhythmically compared, and perhaps with predictions of the number of breaks between short, middle, long words, then AIC's choice, the two-step memory model, appears best. If, on the other hand, one wished to exhibit the one most probable background model that generated a long poem's rhythmic structure, then the BIC's choice, the one-step memory model, is tentatively the best.

We have limited discussion here to full models, of dimension 2, 6, 18, 54, and so on; the fact that AIC and BIC give somewhat conflicting advice is also an inspiration for digging a bit deeper, singling out for scrutiny second-order models with fewer parameters than the maximally allowed 18. Some relevant models and alternative estimation methods are discussed in Hjort and Varin (2008). Statistical analysis of large text collections using dependence models, also for literature and poetry, is partly a modern discipline, helped by corpora becoming electronically available. It is, however, worthwhile pointing out that the first ever data analysis using Markov models was in 1913, when Markov fitted a one-step Markov chain model to the first 20,000 consonants and vowels of Pushkin's *Yevgenij Onegin*; see the discussion in Hjort and Varin (2008). ∎

3.2 Derivation of the BIC

The 'B' in BIC is for 'Bayesian'. In the examples above, all we needed was a likelihood for the data, without any of the usual ingredients of Bayesian analysis such as a specification of prior probabilities for each of the models. In this section we explain the basic steps of a Bayesian model comparison and arrive at the BIC via a particular type of approximation to Bayesian posterior probabilities for models.

3.2.1 Posterior probability of a model

When different models are possible, a Bayesian procedure selects that model which is a posteriori most likely. This model can be identified by calculating the posterior probability of each model and selecting the model with the biggest posterior probability. Let the models be denoted $M_1, \ldots, M_k$, and use y as notation for the vector of observed data $y_1, \ldots, y_n$.

Bayes' theorem provides the posterior probability of the models as

$$P(M_j \mid y) = \frac{P(M_j)}{f(y)} \int_{\Theta_j} f(y \mid M_j, \theta_j) \pi(\theta_j \mid M_j) \, \mathrm{d}\theta_j, \qquad (3.2)$$

where Θ_j is the parameter space to which θ_j belongs. In this expression

- $f(y \mid M_j, \theta_j) = \mathcal{L}_{n,j}(\theta_j)$ is the likelihood of the data, given the jth model and its parameter;
- $\pi(\theta_j \mid M_j)$ is the prior density of θ_j given model M_j;
- $P(M_j)$ is the prior probability of model M_j; and
- $f(y)$ is the unconditional likelihood of the data.

The latter is computed via $f(y) = \sum_{j=1}^{k} P(M_j) \lambda_{n,j}(y)$, where

$$\lambda_{n,j}(y) = \int_{\Theta_j} \mathcal{L}_{n,j}(\theta_j) \pi(\theta_j \mid M_j) \, \mathrm{d}\theta_j \qquad (3.3)$$

is the marginal likelihood or marginal density for model j, with θ_j integrated out with respect to the prior in question, over the appropriate parameter space Θ_j. In the comparison of posterior probabilities $P(M_j \mid y)$ across different models, the $f(y)$ is not important since it is constant across models. Hence the crucial aspect is to evaluate the $\lambda_{n,j}(y)$, exactly or approximately.

Let us define

$$\mathrm{BIC}_{n,j}^{\mathrm{exact}} = 2 \log \lambda_{n,j}(y), \qquad (3.4)$$

with which

$$P(M_j \mid y) = \frac{P(M_j) \exp\left(\frac{1}{2} \mathrm{BIC}_{n,j}^{\mathrm{exact}}\right)}{\sum_{j'=1}^{k} P(M_{j'}) \exp\left(\frac{1}{2} \mathrm{BIC}_{n,j'}^{\mathrm{exact}}\right)}. \qquad (3.5)$$

These 'exact BIC values' are in fact quite rarely computed in practice, since they are difficult to reach numerically. Moreover, this approach requires the specification of priors for all models and for all parameters in the models. The BIC expression that will be derived in Section 3.2.2 is an effective and practical approximation to the exact BIC.

3.2.2 BIC, BIC*, BIC^{exact}

To show how the approximation (3.1) emerges, we introduce the basic Laplace approximation. We wish to have an approximation of the integral $\lambda_{n,j}(y)$, which we now write as

$$\lambda_{n,j}(y) = \int_\Theta \exp\{nh_{n,j}(\theta)\}\pi(\theta \mid M_j)\,d\theta,$$

with $h_{n,j}(\theta) = n^{-1}\ell_{n,j}(\theta)$ and p the length of θ. The basic Laplace approximation method works precisely for such integrals, and states that

$$\int_\Theta \exp\{nh(\theta)\}g(\theta)\,d\theta = \left(\frac{2\pi}{n}\right)^{p/2}\exp\{nh(\theta_0)\}\{g(\theta_0)|J(\theta_0)|^{-1/2} + O(n^{-1})\},$$

where θ_0 is the value that maximises the function $h(\cdot)$ and $J(\theta_0)$ is the Hessian matrix $-\partial^2 h(\theta)/\partial\theta\partial\theta^t$ evaluated at θ_0. We note that the implied approximation becomes exact when h is a negative quadratic form (as with a Gaussian log-likelihood) and g is constant. In our case, we have $h(\theta) = n^{-1}\ell_{n,j}(\theta)$ and that its maximiser equals the maximum likelihood estimator $\widehat{\theta}_j$ for model M_j. Hence, with $J_{n,j}(\widehat{\theta}_j)$ as in (2.8),

$$\lambda_{n,j}(y) \approx \mathcal{L}_{n,j}(\widehat{\theta})(2\pi)^{p/2}n^{-p/2}|J_{n,j}(\widehat{\theta}_j)|^{-1/2}\pi(\widehat{\theta}_j \mid M_j). \tag{3.6}$$

Going back to (3.2) and (3.3), this leads to several possible approximations to each $\lambda_{n,j}(y)$. The first of these is obtained by taking the approximation obtained in the right-hand side of (3.6). After taking the logarithm and multiplying by two we arrive at the approximation which we denote by $\text{BIC}^*_{n,j}$. In other words, $2\log\lambda_{n,j}(y)$ is close to

$$\text{BIC}^*_{n,j} = 2\ell_{n,j}(\widehat{\theta}_j) - p_j\log n + p_j\log(2\pi) - \log|J_{n,j}(\widehat{\theta}_j)| + 2\log\pi_j(\widehat{\theta}_j), \tag{3.7}$$

where $p_j = \text{length}(\theta_j)$. The dominant terms of (3.7) are the two first ones, of sizes respectively $O_P(n)$ and $\log n$, while the others are $O_P(1)$. Ignoring these lower-order terms, then, gives a simpler approximation that we recognise as the BIC, that is,

$$2\log\lambda_{n,j}(y) \approx \text{BIC}_{n,j} = 2\ell_{n,j,\max} - p_j\log n, \tag{3.8}$$

or

$$P(M_j \mid y) \approx \frac{P(M_j)\exp\left(\frac{1}{2}\text{BIC}_{n,j}\right)}{\sum_{j'=1}^{k} P(M_{j'})\exp\left(\frac{1}{2}\text{BIC}_{n,j'}\right)}.$$

The above derivation only requires that the maximum likelihood estimator is an interior point of the parameter space and that the log-likelihood function and the prior densities are twice differentiable. The result originally obtained by Schwarz (1978) assumed rather stronger conditions; in particular, the models he worked with were taken to belong to an exponential family.

Note that the specification of the prior completely disappears in the formula of BIC. No prior information is needed to obtain BIC values, only the maximised log-likelihood function is used. In large samples, the BIC provides an easy to calculate alternative to the actual calculation of the marginal likelihoods, or Bayes factors. For two models M_1 and M_2, the Bayes factor is equal to the posterior odds divided by the prior odds,

$$\frac{P(M_2 \mid y)/P(M_1 \mid y)}{P(M_2)/P(M_1)} = \frac{\lambda_{n,2}(y)}{\lambda_{n,1}(y)}.$$

This can be used for a pairwise comparison of models. Note how the Bayes factor resembles a likelihood ratio. The difference with a true likelihood ratio is that here the unconditional likelihood of the data is used (where the parameter is integrated out), whereas for a likelihood ratio we use the maximised likelihood values. Efron and Gous (2001) discuss and compare different ways to interpret the value of a Bayes factor.

We will return to the Bayesian model selection theme in Section 7.7, where Bayesian model averaging is discussed. The book Lahiri (2001) contains extensive overview papers (with discussion) on the theme of Bayesian model selection; we refer to the papers of Chipman *et al.* (2001), Berger and Pericchi (2001) and Efron and Gous (2001) for more information and many useful references.

3.2.3 A robust version of the BIC using M-estimators

Robust versions of the BIC have been proposed in the literature, partly resembling similar suggestions for robustified AIC scores, discussed in Section 2.10. In a sense robustifying AIC scores is a conceptually more sensible task, since the basic arguments are frequentist in nature and relate directly to performance of estimators for expected distances between models. The BIC, on the other hand, is Bayesian, at least in spirit and construction, and a Bayesian with a model, a prior and a loss function is in principle not in need of robustification of his estimators. If sensitivity to outlying data is a concern, then the model itself might be adjusted to take this into account.

To briefly illustrate just one such option, in connection with the linear regression model, suppose the model uses densities proportional to $\sigma^{-1} \exp\{-\tau((y_i - x_i^t\beta)/\sigma)\}$, with a $\tau(u)$ function that corresponds to more robust analysis than the usual Gaussian form, which corresponds to $\tau(u) = \frac{1}{2}u^2$. One choice is $\tau(u) = |u|$ and is related to the double exponential model. Another choice that leads to maximum likelihood estimators identical to those of the optimal Huber type has $\tau(u)$ equal to $\frac{1}{2}u^2$ for $|u| \leq c$ and to $c(|u| - \frac{1}{2}c)$ for $|u| \geq c$. Inside such a family, the likelihood theory as well as the Laplace approximations still work, and lead to

$$\text{BIC} = -2 \sum_{i=1}^{n} \left\{ \tau\left(\frac{y_i - x_i^t\widehat{\beta}}{\widehat{\sigma}} \right) + \log\widehat{\sigma} \right\} - (p+1)\log n,$$

with p the number of β_j coefficients. This BIC score is then computed for each candidate subset of regressors. For some discussion of similar proposals, see Machado (1993).

3.3 Who wrote 'The Quiet Don'?

The question about the authorship of 'The Quiet Don' (see the introduction in Section 1.3) is formulated in terms of selecting one of three models:

M_1: Text corpora Sh and QD come from the same statistical distribution, while Kr represents another;

M_2: Sh is not statistically compatible with Kr and QD, which are however coming from the same distribution;

M_3: Sh, Kr, QD represent three statistically different corpora.

We write θ_{Sh}, θ_{Kr}, θ_{QD} for the three parameter vectors (p, ξ, a, b), for respectively Sh, Kr, QD. Model M_1 states that $\theta_{Sh} = \theta_{QD}$ while θ_{Kr} is different; model M_2 claims that $\theta_{Kr} = \theta_{QD}$ while θ_{Sh} is different; and finally model M_3 allows the possibility that the three parameter vectors are different.

For the BIC-related analysis to follow we use parameter estimates based on the raw data for each of Sh, Kr and QD separately, i.e. the real sentence counts, not only in their binned form. These parameter values are found numerically using `nlm` in R:

	$\widehat{\theta}_{Sh}$	se	$\widehat{\theta}_{Kr}$	se	$\widehat{\theta}_{QD}$	se
p	0.184	0.021	0.057	0.023	0.173	0.022
ξ	9.099	0.299	9.844	0.918	9.454	0.367
a	2.093	0.085	2.338	0.092	2.114	0.090
b	0.163	0.007	0.178	0.008	0.161	0.007

The standard deviations (se) are obtained from the estimated inverse Fisher information matrix, as per the theory surveyed in Section 2.2.

Let in general $P(M_1)$, $P(M_2)$, $P(M_3)$ be any prior probabilities for the three possibilities; Solzhenitsyn would take $p(M_1)$ rather low and $p(M_2)$ rather high, for example, whereas more neutral observers might start with the three probabilities equal to $1/3$. Let $\mathcal{L}_1(\theta_1)$, $\mathcal{L}_2(\theta_2)$, $\mathcal{L}_3(\theta_3)$ be the three likelihoods in question and denote by π_1, π_2, π_3 any priors used for $(\theta_{Sh}, \theta_{Kr}, \theta_{QD}) = (\theta_1, \theta_2, \theta_3)$. Under M_1, there is one prior $\pi_{1,3}$ for $\theta_1 = \theta_3$, and similarly there is one prior $\pi_{2,3}$ for $\theta_2 = \theta_3$ under M_2. Following the general set-up for Bayesian model selection, we have

$$P(M_1 \mid \text{data}) = P(M_1)\lambda_1 / \{P(M_1)\lambda_1 + P(M_2)\lambda_2 + P(M_3)\lambda_3\},$$
$$P(M_2 \mid \text{data}) = P(M_2)\lambda_2 / \{P(M_1)\lambda_1 + P(M_2)\lambda_2 + P(M_3)\lambda_3\}, \qquad (3.9)$$
$$P(M_3 \mid \text{data}) = P(M_3)\lambda_3 / \{P(M_1)\lambda_1 + P(M_2)\lambda_2 + P(M_3)\lambda_3\},$$

in terms of marginal observed likelihoods

$$\lambda_1 = \int \{\mathcal{L}_1(\theta)\mathcal{L}_3(\theta)\}\mathcal{L}_2(\theta_2)\pi_{1,3}(\theta)\pi_2(\theta_2)\,d\theta\,d\theta_2,$$

$$\lambda_2 = \int \{\mathcal{L}_2(\theta)\mathcal{L}_3(\theta)\}\mathcal{L}_1(\theta_1)\pi_{2,3}(\theta)\pi_1(\theta_1)\,d\theta\,d\theta_1,$$

$$\lambda_3 = \int \mathcal{L}_1(\theta_1)\mathcal{L}_2(\theta_2)\mathcal{L}_3(\theta_3)\pi_1(\theta_1)\pi_2(\theta_2)\pi_3(\theta_3)\,d\theta_1\,d\theta_2\,d\theta_3.$$

The integrals are respectively 8-, 8- and 12-dimensional.

Let now $n_{\text{Sh}} = n_1, n_{\text{Kr}} = n_2$ and $n_{\text{QD}} = n_3$. Applying the methods of Section 3.2 leads via (3.6) to

$$\lambda_1 \doteq \mathcal{L}_{1,3}(\widehat{\theta}_{1,3})(2\pi)^{4/2}(n_1+n_3)^{-4/2}|J_{1,3}|^{-1/2}\pi_{1,3}(\widehat{\theta}_{1,3})$$
$$\times \mathcal{L}_2(\widehat{\theta}_2)(2\pi)^{4/2}n_2^{-4/2}|J_2|^{-1/2}\pi_2(\widehat{\theta}_2),$$

$$\lambda_2 \doteq \mathcal{L}_{2,3}(\widehat{\theta}_{2,3})(2\pi)^{4/2}(n_2+n_3)^{-4/2}|J_{2,3}|^{-1/2}\pi_{2,3}(\widehat{\theta}_{2,3})$$
$$\times \mathcal{L}_1(\widehat{\theta}_1)(2\pi)^{4/2}n_1^{-4/2}|J_1|^{-1/2}\pi_1(\widehat{\theta}_1),$$

$$\lambda_3 \doteq \prod_{j=1,2,3} \mathcal{L}_j(\widehat{\theta}_j)(2\pi)^{4/2}n_j^{-4/2}|J_j|^{-1/2}\pi_j(\widehat{\theta}_j).$$

One may show that the approximations are quite accurate in this situation. To proceed further, we argue that there should be no real differences between the priors involved. These all in principle relate to prior assessment of the (p, ξ, a, b) parameters of the three probability distributions. For a neutral investigation these ought to be taken, if not fully equal, then fairly close. This way we ensure that the data (and Sholokhov and Kriukov) speak for themselves. Furthermore, we observe that the three parameter estimates are rather close, and any differences in the $\pi_j(\widehat{\theta}_j)$ terms will be dominated by what goes on with the growing likelihoods $\mathcal{L}_j(\widehat{\theta}_j)$. We may hence drop these terms from the model comparisons. Taking two times the logarithm of the remaining factors, then, leads to

$$\text{BIC}_1^* = 2(\ell_{1,3,\max} + \ell_{2,\max}) - 4\log(n_1+n_3) - 4\log n_2$$
$$- \log|J_{1,3}| - \log|J_2| + 8\log(2\pi),$$

$$\text{BIC}_2^* = 2(\ell_{2,3,\max} + \ell_{1,\max}) - 4\log(n_2+n_3) - 4\log n_1$$
$$- \log|J_{2,3}| - \log|J_1| + 8\log(2\pi),$$

$$\text{BIC}_3^* = 2(\ell_{1,\max} + \ell_{2,\max} + \ell_{3,\max}) - 4\log n_1 - 4\log n_2 - 4\log n_3$$
$$- \log|J_1| - \log|J_2| - \log|J_3| + 12\log(2\pi).$$

Calculations to find maximum likelihood estimators for the common θ of Sh and QD under M_1, as well as for the common θ of Kr and QD under M_2, lead finally to

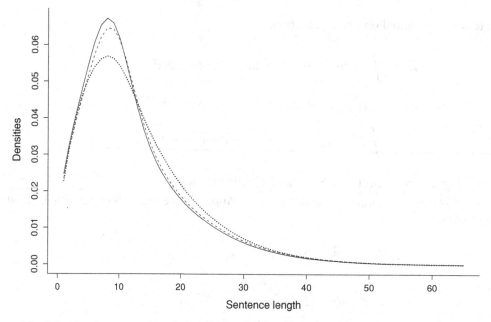

Fig. 3.2. Fitted sentence length distribution for the three text corpora Sh (solid line), QD (broken line) and Kr (dotted line), using the four-parameter model (1.1). The analysis shows that the Kr curve is sufficiently distant from the other two.

	M_1	M_2	M_3
AIC	−79490.8	−79504.4	−79494.5
BIC*	−79515.5	−79530.6	−79528.6
ΔAIC	0.0	−13.7	− 3.8
ΔBIC*	0.0	−15.1	−13.1

The ΔAIC and ΔBIC* scores are differences between the AIC and BIC* values, respectively, subtracting in each case the biggest value. For model selection and for posterior model probabilities via (3.5) and its approximations, only differences count.

We may conclude that the sentence length data speak very strongly in the Nobel laureate's favour, and dismiss D's allegations as speculations. Computing posterior model probabilities via (3.9) gives numbers very close to zero for M_2 and M_3 and very close to one for M_1. Using (3.5) with equal prior probabilities we actually find 0.998 for Sholokhov and the remaining 0.002 shared between Kriukov and the neutral model that the three corpora are different. Even Solzhenitsyn, starting perhaps with $P(M_1) = 0.05$ and $P(M_2) = 0.95$, will be forced to revise his M_1-probability to 0.99 and down-scale his M_2-probability to 0.01. This might sound surprisingly clear-cut, in view of the relative similarity of the distributions portrayed in Figure 1.3. The reason lies with the large sample sizes, which increases detection power. The picture is clearer in Figure 3.2, which shows fitted distributions for Sh and QD in strong agreement and different from Kr. For further analysis and a broader discussion, see Hjort (2007a).

Classification problems come in many forms and by different names, including discriminant analysis, pattern recognition and machine learning; see e.g. Ripley (1996). Arguments akin to those above might be used to give a quite general recipe for classification, at least for cases where the class densities in question are modelled parametrically.

3.4 The BIC and AIC for hazard regression models

Models for survival and event history data are often specified in terms of hazard rate functions instead of densities or cumulative distribution functions. Suppose T is a random failure time variable, with density $f(t)$ for $t \geq 0$. Its survival function is $S(t) = P(T \geq t)$, which is 1 minus the cumulative distribution function $F(t)$. The hazard rate function $h(\cdot)$ is defined as

$$h(t) = \lim_{\varepsilon \to 0} P(\text{die in } [t, t + \varepsilon] \mid \text{survived up to } t)/\varepsilon = \frac{f(t)}{S(t)} \quad \text{for } t \geq 0.$$

Since the right-hand side is the derivative of $-\log\{1 - F(t)\}$, it follows by integration that $F(t) = 1 - \exp\{-H(t)\}$, with cumulative hazard rate $H(t) = \int_0^t h(s)\,ds$. Thus we may find the hazard rate from the density and vice versa. Below we first consider both semiparametric and purely parametric regression models for hazard rates.

In a Cox proportional hazards regression model (Cox, 1972) the hazard rate at time t, for an individual with covariates x_i, z_i, takes the form

$$h(t \mid x_i, z_i) = h_0(t) \exp(x_i^t \beta + z_i^t \gamma).$$

As usual we use x_i to indicate protected but z_i to mean open or nonprotected covariates. The baseline hazard function $h_0(t)$ is assumed to be continuous and positive over the range of lifetimes of interest, but is otherwise left unspecified. This makes the model semiparametric, with a parametric part $\exp(x_i^t \beta + z_i^t \gamma)$ and a nonparametric baseline hazard $h_0(t)$.

A complication with life- and event-time data is that the studied event has not always taken place before the end of the study, in which case the observation is called censored. The complication is both technical (how to construct methods that properly deal with partial information) and statistical (there is a loss in information content and hence in precision of inference). Define the indicator $\zeta_i = 1$ if the event took place before the end of the study ('observed') and $\zeta_i = 0$ if otherwise ('censored'), and introduce the counting process $N_i(t) = I\{t_i \leq t, \zeta_i = 1\}$ which registers whether for subject i the (noncensored) event took place at or before time t. The jump $dN_i(t) = I\{t_i \in [t, t + dt], \zeta_i = 1\}$ is equal to 1 precisely when the event is observed for individual i. Define also the at-risk process $Y_i(t) = I\{t_i \geq t\}$, indicating for each time t whether subject i is still at risk (neither the event nor censoring has yet taken place). The log-partial likelihood $\ell_n^p(\beta, \gamma)$ in a Cox proportional hazards regression model is defined as

$$\sum_{i=1}^{n} \int_{0}^{t_{\text{up}}} \left[x_i^t \beta + z_i^t \gamma - \log \left\{ \sum_{i=1}^{n} Y_i(u) \exp(x_i^t \beta + z_i^t \gamma) \right\} \right] dN_i(u), \qquad (3.10)$$

with t_{up} an upper limit for the time observation period. The maximum partial likelihood estimators of (β, γ), also referred to as the Cox estimators, are the values $(\widehat{\beta}, \widehat{\gamma})$ that maximise (3.10). A full background regarding both mathematical details and motivation for the partial likelihood may be found in, for example, Andersen *et al.* (1993).

Since the BIC is defined through a full likelihood function, it is not immediately applicable to Cox regression models where the partial likelihood is used. However, we may *define*

$$\text{BIC}_{n,S} = 2\ell_{n,S}^{p}(\widehat{\beta}_S, \widehat{\gamma}_S) - (p + |S|) \log n,$$

where S is a subset of $\{1, \ldots, q\}$, indicating a model with only parameters γ_j included for which j is in S. Here $|S|$ is the cardinality of S. All of $x_{i,1}, \ldots, x_{i,p}$ are included in all models. Similarly we define

$$\text{AIC}_{n,S} = 2\ell_{n,S}^{p}(\widehat{\beta}_S, \widehat{\gamma}_S) - 2(p + |S|).$$

The model indexed by S having the largest value of $\text{AIC}_{n,S}$, respectively $\text{BIC}_{n,S}$, is selected as the best model. The AIC and BIC theory does not automatically support the use of the two formulae above, since we have defined them in terms of the profile likelihood rather than the full likelihood. Model selection may nevertheless be carried out according to the two proposals, and some theory for their performance is covered in Hjort and Claeskens (2006), along with development of competing selection strategies. Their results indicate that the AIC and BIC as defined above remain relevant and correct as long as inference questions are limited to statements pertaining to the regression coefficients only. Quantities like median remaining survival time, survival curves for given individuals, etc., are instances where inference needs to take both regression coefficients as well as the baseline hazard into account.

Example 3.9 The Danish melanoma study: AIC and BIC variable selection

This is the set of data on skin cancer survival described in Andersen *et al.* (1993). There were 205 patients with malignant melanoma in this study who had surgery to remove the tumour. These patients were followed after operation over the time period 1962–1977. Several covariates are of potential interest for studying the survival probabilities. In all proportional hazard regression models that we build, we use x_1 as indicator for gender (woman = 1, man = 2). One reason to include x_1 in all of our models is that gender differences are important in cancer studies and we wish to see the effect in the model. Andersen *et al.* (1993) found that men tend to have higher hazard than women. Apart from keeping x_1 inside each model, we now carry out a search among all submodels for the following list of potential variables:

Table 3.6. *Danish melanoma study. The five
best values of the model selection criteria
AIC and BIC, with the variables in the
selected model.*

vars	AIC	vars	BIC
1,2,3,4,6	−529.08	1,4	−542.98
2,3,4,6	−530.00	4	−544.88
3,4,5,6	−530.12	1,3,4	−545.42
2,3,4,5,6	−530.23	1,4,6	−546.07
1,3,4,6	−530.51	3,4	−546.11

- z_1, thickness of the tumour, more precisely $z_1 = (z_1^0 - 292)/100$ where z_1^0 is the real thickness, in 1/100 mm, and the average value 292 is subtracted;
- z_2, infection infiltration level, which is a measure of resistance against the tumour, from high resistance 1 to low resistance 4;
- z_3, indicator of epithelioid cells (present = 1, non-present = 2);
- z_4, indicator of ulceration (present = 1, non-present = 2);
- z_5, invasion depth (at levels 1, 2, 3);
- z_6, age of the patient at the operation (in years).

Patients who died of other causes than malignant melanoma or who were still alive in 1977 are treated as censored observations. Variables z_1 and z_6 are continuous, x_1, z_3 and z_4 are binary, whereas z_2 and z_5 are ordered categorical variables. We model time to death (in days), from time of operation, via the proportional hazard regression model (3.10), with one β parameter associated with x_1 and actually nine γ parameters associated with $z_1, \ldots, z_6$. This is since we represent the information about z_2, which takes values 1, 2, 3, 4, via three indicator variables, and similarly the information about z_5, which takes values 1, 2, 3, via two indicator variables.

Table 3.6 displays the five highest candidate models, along with their AIC and BIC values, having searched through all $2^6 = 64$ candidate models. In the table, '1,3,4' indicates the model that includes variables z_1, z_3, z_4 but not the others, etc. Note that the values are sorted in importance per criterion. The AIC method selects a model with the five variables z_1, z_2, z_3, z_4, z_6; only invasion depth z_5 is not selected. The BIC, on the other hand, selects only variables z_1 (tumour thickness) and z_4 (ulceration). Note that variable z_4 is present in all the five best BIC models. This example is another illustration of a situation where the BIC tends to select simpler models than AIC.

Whether the AIC or BIC advice ought to be followed depends on the aims of the statistical analysis. As we shall see in Chapter 4, a reasonable rule of thumb is that AIC is more associated with (average) prediction and (average) precision of estimates whereas the BIC is more linked to the task of searching for all non-null regression coefficients.

See also Example 6.12, where focussed information criteria are used to select models optimal for estimation of various given quantities. ∎

The proportional hazards assumption underlying the Cox model is sometimes questionable in biostatistical applications. The illustration we give below is one instance where a fully parametric hazard regression family fits the data very well, but where the resulting hazards for different individuals do not conform to the proportionality condition.

Assume in general terms that the hazard rate for individual i is of a certain parametric form $h_i(t) = h(t \mid x_i, \theta)$. The log-likelihood function for a survival data set of triples (t_i, x_i, ζ_i) may be expressed in several ways, for example as

$$\ell_n(\theta) = \sum_{i=1}^{n} \{\zeta_i \log f(t_i \mid x_i, \theta) + (1 - \zeta_i) \log \mathcal{S}(t_i \mid x_i, \theta)\},$$

in terms of the survival function $\mathcal{S}(t \mid x_i, \theta) = \exp\{-H(t \mid x_i, \theta)\}$ and density $f(t \mid x_i, \theta) = h(t \mid x_i, \theta)\mathcal{S}(t \mid x_i, \theta)$, with $H(t \mid x_i, \theta)$ the cumulative hazard. One may extend first of all the general likelihood theory to such parametric hazard regression models for survival and event history data, where there are distance measures between true hazards and modelled hazards that appropriately generalise the Kullback–Leibler divergence; see Hjort (1992a). Secondly, the essentials of the theory surveyed in Chapters 2 and 3 about the AIC, BIC, TIC and cross-validation may also be extended, with the appropriate modifications. The influence functions of hazard estimator functionals play a vital role here. We briefly illustrate the resulting methods in the following real data example.

Example 3.10 Survival analysis via Gamma process crossings*

We shall use Gamma process level crossing models to present an alternative analysis of a classic survival data set on carcinoma of the oropharynx. The data are given in Kalbfleisch and Prentice (2002, p. 378), and are for our purposes of the form $(t_i, \zeta_i, x_{i,1}, x_{i,2}, x_{i,3}, x_{i,4})$ for $i = 1, \ldots, n = 193$ patients, where t_i is the survival time, non-censored in case $\zeta_i = 1$ but censored in case $\zeta_i = 0$, and the four covariates are

- x_1: gender, 1 for male and 2 for female;
- x_2: condition, 1 for no disability, 2 for restricted work, 3 for requires assistance with self-care, and 4 for confined to bed;
- x_3: T-stage, an index of size and infiltration of tumour, ranging from 1 (a small tumour) to 4 (a massive invasive tumour);
- x_4: N-stage, an index of lymph node metastasis, ranging from 0 (no evidence of metastasis) to 3 (multiple positive or fixed positive nodes).

The task is to understand how the covariates influence different aspects of the survival mechanisms involved.

Aalen and Gjessing (2001) present an analysis of these data, in terms of a model for how quickly certain individual Gaussian processes with negative trends move from

Table 3.7. *Analysis of the oropharynx cancer survival data via the three parametric hazard rate models (i) the Aalen–Gjessing model; (ii) the Aalen–Gjessing model with exponential link; (iii) the Gamma process threshold crossing model. The table lists parameter estimates (with standard errors), along with attained log-likelihood maxima and AIC, TIC, BIC scores. The first five parameters are intercept and regression coefficients for the four covariates $x_1, \ldots, x_4$, while the two last lines relate to parameters (μ, τ^2) for the Aalen–Gjessing models and (a, κ) for the Gamma process model.*

	Model 1	Model 2	Model 3
ℓ_{max}	−393.508	−391.425	−386.585
AIC	−801.015	−796.851	−787.170
TIC	−805.928	−801.226	−788.463
BIC	−823.854	−819.689	−810.009
	estimates	estimates	estimates
−	5.730 (0.686)	2.554 (0.303)	0.513 (0.212)
x_1	0.101 (0.199)	0.010 (0.112)	0.087 (0.085)
x_2	−0.884 (0.128)	−0.626 (0.091)	−0.457 (0.087)
x_3	−0.548 (0.132)	−0.190 (0.052)	−0.127 (0.046)
x_4	−0.224 (0.090)	−0.120 (0.040)	−0.065 (0.031)
	0.341 (0.074)	0.386 (0.084)	6.972 (1.181)
	0.061 (0.053)	0.091 (0.066)	0.252 (0.034)

positive start positions $c_1, \ldots, c_n$ to zero. Their model takes $c_i = x_i^t \beta = \beta_0 + \beta_1 x_{i,1} + \cdots + \beta_4 x_{i,4}$ for $i = 1, \ldots, n$, where we write $x_i = (1, x_{i,1}, \ldots, x_{i,4})^t$ in concise vector notation. The model further involves, in addition to the $p = 5$ regression coefficients of c_i, a drift parameter μ and a variance parameter τ^2 associated with the underlying processes. Parameter estimates, along with standard errors (estimated standard deviations) and Wald test ratios z are given in Table 3.7. Other types of time-to-failure model associated with Gaussian processes are surveyed in Lee and Whitmore (2007).

Here we model the random survival times $T_1, \ldots, T_n$ as level crossing times $T_i = \min\{t: Z_i(t) \geq ac_i\}$ for $i = 1, \ldots, n$, where $Z_1, \ldots, Z_n$ are independent nondecreasing Gamma processes with independent increments, of the type

$$Z_i(t) \sim \text{Gamma}(aM(t), 1) \quad \text{with } M(t) = 1 - \exp(-\kappa t).$$

Different individuals have the same type of Gamma damage processes, but have different tolerance thresholds, which we model as

$$c_i = \exp(x_i^t \beta) = \exp(\beta_0 + \beta_1 x_{i,1} + \beta_2 x_{i,2} + \beta_3 x_{i,3} + \beta_4 x_{i,4}),$$

in terms of the four covariates x_1, x_2, x_3, x_4. The survival functions hence take the form $\mathcal{S}_i(t) = \mathrm{P}(Z_i(t) \leq ac_i) = G(ac_i, aM(t), 1)$, and define a model with $5 + 2 = 7$ parameters. Note that larger c_i values indicate better health conditions and life prospects, since the Gamma processes need longer time to cross a high c_i threshold than a low threshold.

Table 3.7 summarises analysis of three models, in terms of parameter estimates and standard errors (estimated standard deviations), along with attained log-likelihood maxima and AIC, TIC and BIC scores. The TIC or model-robust AIC score uses an appropriate generalisation of the penalty factor $\mathrm{Tr}(\widehat{J}^{-1}\widehat{K})$ for hazard rate models. The models are (1) the one used in Aalen and Gjessing (2001), with a linear link $c_i = x_i^t \beta$ for their start position parameters c_i, featuring coefficients $\beta_0, \ldots, \beta_5$ along with diffusion process parameters μ, τ^2; (2) an alternative model of the same basic type, but with an exponential link $c_i = \exp(x_i^t \beta)$ instead; and (3) the Gamma process level crossing model with threshold parameters $c_i = \exp(x_i^t \beta)$, concentration parameter a and transformation parameter κ. [For easier comparison with other published analyses we have used the data precisely as given in Kalbfleisch and Prentice (2002, p. 378), even though these may contain an error; see the data overview starting on page 290. We also report that we find the same parameter estimates as in Aalen and Gjessing (2001) for model (1), but that the standard errors they report appear to be in error.]

Our conclusions are first that the Aalen–Gjessing model with exponential link works rather better than in their analysis with a linear link, and second that the Gamma process level crossing model is clearly superior. Of secondary importance is that the subset model that uses only x_2, x_3, x_4, bypassing gender information x_1, works even better. We also point out that variations of model (3), in which individuals have damage processes working at different time scales, may work even better. Specifically, the model that lets $Z_i(t)$ be a Gamma process with parameters $aM_i(t)$, with $M_i(t) = 1 - \exp(-\kappa_i t)$ and $\kappa_i = \exp(\varepsilon v_i)$, leads via this one extra parameter ε to better AIC and BIC scores, for the choice v_i equal to $x_2 + x_3 + x_4$ minus its mean value. The point is that different patients have different 'onset times', moving at different speeds towards their supreme levels $Z_i(\infty)$. Such a model, with $5 + 3 = 8$ parameters, allows crossing hazards and survival curves, unlike the Cox model. ∎

3.5 The deviance information criterion

We have seen that the BIC is related to posterior probabilities for models and to marginal likelihoods. From the Bayesian perspective there are several alternatives for selecting a model. One of these is the deviance information criterion, the DIC (Spiegelhalter *et al.*, 2002). It is in frequent use in settings with Bayesian analysis of many parameters in complex models, where its computation typically is an easy consequence of output from Markov chain Monte Carlo simulations.

Deviances are most often discussed and used in regression situations where there is a well-defined 'saturated model', typically containing as many parameters as there are data points, thus creating a perfect fit. For most purposes what matters is differences of deviances, not so much the deviances themselves. We define the deviance, for a given model and given data y, as

$$D(y, \theta) = -2 \log f(y, \theta) = -2\ell_n(\theta),$$

involving the log-likelihood in some p-dimensional parameter. The deviance difference is

$$\mathrm{dd}(y, \theta_0, \widehat{\theta}) = D(y, \theta_0) - D(y, \widehat{\theta}) = 2\{\ell_n(\widehat{\theta}) - \ell_n(\theta_0)\}, \tag{3.11}$$

where $\widehat{\theta}$ could be the maximum likelihood estimator and θ_0 the underlying true parameter, or least false parameter in cases where the model used is not fully correct. Via a second-order Taylor expansion,

$$\mathrm{dd}(Y, \theta_0, \widehat{\theta}) \doteq_d n(\widehat{\theta} - \theta_0)^{\mathrm{t}} J_n(\widehat{\theta} - \theta_0) \xrightarrow{d} (U')^{\mathrm{t}} J^{-1} U', \tag{3.12}$$

where $U' \sim \mathrm{N}_p(0, K)$; the limit has expected value $p^* = \mathrm{Tr}(J^{-1}K)$. Again, if the model used is correct, then $J = K$ and $p^* = p$. This is one component of the story that leads to AIC; indeed one of its versions is $2\ell_n(\widehat{\theta}) - 2p^* = -\{D(y, \widehat{\theta}) + 2p^*\}$, cf. Section 2.5.

Spiegelhalter *et al.* (2002) consider a Bayesian version of the deviance difference, namely

$$\mathrm{dd}(y, \theta, \bar{\theta}) = D(y, \theta) - D(y, \bar{\theta}) = 2\{\ell_n(\bar{\theta}) - \ell_n(\theta)\}. \tag{3.13}$$

This is rather different, in interpretation and in execution, from (3.11). Here θ is seen as a random parameter, having arisen via a prior density $\pi(\theta)$; also, $\bar{\theta} = \mathrm{E}(\theta \mid \mathrm{data}) = \int \theta \pi(\theta \mid \mathrm{data}) \, \mathrm{d}\theta$ is the typical Bayes estimate, the posterior mean. So, in (3.13) data y and estimate $\bar{\theta}$ are fixed, while θ has its posterior distribution. Spiegelhalter *et al.* (2002) define

$$p_D = \mathrm{E}\{\mathrm{dd}(y, \theta, \bar{\theta}) \mid \mathrm{data}\} = \int \mathrm{dd}(y, \theta, \bar{\theta}) \pi(\theta \mid \mathrm{data}) \, \mathrm{d}\theta, \tag{3.14}$$

and then

$$\mathrm{DIC} = D(y, \bar{\theta}) + 2p_D, \tag{3.15}$$

intended as a Bayesian measure of fit. The p_D acts as a penalty term to the fit, and is sometimes interpreted as 'the effective number of parameters' in the model. Models with smaller values of the DIC are preferred.

We note that it may be hard to obtain explicit formulae for the DIC, but that its computation in practice is often easy in situations where analysis involves simulating samples of θ from the posterior, as in a fair portion of applied Bayesian contexts. One then computes $\bar{\theta}$ as the observed mean of a large number of simulated parameter vectors, and similarly p_D as the observed mean of $\mathrm{dd}(y, \theta, \bar{\theta})$, at virtually no extra computational cost. Note also that

$$p_D = \text{average}(D(\theta)) - D(\text{average}(\theta)) = \overline{D(y, \theta)} - D(y, \bar{\theta})$$

indicating posterior averages over a large number of simulated θs, and

$$\text{DIC} = \overline{D(y, \theta)} + p_D,$$

where $\overline{D(y, \theta)}$ is the posterior mean of $D(y, \theta)$.

It is already apparent that the DIC shares more similarities with the frequentist criterion AIC than with the Bayesian BIC. There is even a parallel result to (3.12), which we now describe. From (3.13), with a second-order Taylor expansion in θ around $\bar{\theta}$,

$$\text{dd}(y, \theta, \bar{\theta}) = -(\theta - \bar{\theta})^{\text{t}} \frac{\partial^2 \ell_n(\tilde{\theta})}{\partial \theta \partial \theta^{\text{t}}} (\theta - \bar{\theta}) = n(\theta - \bar{\theta})^{\text{t}} J_n(\tilde{\theta})(\theta - \bar{\theta}),$$

where $\tilde{\theta}$ is between the Bayes estimate $\bar{\theta}$ and the random θ. But a classic result about Bayesian parametric inference says that when information in data increases, compared to the dimension of the model, then $\theta \mid \text{data}$ is approximately distributed as a $N_p(\bar{\theta}, J_n(\bar{\theta})^{-1}/n)$. This is a version of the Bernshteĭn–von Mises theorem. This now leads to

$$\text{dd}(y, \theta, \bar{\theta}) \mid \text{data} \xrightarrow{d} (U')^{\text{t}} J^{-1} U', \tag{3.16}$$

the very same distributional limit as with (3.12), in spite of being concerned with a rather different set-up, with fixed data and random parameters. The consequence is that, for regular parametric families, and with data information increasing in comparison to the complexity of the model, p_D tends in probability to $p^* = \text{Tr}(J^{-1}K)$, and the model-robust AIC of (2.20) and the DIC will be in agreement. Spiegelhalter *et al.* (2002) give some heuristics for statements that with our notation mean $p_D \approx p^*$, but the stronger fact that (3.16) holds is apparently not noted in their article or among the ensuing discussion contributions. Thus an eclectic view on the DIC is that it works well from the frequentist point of view, by emulating the model-robust AIC; for an example of this phenomenon, see Exercise 3.10.

Software for computing the DIC is included in WinBUGS, for example, for broad classes of models.

Example 3.11 DIC selection among binomial models
Dobson (2002, Chapter 7) gives data on 50 years survival after graduation for students at the Adelaide University in Australia, sorted by year of graduation, by field of study, and by gender. For this particular illustration we shall consider a modest subset of these data, relating to students who graduated in the year 1940, and where the proportions of those who survived until 1990 were 11/25, 12/19, 15/18, 6/7, for the four categories (men, arts), (men, science), (women, arts), (women, science), respectively. We view the data as realisations of four independent binomial experiments, say $Y_j \sim \text{Bin}(m_j, \theta_j)$

for $j = 1, 2, 3, 4$, with $\theta_1, \theta_2, \theta_3, \theta_4$ representing the chance of surviving 50 years after graduation for persons in the four categories just described.

We shall evaluate and compare three models. Model M_1 takes $\theta_1 = \theta_2 = \theta_3 = \theta_4$; model M_2 takes $\theta_1 = \theta_2$ and $\theta_3 = \theta_4$; while model M_3 operates with four different probabilities. Thus the number of unknown probabilities used in the three models are respectively one, two, four. For each of the three models, we use a prior for the unknown probabilities of the type Beta($\frac{1}{2}c, \frac{1}{2}c$), where c may be 2 (corresponding to a uniform prior on the unit interval) or 1 (corresponding to the Jeffreys prior for an unknown binomial probability); the numerical illustration below uses $c = 1$. The likelihood is

$$\mathcal{L}(\theta_1, \theta_2, \theta_3, \theta_4) = A \prod_{j=1}^{4} \theta_j^{y_j}(1 - \theta_j)^{m_j - y_j} \quad \text{with } A = \prod_{j=1}^{4} \binom{m_j}{y_j}. \tag{3.17}$$

To compute the DIC as in (3.15), we need for each candidate model the posterior distribution of the four θ_j. Under model M_1, the common θ is Beta with parameters $(\frac{1}{2}c + \sum_{j=1}^{4} y_j, \frac{1}{2}c + \sum_{j=1}^{4}(m_j - y_j))$; under model M_2, $\theta_1 = \theta_2$ is Beta with $(\frac{1}{2}c + \sum_{j=1,2} y_j, \frac{1}{2}c + \sum_{j=1,2}(m_j - y_j))$ while $\theta_3 = \theta_4$ is Beta with $(\frac{1}{2}c + \sum_{j=3,4} y_j, \frac{1}{2}c + \sum_{j=3,4}(m_j - y_j))$; and finally under model M_3, the four θ_js are independent and Beta with parameters $(\frac{1}{2}c + y_j, \frac{1}{2}c + m_j - y_j)$. This in particular leads to Bayes estimates (posterior means)

$$\bar{\theta}_j = \left(\tfrac{1}{2}c + \sum_{j=1}^{4} y_j\right) \Big/ \left(c + \sum_{j=1}^{4} m_j\right) \quad \text{for } j = 1, 2, 3, 4$$

for model M_1; then

$$\bar{\theta}_j = \frac{\frac{1}{2}c + \sum_{j=1,2} y_j}{c + \sum_{j=1,2} m_j} \text{ for } j = 1, 2 \quad \text{and} \quad \bar{\theta}_j = \frac{\frac{1}{2}c + \sum_{j=3,4} y_j}{c + \sum_{j=3,4} m_j} \text{ for } j = 3, 4$$

for model M_2; and finally $\bar{\theta}_j = (\frac{1}{2}c + y_j)/(c + m_j)$ for $j = 1, 2, 3, 4$ for model M_3.

We are now in a position to compute the DIC score for the three models. We first need $D(y, \bar{\theta})$, with the Bayes estimates $\bar{\theta}_j$ just described inserted into $D(y, \theta) = -2 \sum_{j=1}^{4}\{y_j \log \theta_j + (m_j - y_j) \log(1 - \theta_j)\}$, where we choose to ignore the $-2 \log A$ term that is common to each model. Secondly, we need for each candidate model to consider

$$\text{dd}(y, \theta, \bar{\theta}) = 2 \sum_{j=1}^{4} \left\{ y_j \log \frac{\bar{\theta}_j}{\theta_j} + (m_j - y_j) \log \frac{1 - \bar{\theta}_j}{1 - \theta_j} \right\},$$

and find its posterior mean p_D, where data and the $\bar{\theta}_j$s are fixed, but where the θ_js have their posterior distribution. We find p_D for models M_1, M_2, M_3 by simulating a million θ_s vectors from the appropriate posterior distribution and then computing the mean of the recorded $\text{dd}(y, \theta_s, \bar{\theta})$; numerical integration is also easily carried out for this illustration.

Table 3.8. *Fifty years after graduation in 1940, for the four categories (men, arts),*
(men, science), (women, arts), (women, science), evaluated via three models: number
of free parameters; estimates of survival; $2\ell_{max}$; $D(y, \bar{\theta})$; p_D; AIC; DIC; and ranking.

	θ_1	θ_2	θ_3	θ_4	$2\ell_{max}$	$D(y, \bar{\theta})$	p_D	AIC	DIC
1	0.636	0.636	0.636	0.636	−90.354	90.355	0.991	−92.354	92.336 (3)
2	0.522	0.522	0.827	0.827	−82.889	82.920	1.926	−86.889	86.772 (1)
4	0.442	0.625	0.816	0.813	−81.267	81.406	3.660	−89.266	88.727 (2)

This analysis leads to results summarised in Table 3.8. We see that DIC and AIC agree
well, also about the ranking of models, with M_2 suggested as the best model – men and
women have different 50 years survival chances, but differences between arts and science
students, if any, are not clear enough to be taken into account, for this particular data
set that uses only 1940 students. Note that $D(y, \bar{\theta})$ is close to but not quite identical to
$-2\,\ell_{max}$, since the Bayes estimates (with Jeffreys priors) are close to but not equal to the
maximum likelihood estimators. Note further that the p_D numbers are close to but some
distance away from the model dimensions 1, 2, 4.

This simple illustration has used the data stemming from graduation year 1940 only. If
we supplement these data with corresponding information from graduation years 1938,
1939, 1941, 1942, and assume the probabilities for these classes are exactly as for 1940,
then preference shifts to the largest model M_3, for both AIC and DIC. Intermediate models
may also be formed and studied, concerning the relative proximity of these five times
four survival probabilities, and again the DIC, along with Bayesian model averaging,
may be used; see also Example 7.9. For some further calculations involving J and K
matrices and corrections to AIC, see Exercise 3.10. ■

3.6 Minimum description length

The principle of the minimum description length (MDL) stems from areas associated
with communication theory and computer science, and is at the outset rather different
from, for example, the BIC; indeed one may sometimes define and interpret the MDL
algorithm without connections to probability or statistical models as such. We shall see
that there are connections to the BIC, however.

The basic motivation of the MDL is to measure the *complexity* of models, after which
one selects the least complex candidate model. Kolmogorov (1965) defined the complexity of a sequence of numbers (the data set) as the length of the shortest computer
program that reproduces the sequence. The invariance theorem by Solomonoff (1964a,b)
states that the programming language is basically irrelevant. The Kolmogorov complexity, though, cannot be computed in practice. Rissanen (1987) introduced the *stochastic*
complexity of a data set with respect to a class of models as the shortest code length

(or description length) when coding is performed using the given model class. A model inside this class, which gives a good fit to the data, will give a short code length. We will not go into the coding aspects of data. Instead, we immediately make the link to probability distributions. As a consequence of the Kraft–McMillan inequality (see, for example, Section 5.5 of Cover and Thomas, 1991) it follows that for every code defined on a finite alphabet A, there exists a probability distribution P such that for all data sets $x \in A$, $P(x) = 2^{-L(x)}$, where L is the code length. And for every probability distribution P defined over a finite set A, there exists a code C such that its code length equals the smallest integer greater than or equal to $-\log P(x)$. Hence, minimising the description length corresponds to maximising the probability.

The MDL searches for a model in the set of models of interest that minimises the sum of the description length of the model, plus the description length of the data when fit with this model. Below we explicitly include the dependence of the likelihood on the observed data $y = (y_1, \ldots, y_n)$ and let $\widehat{\theta}_j(y)$ denote the dependence of the estimate on the data;

$$M_{\mathrm{MDL}} = \arg \min_{M_j} \big\{ - \ell_{n,j}(\widehat{\theta}_j(y)) + \text{code length}(M_j) \big\}.$$

We see that the MDL takes the complexity of the model itself into account, in a manner similar to that of the AIC and BIC methods, but with code length as penalty term. The determination of the code length of the model, or an approximation to it, is not unique. For a k-dimensional parameter which is estimated with $1/\sqrt{n}$ consistency, the complexity or code length can be argued to be approximately $-k \log(1/\sqrt{n}) = \frac{1}{2} k \log n$. In such cases,

$$M_{\mathrm{MDL}} = \arg \min_{M_j} \big\{ - \ell_{n,j}(\widehat{\theta}_j(y)) + \tfrac{1}{2} p_j \log n \big\},$$

with p_j the parameter length in model M_j; the resulting selection mechanism coincides with the BIC.

This equivalence does not hold in general. Rissanen (1996) obtained a more general approximation starting from a normalised maximum likelihood distribution

$$\mathcal{L}_{n,j}(\widehat{\theta}_j(y))/B_{n,j} = \mathcal{L}_{n,j}(\widehat{\theta}_j(y)) \Big/ \int_{A_{n,j}} \mathcal{L}_{n,j}(\widehat{\theta}_j(z_1, \ldots, z_n)) \, dz_1 \cdots dz_n,$$

with $A_{n,j}$ the set of data sets $z = (z_1, \ldots, z_n)$ for which $\widehat{\theta}_j(z)$ belongs to the interior $\mathrm{int}(\Omega_j)$ of the parameter space Ω_j for model j. The normalising constant can also be rewritten as

$$B_{n,j} = \int_{\theta \in \mathrm{int}(\Omega_j)} \int_{A'_{n,j}(\theta)} \mathcal{L}_{n,j}(\widehat{\theta}_j(z_1, \ldots, z_n)) \, dz_1 \cdots dz_n \, d\theta,$$

where $A'_{n,j}(\theta) = \{z_1, \ldots, z_n : \widehat{\theta}(z) = \theta\}$. Rissanen (1996) showed that if the estimator $\widehat{\theta}(y)$ satisfies a central limit theorem, and the parameter region is compact (bounded and closed), then

$$\log B_{n,j} = \tfrac{1}{2} p_j \log\left(\frac{n}{2\pi}\right) + \log \int_{\Omega_j} |J_j(\theta)|^{1/2}\, d\theta + o_P(1),$$

where $J_j(\theta)$ is the limiting Fisher information matrix, computed under the assumption that the model M_j is correct, cf. (2.8). It is also assumed that J_j is a continuous function of θ. This leads to a refined MDL criterion, selecting the model that minimises

$$-\ell_{n,j}(\widehat{\theta}_j(y)) + \tfrac{1}{2} p_j \log\left(\frac{n}{2\pi}\right) + \log \int_{\Omega_j} |J_j(\theta)|^{1/2}\, d\theta.$$

When comparing this to the more accurate BIC approximation $\mathrm{BIC}^*_{n,j}$ in (3.7), we see that with Jeffreys prior $\pi(\widehat{\theta}_j) = |J_j(\widehat{\theta}_j)|^{1/2} / \int |J_j(\theta_j)|^{1/2}\, d\theta_j$, there is exact agreement. When the parameter spaces are not compact the integral is typically infinite, and more complicated approximations hold. With priors different from the Jeffreys one, the (exact) Bayesian approach is no longer equivalent to the MDL. See also Myung *et al.* (2006) and Grünwald (2007) for a broader discussion.

3.7 Notes on the literature

There is much more to say about Bayesian model selection, the use of Bayes factors, intrinsic Bayes factors, and so on. For an overview of Bayesian model selection (and model averaging), see Wasserman (2000). A thorough review and discussion of Bayes factors can be found in Kass and Raftery (1995). The relationship with the Bayesian information criterion is clearly expressed in Kass and Wasserman (1995) and Kass and Vaidyanathan (1992). Ways of post-processing posterior predictive p-values for Bayesian model selection and criticism are developed in Hjort *et al.* (2006). Further relevant references are van der Linde (2004, 2005), and Lu *et al.* (2005).

Model selection in Cox regression models has received far less attention in the literature than its linear regression counterpart. Fan and Li (2002) introduced a penalised log-partial likelihood method, with a penalty called the smoothly clipped absolute deviation. There are two unknown parameters involved in this strategy, one is set fixed at a predetermined value, while the other is chosen via an approximation to generalised cross-validation. Bunea and McKeague (2005) studied a penalised likelihood where the penalty depends on the number of variables as well as on the number of parameters in the sieve construction to estimate $h_0(\cdot)$. The methods of Fan and Li (2002) and Bunea and McKeague (2005) have good properties regarding model consistency, but as we shall see in Chapter 4 this is associated with negative side-effects for associated risk functions. Augustin *et al.* (2005) study the problem of incorporating model selection uncertainty for survival data. Their model average estimators use weights estimated from bootstrap resampling. The lasso method for variable selection in the Cox model is studied by Tibshirani (1997). Especially when the number of variables q is large in comparison with the sample size, this method is of particular value, as are similar L_1-based methods of Efron *et al.* (2004). Methods presented in Section 3.4 are partly based on Hjort and Claeskens (2006). More general relative risk functions, different from the simple exponential, may also be studied, to generate important variations on the traditional Cox model, see DeBlasi and Hjort (2007).

A review paper on the MDL is Hansen and Yu (2001), and much more can be found in a recent tutorial provided by Grünwald (2005). De Rooij and Grünwald (2006) perform an empirical study of the MDL with infinite parametric complexity. Qian (1999) developed a stochastic complexity criterion for robust linear regression. The MDL is used by Antoniadis *et al.* (1997) for application in wavelet estimation.

Exercises

3.1 *Exponential, Weibull or Gamma for nerve impulse data:* Consider the nerve impulse data of Hand *et al.* (1994, case #166), comprising 799 reaction times ranging from 0.01 to 1.38 seconds. Use both AIC and the BIC to select among the exponential, the Weibull, and the Gamma models. Attempt also to find even better models for these data.

3.2 *Stackloss:* Consider again the stackloss data of Exercise 2.7. Perform a similar model selection search, now using the BIC.

3.3 *Switzerland in 1888:* Consider again the data on Swiss fertility measures of Exercise 2.8. Perform a similar model selection search, now using the BIC.

3.4 *The Ladies Adelskalenderen:* Refer to the data of Exercise 2.9. Perform a model selection search using the BIC. Compare the results with those obtained by using AIC.

3.5 *AIC and BIC with censoring:* Assume that survival data of the form (t_i, ζ_i) are available for $i = 1, \ldots, n$, with ζ_i an indicator for non-censoring. Generalise the formulae of Examples 2.5 and 3.1 to the present situation, showing how AIC and BIC may be used to decide between the exponential and the Weibull distribution, even when some of the failure times may be censored.

3.6 *Deviance in linear regression:* For normally distributed data in a linear regression model, verify that the deviance is equal to the residual sum of squares.

3.7 *Sometimes AIC = BIC:* Why are the AIC and BIC differences in Example 3.6 identical?

3.8 *The Raven:* In Example 3.8 we sorted words into categories short, middle, long if their lengths were respectively inside the 1–3, 4–5, 6–above cells. Use the same poem to experiment with other length categories and with four length categories rather than only three. Investigate whether AIC and the BIC select different memory orders.

3.9 *DIC for blood groups:* For Landsteiner's blood group data, see Example 3.6, place reasonable priors on respectively (a, b) and (p, q), and carry out DIC analysis for comparing the two candidate models.

3.10 *J and K calculations for binomial models:* Consider the situation of Example 3.11. This exercise provides the details associated with *J*- and *K*-type calculations, that are also required in order to use the model-robust AIC, for example.

(a) For model M_1, show that

$$J = \frac{\sum_{j=1}^{4} m_j \theta_j}{\theta^2} + \frac{\sum_{j=1}^{4} m_j (1 - \theta_j)}{(1 - \theta)^2} \quad \text{and} \quad K = \frac{\sum_{j=1}^{4} m_j \theta_j (1 - \theta_j)}{\theta^2 (1 - \theta)^2},$$

with the θ in the denominator denoting the under M_1 assumed common probability (i.e. the implied least false parameter value). Compute model-robust estimates of these via $\widehat{\theta}_j = y_j / m_j$ and $\widehat{\theta} = \sum_{j=1}^{4} y_j / \sum_{j=1}^{4} m_j$.

(b) For model M_2, show that J and K are diagonal 2×2 matrices, with elements respectively

$$\left(\frac{\sum_{j=1,2} m_j \theta_j}{\theta_a^2} + \frac{\sum_{j=1,2} m_j (1 - \theta_j)}{(1 - \theta_a)^2}, \frac{\sum_{j=3,4} m_j \theta_j}{\theta_b^2} + \frac{\sum_{j=3,4} m_j (1 - \theta_j)}{(1 - \theta_b)^2} \right)$$

and

$$\left(\frac{\sum_{j=1,2} m_j \theta_j (1 - \theta_j)}{\theta_a^2 (1 - \theta_a)^2}, \frac{\sum_{j=3,4} m_j \theta_j (1 - \theta_j)}{\theta_b^2 (1 - \theta_b)^2} \right),$$

with θ_a denoting the assumed common value of θ_1 and θ_2 and similarly θ_b the assumed common value of θ_3 and θ_4. Find estimates of J and K by inserting $\widehat{\theta}_j$ along with $(y_1 + y_2)/(m_1 + m_2)$ and $(y_3 + y_4)/(m_3 + m_4)$ for θ_a and θ_b.

(c) Considering finally the largest model M_3, show that J and K are both equal to the diagonal matrix with elements $m_j / \{\theta_j (1 - \theta_j)\}$. Implementing these findings, one arrives at $\text{Tr}(\widehat{J}^{-1} \widehat{K})$ numbers equal to 0.874, 1.963, 4.000 for models 1, 2, 3. These are in good agreement with the observed p_D numbers in the DIC strategy, as per theory.

4

A comparison of some selection methods

In this chapter we compare some information criteria with respect to consistency and efficiency, which are classical themes in model selection. The comparison is driven by a study of the 'penalty' applied to the maximised log-likelihood value, in a framework with increasing sample size. AIC is not strongly consistent, though it is efficient, while the opposite is true for the BIC. We also introduce Hannan and Quinn's criterion, which has properties similar to those of the BIC, while Mallows's C_p and Akaike's FPE behave like AIC.

4.1 Comparing selectors: consistency, efficiency and parsimony

If we make the assumption that there exists one true model that generated the data and that this model is one of the candidate models, we would want the model selection method to identify this true model. This is related to consistency. A model selection method is weakly consistent if, with probability tending to one as the sample size tends to infinity, the selection method is able to select the true model from the candidate models. Strong consistency is obtained when the selection of the true model happens almost surely. Often, we do not wish to make the assumption that the true model is amongst the candidate models. If instead we are willing to assume that there is a candidate model that is closest in Kullback–Leibler distance to the true model, we can state weak consistency as the property that, with probability tending to one, the model selection method picks such a closest model. In this chapter we explain why some information criteria have this property, and others do not. A different nice property that we might want an information criterion to possess is that it behaves 'almost as well', in terms of mean squared error, or expected squared prediction error, as the theoretically best model for the chosen type of squared error loss. Such a model selection method is called efficient. In Section 4.5 we give a more precise statement of the efficiency of an information criterion, and identify several information criteria that are efficient.

At the end of the chapter we will explain that consistency and efficiency cannot occur together, thus, in particular, a consistent criterion can never be efficient.

Several information criteria have a common form, which is illustrated by AIC, see
(2.1), and the BIC, see (3.1):

$$\text{AIC}\{f(\cdot;\theta)\} = 2\ell_n(\widehat{\theta}) - 2\,\text{length}(\theta),$$
$$\text{BIC}\{f(\cdot;\theta)\} = 2\ell_n(\widehat{\theta}) - (\log n)\,\text{length}(\theta).$$

Both criteria are constructed as twice the maximised log-likelihood value minus a penalty
for the complexity of the model. BIC's penalty is larger than that of AIC, for all n at
least 8. This shows that the BIC more strongly discourages choosing models with many
parameters.

For the kth model in our selection list ($k = 1, \ldots, K$), denote the parameter vector by
θ_k and the density function for the ith observation by $f_{k,i}$. For the regression situation,
$f_{k,i}(y_i, \theta_k) = f_k(y_i \mid x_i, \theta_k)$. This density function depends on the value of the covariate
x_i for the ith observation. For the i.i.d. case there is no such dependence and $f_{k,i} = f_k$,
for all observations. In its most general form, the data are not assumed to be independent.

Both information criteria AIC and BIC take the form

$$\text{IC}(M_k) = 2 \sum_{i=1}^{n} \log f_{k,i}(Y_i, \widehat{\theta}_k) - c_{n,k},$$

where $c_{n,k} > 0$ is the penalty for model M_k, for example, $c_{n,k} = 2\,\text{length}(\theta_k)$ for AIC.
The better model has the larger value of IC. The factor 2 is not really needed, though it
is included here for historical reasons. Other examples of criteria which take this form
are AIC_c, TIC and BIC^*.

A parsimonious model. One underlying purpose of model selection is to use the information criterion to select the model that is closest to the (unknown) but true model.
The true data-generating density is denoted by $g(\cdot)$. The Kullback–Leibler distance, see
equation (2.2), can be used to measure the distance from the true density to the model
density. If there are two or more models that minimise the Kullback–Leibler distance,
we wish to select that model which has the fewest parameters. This is called the most
parsimonious model.

Sin and White (1996) give a general treatment on asymptotic properties of information
criteria of the form $\text{IC}(M_k)$. We refer to that paper for precise assumptions on the models
and for proofs of the results phrased in this section.

Theorem 4.1 Weak consistency. *Suppose that amongst the models under consideration
there is exactly one model M_{k_0} which reaches the minimum Kullback–Leibler distance.
That is, for this model it holds that*

$$\liminf_{n \to \infty} \min_{k \neq k_0} n^{-1} \sum_{i=1}^{n} \{\text{KL}(g, f_{k,i}) - \text{KL}(g, f_{k_0,i})\} > 0.$$

*Let the strictly positive penalty be such that $c_{n,k} = o_p(n)$. Then, with probability going
to one, the information criterion selects this closest model M_{k_0} as the best model.*

Thus, in order to pick the Kullback–Leibler best model with probability going to one, or in other words, to have weak consistency of the criterion, the condition on the penalty is that when it is divided by n, it should tend to zero for growing sample size. Note that for the result on weak consistency to hold it is possible to have one of the $c_{n,k} = 0$, while all others are strictly positive.

As a consequence of the theorem, we immediately obtain the weak consistency of the BIC where $c_{n,k} = (\log n)\,\text{length}(\theta_k)$. The fixed penalty $c_{n,k} = 2\,\text{length}(\theta_k)$ of AIC also satisfies the condition, and hence AIC is weakly consistent under these assumptions.

Suppose now that amongst the models under consideration there are two or more models which reach the minimum Kullback–Leibler distance. Parsimony means that amongst these models the model with the fewest number of parameters, that is, the 'simplest model', is chosen. In the literature, this parsimony property is sometimes referred to as consistency, hereby ignoring the situation where there is a unique closest model. Consistency can be obtained under two types of technical condition. Denote by $\mathcal{J}$ the set of indices of the models which all reach the minimum Kullback–Leibler distance to the true model, and denote by $\mathcal{J}_0$ the subset of $\mathcal{J}$ containing models with the smallest dimension (there can be more than one such smallest model).

Theorem 4.2 Consistency. *Assume either set of conditions (a) or (b).*

(a) *Assume that for all $k_0 \neq \ell_0 \in \mathcal{J}$:*

$$\limsup_{n\to\infty} n^{-1/2} \sum_{i=1}^{n} \{\text{KL}(g, f_{k_0,i}) - \text{KL}(g, f_{\ell_0,i})\} < \infty.$$

For any index j_0 in $\mathcal{J}_0$, and for any index $\ell \in \mathcal{J}\backslash\mathcal{J}_0$, let the penalty be such that $\text{P}\{(c_{n,\ell} - c_{n,j_0})/\sqrt{n} \to \infty\} = 1$.

(b) *Assume that for all $k_0 \neq \ell_0 \in \mathcal{J}$, the log-likelihood ratio*

$$\sum_{i=1}^{n} \log \frac{f_{k_0,i}(Y_i, \theta_{k_0}^*)}{f_{\ell_0,i}(Y_i, \theta_{\ell_0}^*)} = O_P(1),$$

and that for any j_0 in $\mathcal{J}_0$ and $\ell \in \mathcal{J}\backslash\mathcal{J}_0$, $\text{P}(c_{n,\ell} - c_{n,j_0} \to \infty) = 1$.

Then, with probability tending to one the information criterion will pick such a smallest model:

$$\lim_{n\to\infty} \text{P}\left\{ \min_{\ell\in\mathcal{J}\backslash\mathcal{J}_0} \left(\text{IC}(M_{j_0}) - \text{IC}(M_\ell)\right) > 0 \right\} = 1.$$

Part (b) requires boundedness in distribution of the log-likelihood ratio statistic. The asymptotic distribution of such statistic is studied in large generality by Vuong (1989). The most well-known situation is when the models are nested. In this case it is known that twice the maximised log-likelihood ratio value follows asymptotically a chi-squared distribution with degrees of freedom equal to the difference in number of parameters of the two models, when the smallest model is true. Since the limiting distribution is a chi-squared distribution, this implies that it is bounded in probability (that is, $O_P(1)$), and hence the condition in (b) is satisfied.

The BIC penalty $c_{n,k} = (\log n)\text{length}(\theta_k)$ obviously satisfies the penalty assumption in (b), but the AIC penalty fails. Likewise, the AIC_c and TIC penalties fail this assumption.

In fact, criteria with a fixed penalty, not depending on sample size, do not satisfy either penalty condition in Theorem 4.2. This implies that, for example, AIC will not necessarily choose the most parsimonious model, there is a probability of overfitting. This means that the criterion might pick a model that has more parameters than actually needed. Hence with such criteria, for which the assumptions of Theorem 4.2 do not hold, there is a probability of selecting too many parameters when there are two or more models which have minimal Kullback–Leibler distance to the true model. We return to the topic of overfitting in Section 8.3.

4.2 Prototype example: choosing between two normal models

A rather simple special case of the general model selection framework is the following. Observations $Y_1, \ldots, Y_n$ are i.i.d. from the normal density $N(\mu, 1)$, with two models considered: model M_0 assumes that $\mu = 0$, while model M_1 remains 'general' and simply says that $\mu \in \mathbb{R}$. We investigate consequences of using different penalty parameters $c_{n,k} = d_n \text{length}(\theta_k)$. The values of the information criteria are

$$\text{IC}_0 = 2 \max_{\mu}\{\ell_n(\mu): M_0\} - d_n \cdot 0,$$
$$\text{IC}_1 = 2 \max_{\mu}\{\ell_n(\mu): M_1\} - d_n \cdot 1.$$

The log-likelihood function here is that of a normal model with unknown mean μ and known variance equal to one,

$$\ell_n(\mu) = -\tfrac{1}{2}\sum_{i=1}^{n}(Y_i - \mu)^2 - \tfrac{1}{2}n\log(2\pi) = -\tfrac{1}{2}n(\bar{Y} - \mu)^2 - \tfrac{1}{2}n\{\widehat{\sigma}^2 + \log(2\pi)\}.$$

The right-hand side of this equation is obtained by writing $Y_i - \mu = Y_i - \bar{Y} + \bar{Y} - \mu$, where $\bar{Y}$ as usual denotes the average and $\widehat{\sigma}^2 = n^{-1}\sum_{i=1}^{n}(Y_i - \bar{Y})^2$. Apart from the additive terms $-\tfrac{1}{2}n(\widehat{\sigma}^2 + \log(2\pi))$, not depending on the models,

$$\text{IC}_0 = -\tfrac{1}{2}n\bar{Y}^2 \quad \text{and} \quad \text{IC}_1 = -d_n,$$

showing that

$$\text{selected model} = \begin{cases} M_1 & \text{if } |\sqrt{n}\bar{Y}| \geq d_n^{1/2}, \\ M_0 & \text{if } |\sqrt{n}\bar{Y}| < d_n^{1/2}. \end{cases} \tag{4.1}$$

Secondly, the resulting estimator of the mean parameter is

$$\widehat{\mu} = \begin{cases} \bar{Y} & \text{if } |\sqrt{n}\bar{Y}| \geq d_n^{1/2}, \\ 0 & \text{if } |\sqrt{n}\bar{Y}| < d_n^{1/2}. \end{cases} \tag{4.2}$$

Primary candidates for the penalty factor include $d_n = 2$ for AIC and $d_n = \log n$ for the BIC. We now investigate consequences of such and similar choices, in the large-sample framework where n increases.

We first apply Theorem 4.1. Under the assumption that the biggest model is the true model, the true density g equals the $N(\mu, 1)$ density. Obviously, for model M_1 the Kullback–Leibler distance to the true model is equal to zero. For model M_0, this distance is equal to $\frac{1}{2}\mu^2$, which shows that the assumption of Theorem 4.1 on the KL-distances holds. In particular, both AIC and the BIC will consistently select the wide model, the model with the most variables, as the best one. However, if the true model is model M_0, then $\mu = 0$ and the difference in Kullback–Leibler values equals zero. Both models have the same Kullback–Leibler distance. For such a situation Theorem 4.1 is not applicable, instead we use Theorem 4.2. Since the two models with distributions $N(0, 1)$ and $N(\mu, 1)$ are nested, the limit of the log-likelihood ratio statistic, with μ estimated by its maximum likelihood estimator, has a χ_1^2 distribution if the $N(0, 1)$ model is true, and hence we need the additional requirement that the penalty diverges to infinity for growing sample size n in order to select the most parsimonious model with probability one. Only the BIC leads to consistent model selection in this case, AIC does not.

The model selection probabilities for the AIC scheme are easily found here. With Z standard normal, they are given by

$$P_n(M_1 \mid \mu) = P(|\sqrt{n}\mu + Z| \leq \sqrt{2}).$$

Figure 4.1 shows the probability $P_n(M_1 \mid \mu)$ for the AIC and BIC schemes, for a high number of data points, $n = 1000$. It illustrates that the practical difference between the two methods might not be very big, and that also AIC has detection probabilities of basically the same shape as with the BIC. We note that $P_n(M_1 \mid 0) = P(\chi_1^2 \geq d_n)$, which is 0.157 for the AIC method and which goes to zero for the BIC.

Next consider the risk performance of $\widehat{\mu}$, which we take to be the mean squared error multiplied by the sample size; this scaling is natural since variances of regular estimators are $O(1/n)$. We find

$$r_n(\mu) = n \, E_\mu\{(\widehat{\mu} - \mu)^2\} = E\big[\{(\sqrt{n}\mu + Z)I\{|\sqrt{n}\mu + Z| \geq d_n^{1/2}\} - \sqrt{n}\mu\}^2\big],$$

which can be computed and plotted via numerical integration or via an explicit formula. In fact, as seen via Exercise 4.3,

$$r_n(\mu) = 1 - \int_{\text{low}_n}^{\text{up}_n} z^2\phi(z)\,dz + n\mu^2\{\Phi(\text{up}_n) - \Phi(\text{low}_n)\}, \tag{4.3}$$

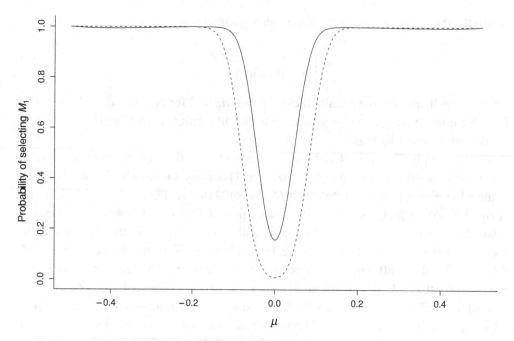

Fig. 4.1. The probability that model M_1 is selected, for the AIC (full line) and the BIC method (dashed line), for $n = 1000$ data points.

where

$$\text{low}_n = -d_n^{1/2} - \sqrt{n}\mu \quad \text{and} \quad \text{up}_n = d_n^{1/2} - \sqrt{n}\mu.$$

Risk functions are plotted in Figure 4.2 for three information criteria: the AIC with $d_n = 2$, the BIC with $d_n = \log n$, and the scheme that uses $d_n = \sqrt{n}$. Here AIC does rather better than the BIC; in particular, its risk function is bounded (with maximal value 1.647, for all n, actually), whereas $r_n(\mu)$ for the BIC exhibits much higher maximal risk. We show in fact below that its max-risk $r_n = \max r_n(\mu)$ is unbounded, diverging to infinity as n increases.

The information criterion with $d_n = \sqrt{n}$ is included in Figure 4.2, and in our discussion, since it corresponds to the somewhat famous estimator

$$\widehat{\mu}_H = \begin{cases} \bar{Y} & \text{if } |\bar{Y}| \geq n^{-1/4}, \\ 0 & \text{if } |\bar{Y}| < n^{-1/4}. \end{cases}$$

The importance of this estimator, which stems from J. L. Hodges Jr., cf. Le Cam (1953), is not that it is meant for practical use, but that it exhibits what is known as 'superefficiency' at zero:

$$\sqrt{n}(\widehat{\mu}_H - \mu) \xrightarrow{d} \begin{cases} \mathrm{N}(0, 1) & \text{for } \mu \neq 0, \\ \mathrm{N}(0, 0) & \text{for } \mu = 0. \end{cases} \tag{4.4}$$

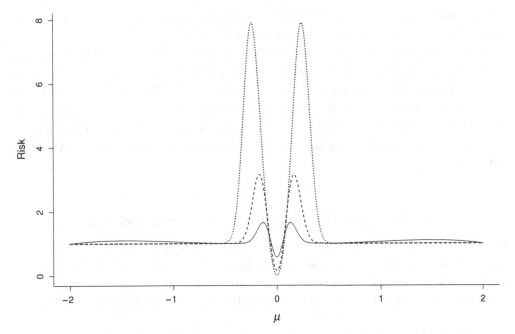

Fig. 4.2. Risk functions for estimators $\widehat{\mu}$ of 4.2, for $n = 200$, for penalties d_n equal to 2 (AIC, full line), $\log n$ (BIC, dashed line), $\sqrt{n}$ (Hodges, dotted line), with maximal risks equal to 1.647, 3.155, 7.901 respectively. The conservative estimator $\widehat{\mu} = \bar{Y}$ has constant risk function equal to 1.

Here the limit constant 0 is written as N(0, 0), to emphasise more clearly the surprising result that the variance of the limit distribution is $v(\mu) = 1$ for $\mu \neq 0$ but $v(\mu) = 0$ for $\mu = 0$. The result is not only surprising but also alarming in that it appears to clash with the Cramér–Rao inequality and with results about the asymptotic optimality of maximum likelihood estimators. At the same time, there is the unpleasant behaviour of the risk function close to zero for large n, as illustrated in Figure 4.2. Thus there are crucial differences between pointwise and uniform approximations of the risk functions; result (4.4) is mathematically true, but only in a restricted pointwise sense.

It can be shown that consistency at the null model (an attractive property) actually implies unlimited max-risk with growing n (an unattractive property). To see this, consider the risk expression (4.3) at points $\mu = \delta/\sqrt{n}$, where

$$r_n(\delta/\sqrt{n}) \geq \delta^2 \big\{ \Phi\big(\delta + d_n^{1/2}\big) - \Phi\big(\delta - d_n^{1/2}\big) \big\}.$$

At the particular point $\mu = (d_n/n)^{1/2}$, therefore, the risk is bigger than the value $d_n\{\Phi(2d_n^{1/2}) - \frac{1}{2}\}$, which diverges to infinity precisely when $d_n \to \infty$, which happens if and only if $P_n(M_0 \mid \mu = 0) \to 1$. That is, when there is consistency at the null model. As illustrated in Figure 4.2, the phenomenon is more pronounced for the Hodges estimator than for the BIC. The aspect that an estimator-post-BIC carries max risk proportional to $\log n$, and that this max risk is met close to the null model at which consistency is achieved, remains however a somewhat negative property.

We have uncovered a couple of crucial differences between the AIC and BIC methods. (i) The BIC behaves very well from the consistency point of view; with large n it gives a precise indication of which model is correct. This might or might not be relevant in the context in which it is used. (ii) But it pays a price for its null model consistency property: the risk function for the μ estimator exhibits unpleasant behaviour near the null model, and its maximal risk is unbounded with increasing sample size. In this respect AIC fares rather better.

4.3 Strong consistency and the Hannan–Quinn criterion

Weak consistency is defined via weak convergence: it states that with probability converging to one, a most parsimonious model with minimum Kullback–Leibler distance to the true model is selected. The results can be made stronger by showing that under some conditions such a model is selected almost surely. This is called strong consistency.

Theorem 4.3 *Strong consistency. Suppose that amongst the models under consideration there is exactly one model M_{k_0} which reaches the minimum Kullback–Leibler distance, such that*

$$\liminf_{n\to\infty} \min_{k\neq k_0} n^{-1} \sum_{i=1}^{n} \{\mathrm{KL}(g, f_{k,i}) - \mathrm{KL}(g, f_{k_0,i})\} > 0.$$

Let the strictly positive penalty be such that $c_{n,k} = o(n)$ almost surely. Then

$$\mathrm{P}\Big\{ \min_{\ell\neq k_0} \big(\mathrm{IC}(M_{k_0}) - \mathrm{IC}(M_\ell)\big) > 0, \text{ for almost all } n\Big\} = 1.$$

The assumption on the penalty is very weak, and is satisfied for the nonstochastic penalties of AIC and the BIC. This tells us that there are situations where AIC, and the BIC, are strongly consistent selectors of the model that is best in minimising the Kullback–Leibler distance to the true model.

The situation is more complex if there is more than one model reaching minimum Kullback–Leibler distance to the true model. A general treatment of this subject is presented in Sin and White (1996). For details we refer to that paper. The main difference between showing weak and strong consistency is, for the latter, the use of the law of the iterated logarithm instead of the central limit theorem.

Using the notation of Theorem 4.2, if for all $k_0 \neq \ell_0 \in \mathcal{J}$:

$$\limsup_{n\to\infty} \frac{1}{\sqrt{n \log\log n}} \sum_{i=1}^{n} \{\mathrm{KL}(g, f_{k_0,i}) - \mathrm{KL}(g, f_{\ell_0,i})\} \leq 0,$$

then (under some additional assumptions) the requirement on the penalty $c_{n,k}$ to guarantee strong consistency of selection is that

$$P(c_{n,k} \geq a_n \sqrt{n \log \log n} \text{ for almost all } n) = 1,$$

where a_n is a random sequence, almost surely bounded below by a strictly positive number.

If it is rather the situation that for all $k_0 \neq \ell_0 \in \mathcal{J}$, the log-likelihood ratio

$$\sum_{i=1}^{n} \log \frac{f_{k_0,i}(Y_i, \theta_{k_0}^*)}{f_{\ell_0,i}(Y_i, \theta_{\ell_0}^*)} = o(\log \log n) \text{ almost surely,}$$

then the required condition on the penalty is that

$$P(c_{n,k} \geq b_n \log \log n \text{ for almost all } n) = 1,$$

where b_n is a random sequence, almost surely bounded below by a strictly positive number.

For a sequence of strictly nested models, the second condition on the penalty is sufficient (Sin and White, 1996, Corollary 5.3).

The BIC penalty $c_{n,k} = \log n \, \text{length}(\theta_k)$ satisfies these assumptions, leading to strong consistency of model selection, provided the other assumptions hold. The above results indicate that $\log n$ is not the slowest rate by which the penalty can grow to infinity in order to almost surely select the most parsimonious model. The application of the law of the iterated logarithm to ensure strong consistency of selection leads to Hannan and Quinn's (1979) criterion

$$\text{HQ}\{f(\cdot; \theta)\} = 2 \log \mathcal{L}(\widehat{\theta}) - 2c \log \log n \, \text{length}(\theta), \text{ with } c > 1.$$

The criterion was originally derived to determine the order in an autoregressive time series model. Hannan and Quinn (1979) do not give any advice on which value of c to choose. Note that for practical purposes, this choice of penalty might not be that useful. Indeed, even for very large sample sizes the quantity $\log \log n$ remains small, and whatever method is used for determining a threshold value c will be more important than the $\log \log n$ factor in determining the value of $2c \log \log n$.

4.4 Mallows's C_p and its outlier-robust versions

A criterion with behaviour similar to that of AIC for variable selection in regression models is Mallows's (1973)

$$C_p = \text{SSE}_p / \widehat{\sigma}^2 - (n - 2p). \tag{4.5}$$

Here SSE_p is the residual sum of squares $\sum_{i=1}^{n}(y_i - \widehat{y}_i)^2$ in the model with p regression coefficients, and the variance is computed in the largest model. C_p is defined as an estimator of the scaled squared prediction error

$$E\left\{\sum_{i=1}^{n}(\widehat{Y}_i - EY_i)^2\right\}\Big/\sigma^2.$$

If the model with p variables contains no bias, then C_p is close to p. If, on the other hand, there is a large bias, then $C_p > p$. Values close to the corresponding p (but preferably smaller than p) indicate a good model. A plot of C_p versus p often helps in identifying good models.

Ronchetti and Staudte (1994) define an outlier-robust version of Mallows's C_p, as an estimator of $E\{\sum_{i=1}^{n} \widehat{w}_i^2(\widehat{Y}_i - EY_i)^2\}/\sigma^2$, which is a weighted scaled squared prediction error, using the robust weights defined in Section 2.10.3. In detail,

$$RC_p = W_p\widehat{\sigma}^2 - (U_p - V_p),$$

where we need the following definitions. With $\eta(x, \varepsilon)$ as defined in Section 2.10.3, define $M = E\{(\partial/\partial\varepsilon)\eta(x, \varepsilon)xx^t\}$, $W = E(w^2xx^t)$, $R = E\{\eta^2(x, \varepsilon)xx^t\}$, $N = E\{\eta^2(\partial/\partial\varepsilon)\eta xx^t\}$ and $L = E[\{((\partial/\partial\varepsilon)\eta)^2 + 2(\partial/\partial\varepsilon)\eta w - 3w^2\}xx^t]$. Then we let $V_p = \text{Tr}(WM^{-1}RM^{-1})$ and

$$U_p - V_p = \sum_{i=1}^{n} E\{\eta^2(x_i, \varepsilon_i)\} - 2\,\text{Tr}(NM^{-1}) + \text{Tr}(LM^{-1}RM^{-1}).$$

In practice, approximations to U_p and V_p are used, see Ronchetti and Staudte (1994). The robust C_p is, for example, available in the S-Plus library `Robust`.

A different robustification of C_p is arrived at via a weighting scheme similar to that used to obtain the weighted AIC in Section 2.10.1. The weight functions $w(\cdot)$ are here based on smoothed Pearson residuals. This leads to

$$WC_p = \widehat{\sigma}^{-2}\sum_{i=1}^{n} w(y_i - x_i^t\widehat{\beta}, \widehat{\sigma})(y_i - x_i^t\widehat{\beta})^2 - \sum_{i=1}^{n} w(y_i - x_i^t\widehat{\beta}, \widehat{\sigma}) + 2\,p.$$

Agostinelli (2002) shows that without the presence of outliers the WC_p is asymptotically equivalent to C_p. When the weight function $w(\cdot) \equiv 1$, then $WC_p = C_p$.

4.5 Efficiency of a criterion

Judging model selection criteria is not only possible via a study of the selection of the most parsimonious subset of variables (consistency issues). The approach taken in this section is a study of the efficiency of the criterion with respect to a loss function.

Let us consider an example. We wish to select the best set of variables in the regression model

$$Y_i = \beta_0 + \beta_1 x_{i,1} + \cdots + \beta_k x_{i,k} + \varepsilon_i, \quad i = 1, \ldots, n,$$

where $\text{Var}\,\varepsilon_i = \sigma^2$, with the specific purpose of predicting a new (independent) outcome variable $\widehat{Y}_i$ at the observed covariate combination $x_i = (x_{i,1}, \ldots, x_{i,k})^{\mathrm{t}}$, for $i = 1, \ldots, n$. For prediction, the loss is usually taken to be the squared prediction error. This means that we wish to select that set S of covariates x_j, $j \in S$ for which the expected prediction error, conditional on the observed data $\mathcal{Y}_n = (Y_1, \ldots, Y_n)$, is as small as possible. That is, we try to minimise

$$\sum_{i=1}^{n} \mathrm{E}\{(\widehat{Y}_{S,i} - Y_{\mathrm{true},i})^2 | \mathcal{Y}_n\}, \tag{4.6}$$

where it is understood that the $Y_{\mathrm{true},i}$ are independent of $Y_1, \ldots, Y_n$, but otherwise come from the same distribution. The predicted values $\widehat{Y}_{S,i}$ depend on the data $\mathcal{Y}$. The notation $\widehat{Y}_{S,i}$ indicates that the prediction is made using the model containing only the covariates x_j, $j \in S$. Using the independence between the new observations and the original $\mathcal{Y}_n = (Y_1, \ldots, Y_n)$, we can write

$$\sum_{i=1}^{n} \mathrm{E}\{(\widehat{Y}_{S,i} - Y_{\mathrm{true},i})^2 | \mathcal{Y}_n\} = \sum_{i=1}^{n} \mathrm{E}[\{\widehat{Y}_{S,i} - \mathrm{E}(Y_{\mathrm{true},i})\}^2 | \mathcal{Y}_n] + n\sigma^2$$

$$= \sum_{i=1}^{n} (\widehat{\beta}_S - \beta_{\mathrm{true}})^{\mathrm{t}} x_i x_i^{\mathrm{t}} (\widehat{\beta}_S - \beta_{\mathrm{true}}) + n\sigma^2.$$

To keep the notation simple, we use β_S also for the vector with zeros inserted for undefined entries, to make the dimensions of β_S and β_{true} equal. If the model selected via a model selection criterion reaches the minimum of (4.6) as the sample size n tends to infinity, we call the selection criterion *efficient*, conditional on the given set of data. Instead of conditioning on the given sample $\mathcal{Y}_n$, theoretically, we rather work with the (unconditional) expected prediction error

$$L_n(S) = \sum_{i=1}^{n} \mathrm{E}\{(\widehat{Y}_{S,i} - Y_{\mathrm{true},i})^2\}, \tag{4.7}$$

which for the regression situation reads

$$L_n(S) = \mathrm{E}\{(\widehat{\beta}_S - \beta_{\mathrm{true}})^{\mathrm{t}} X^{\mathrm{t}} X (\widehat{\beta}_S - \beta_{\mathrm{true}})\} + n\sigma^2. \tag{4.8}$$

Denote by S^* the index set for which the minimum value of the expected prediction error is attained. Let $\widehat{S}$ be the set of indices in the selected model. The notation $\mathrm{E}_{\widehat{S}}$ denotes that the expectation is taken with respect to all random quantities except for $\widehat{S}$. The criterion used to select $\widehat{S}$ is efficient when

$$\frac{\sum_{i=1}^{n} \mathrm{E}_{\widehat{S}}\{(\widehat{Y}_{\widehat{S},i} - Y_{\mathrm{true},i})^2\}}{\sum_{i=1}^{n} \mathrm{E}\{(\widehat{Y}_{S^*,i} - Y_{\mathrm{true},i})^2\}} = \frac{L_n(\widehat{S})}{L_n(S^*)} \xrightarrow{P} 1, \quad \text{as } n \to \infty.$$

In other words, a model selection criterion is called *efficient* if it selects a model such that the ratio of the expected loss function at the selected model and the expected loss function at its theoretical minimiser converges to one in probability. Note that since we use a selected model $\widehat{S}$, the numerator is a random variable.

When focus is on estimation rather than prediction, we look at the squared estimation error. After taking expectations this leads to studying the mean squared error. We show below, in the context of autoregressive models and normal linear regression, that some model selectors, including AIC, are efficient.

4.6 Efficient order selection in an autoregressive process and the FPE

Historically, the efficiency studies originate from the time series context where a model is sought that asymptotically minimises the mean squared prediction error. Shibata (1980) formulates the problem for the order selection of a linear time series model. Let ε_i ($i = \ldots, -1, 0, 1, \ldots$) be independent and identically distributed $N(0, \sigma^2)$ random variables. Consider the stationary Gaussian process $\{X_i\}$ such that for real-valued coefficients a_j,

$$X_i - (a_1 X_{i-1} + a_2 X_{i-2} + \cdots) = \varepsilon_i.$$

In practice, the time series is truncated at order k, resulting in the kth-order autoregressive model, built from observations $X_1, \ldots, X_n$,

$$X_i - (a_1 X_{i-1} + \cdots + a_k X_{i-k}) = \varepsilon_i, \quad i = 1, \ldots, n. \tag{4.9}$$

In this model we estimate the unknown coefficients $\boldsymbol{a}_k = (a_1, \ldots, a_k)^{\mathrm{t}}$ by $\widehat{\boldsymbol{a}}_k = (\widehat{a}_1(k), \ldots, \widehat{a}_k(k))^{\mathrm{t}}$. The purpose of fitting such an autoregressive model is often to predict the one-step-ahead value Y_{n+1} of a new, independent realisation of the time series, denoted $\{Y_i\}$. The one-step-ahead prediction is built from the model (4.9):

$$\widehat{Y}_{n+1} = \widehat{a}_1(k) Y_n + \widehat{a}_2(k) Y_{n-1} + \cdots + \widehat{a}_k(k) Y_{n-k+1}.$$

Conditional on the original time series $X_1, \ldots, X_n$, the mean squared prediction error of $\widehat{Y}_{n+1}$ equals

$$\begin{aligned}
&\mathrm{E}[(Y_{n+1} - \widehat{Y}_{n+1})^2 | X_1, \ldots, X_n] \\
&= \mathrm{E}[(Y_{n+1} - a_1 Y_{n+j-1} - \cdots - a_k Y_{n+1-k} \\
&\quad - \{(\widehat{a}_1(k) - a_1) Y_{n+1-1} - \cdots - (\widehat{a}_k(k) - a_k) Y_{n+1-k}\})^2 | X_1, \ldots, X_n] \\
&= \sigma^2 + (\widehat{\boldsymbol{a}}_k - \boldsymbol{a})^{\mathrm{t}} \Gamma_k (\widehat{\boldsymbol{a}}_k - \boldsymbol{a}),
\end{aligned} \tag{4.10}$$

where the (i, j)th entry of Γ_k is defined as $\mathrm{E}(Y_i Y_j)$, for $i, j = 1, \ldots, k$. For the equality in the last step, we have used the independence of $\{Y_i\}$ and $X_1, \ldots, X_n$.

Before continuing with the discussion on efficiency, we explain how (4.10) leads to Akaike's final prediction error (Akaike, 1969)

$$\mathrm{FPE}(k) = \widehat{\sigma}_k^2 \frac{n+k}{n-k}, \tag{4.11}$$

where $\widehat{\sigma}_k^2$ is the maximum likelihood estimator of σ^2 in the truncated model of order k, that is, $\widehat{\sigma}_k^2 = \text{SSE}_k/n$. FPE is used as an order selection criterion to determine the best truncation value k, by selecting the k that minimises FPE(k). This criterion is directed towards selecting a model which performs well for prediction. For a modified version, see Akaike (1970).

The expression in (4.10) is a random variable. To compute its expectation with respect to the distribution of $X_1, \ldots, X_n$, we use that for maximum likelihood estimators $\widehat{a}_k$ (see Brockwell and Davis, 1991, Section 8.8),

$$\sqrt{n}(\widehat{a}_k - a_k) \overset{d}{\to} \text{N}(0, \sigma^2 \Gamma_k^{-1}).$$

This implies that for large n we may use the approximation

$$\sigma^2 + \text{E}\{(\widehat{a}_k - a)^{\text{t}} \Gamma_k (\widehat{a}_k - a)\} \approx \sigma^2(1 + k/n).$$

Replacing the true unknown σ^2 by the unbiased estimator $n\widehat{\sigma}_k^2/(n-k)$, leads to the FPE($k$) expression (4.11).

To prove efficiency, some assumptions on the time series are required. We refer to Lee and Karagrigoriou (2001), who provide a set of minimal assumptions in order for efficiency to hold. The main result is the following:

Theorem 4.4 *Under the assumptions as in Lee and Karagrigoriou (2001):*

(a) The criteria $\text{AIC}(k) = -n \log(\widehat{\sigma}_k^2) - 2(k+1)$, *the corrected version* $\text{AIC}_c(k) = -n \log(\widehat{\sigma}_k^2) - n(n+k)/(n-k-2)$, $\text{FPE}(k)$, $S_n(k) = \text{FPE}(k)/n$, *and* $\text{FPE}_\alpha(k) = \widehat{\sigma}_k^2$ $(1 + \alpha k/n)$ *with* $\alpha \neq 2$, *are all asymptotically efficient.*

(b) The criteria $\text{BIC}(k) = \log \widehat{\sigma}_k^2 + k \log n$ *and the Hannan–Quinn criterion* $\text{HQ}(k) = n \log \widehat{\sigma}_k^2 + 2ck \log \log n$ *with* $c > 1$ *are not asymptotically efficient.*

The assumption of predicting the next step of a new and independent series $\{Y_t\}$ is not quite realistic, since in practice the same time series is used for both selecting the order and for prediction. Ing and Wei (2005) develop an efficiency result for AIC-like criteria for what they term same-realisation predictions where the original time series is also used to predict future values. Their main conclusion is that the efficiency result still holds for AIC for the one-series case.

4.7 Efficient selection of regression variables

The onset to studying efficiency in regression variable selection is given by Shibata (1981, 1982) for the situation of a true model containing an infinite or growing number of regression variables. For a normal regression model, we follow the approach of Hurvich and Tsai (1995b) to obtain the efficiency of a number of model selection criteria.

We make the assumption that for all considered index sets S, the corresponding design matrix X_S (containing in its columns those variables x_j with j in S) has full rank $|S|$, where $|S|$ denotes the number of variables in the set S, and that $\max_S |S| = o(n^a)$ for a constant $a \in (0, 1]$. There is another technical requirement that asks for the existence of a constant $b \in [0, 0.5)$ such that for any $c > 0$,

$$\sum_S \exp\left\{ - cn^{-2b} L_n(S)\right\} \to 0, \quad \text{for } n \to \infty.$$

This implies in particular that for all S, $n^{-2b} L_n(S) \to \infty$. This condition is fulfilled when the number of regression variables in the true model is infinite, or for models where the number of variables grows with the sample size. One such example is a sequence of nested models,

$$S_1 = \{1\} \subset S_2 = \{1, 2\} \subset \cdots \subset S_q = \{1, \ldots, q\} \subset \cdots,$$

where $|S_q| = q$. Hurvich and Tsai (1995a) (see also Shibata, 1981) provide some ways to check the technical assumption.

The next theorem states the asymptotic efficiency of the specific model selection criterion

$$\text{IC}(S) = (n + 2|S|)\widehat{\sigma}_S^2, \tag{4.12}$$

where the variance in model S is estimated via $n^{-1}(Y - X_S \widehat{\beta}_S)^t (Y - X_S \widehat{\beta}_S) = \text{SSE}_S/n$. For the proof we refer to Hurvich and Tsai (1995a).

Theorem 4.5 *Let $\widehat{S}_{\text{IC}}$ denote the set selected by minimising criterion (4.12). If the assumptions hold, then, with S^* defined as the minimiser of L_n in (4.7), and $c = \min\{(1 - a)/2, b\}$,*

$$\frac{L_n(\widehat{S}_{\text{IC}})}{L_n(S^*)} - 1 = o_P(n^{-c}).$$

The criterion in (4.12) is for normal regression models closely related to AIC. It is hence no surprise that AIC has a similar behaviour, as the next corollary shows. The same can be said of Mallows's (1973) C_p, see (4.5).

Corollary 4.1 *(a) The criteria AIC, final prediction error FPE, see (4.11), their modifications FPE(S)/n, $\text{AIC}_c(S) = \text{AIC}(S) + 2(|S| + 1)(|S| + 2)/(n - |S| + 2)$ and Mallows's C_p are all asymptotically efficient under the conditions of Theorem 4.5.*

(b) The criteria BIC and the Hannan–Quinn criterion HQ are not asymptotically efficient.

This corollary follows using the method of proof of Shibata (1980, theorem 4.2). Nishii (1984) has shown that FPE and Mallows's C_p are asymptotically equivalent.

4.8 Rates of convergence*

The result in Theorem 4.5, as obtained by Hurvich and Tsai (1995b), is stronger than just showing efficiency. Indeed, it gives the rate by which the ratio of the function L_n

evaluated at the set chosen by the information criterion, and at the optimal set, converges to one; this rate is stated as $o_P(n^{-c})$, where $c = \min\{(1-a)/2, b\}$. For $a = 1$ and $b = 0$, the theorem reduces to the original result of Shibata (1981). In this case we have that $c = 0$ and that the ratio is of the order $o_P(1)$, or in other words, there is convergence to zero in probability. When $c = 0$ we do not know how fast the convergence to zero is. The constants a and b determine how fast the rate of convergence will be. The a value is related to the size of the largest model. The value $0 < a \leq 1$ is such that $\max_S |S| = o(n^a)$. The rate of convergence increases when a decreases from 1 to $1 - 2b$. However, a smaller value of a implies a smaller dimension of the largest model considered. This means that it cannot always be advised to set a to its smallest value $1 - 2b$, since then we are possibly too much restricting the number of parameters in the largest model. The value $0 \leq b < 1/2$ gives us information on the speed by which L_n diverges. From the assumption we learn that b is such that $n^{-2b}L_n(S) \to \infty$.

By changing the values for a and b to have $a \to 0$ and $b \to 1/2$, one obtains rates close to the parametric rate of convergence $o_P(n^{-1/2})$.

For a study on convergence rates of the generalised information criterion for linear models, we refer to Shao (1998). Zhang (1993) obtained rates of convergence for AIC and BIC for normal linear regression.

4.9 Taking the best of both worlds?*

Both AIC and the BIC have good properties, AIC is efficient and the BIC is consistent. A natural question is whether they can be combined. Bozdogan (1987) studies the bias introduced by the maximum likelihood estimators of the parameters and proposes two adjustments to the penalty of AIC. This leads to his 'corrected AIC' of which a first version is defined as $\text{CAIC} = 2\,\ell(\widehat{\theta}) - \text{length}(\theta)(\log n + 1)$. In the notation of Chapter 3, $\text{CAIC} = \text{BIC} - \text{length}(\theta)$. A second version uses the information matrix $J_n(\widehat{\theta})$ and is defined as $\text{CAICF} = 2\,\ell(\widehat{\theta}) - \text{length}(\theta)(\log n + 2) - \log |J_n(\widehat{\theta})|$. Bozdogan (1987) shows that for both corrected versions the probability of overfitting goes to zero asymptotically, similarly as for the BIC, while keeping part of the penalty as in AIC, namely a constant times the dimension of θ.

A deeper question is whether the consistency of the BIC can be combined with the efficiency of AIC. Yang (2005) investigates such questions and comes to a negative answer. More precisely, he investigates a combination of consistency and minimax rate optimality properties in models of the form $Y_i = f(x_i) + \varepsilon_i$. A criterion is minimax rate optimal over a certain class of functions $\mathcal{C}$ when the worst case situation for the risk, that is $\sup_{f \in \mathcal{C}} n^{-1} \sum_{i=1}^{n} \text{E}[\{f(x_i) - f_{\widehat{S}}(x_i, \widehat{\theta}_{\widehat{S}})\}^2]$, converges at the same rate as the best possible worst case risk

$$\inf_{\widehat{f}} \sup_{f \in \mathcal{C}} n^{-1} \sum_{i=1}^{n} \text{E}[\{\widehat{f}(x_i) - f(x_i)\}^2].$$

In the above, the selected model is denoted $f_{\widehat{S}}$, while $\widehat{f}$ is any data-based estimator of f. AIC can be shown to be minimax rate optimal, while the BIC does not have this property. Changing the penalty constant 2 in AIC to some other value takes away this favourable situation. Hence, just changing the penalty constant cannot lead to a criterion that has both properties. In theorem 1 of Yang (2005), he proves the much stronger result that any consistent model selection method cannot be minimax rate optimal. Model averaging (see Chapter 7) is also of no help here; his theorem 2 shows that any model averaging method is not minimax rate optimal if the weights are consistent in that they converge to one in probability for the true model, and to zero for the other models. Theorem 3 of Yang (2005) tells a similar story for Bayesian model averaging.

4.10 Notes on the literature

Results about theoretical properties of model selection methods are scattered in the literature, and often proofs are provided for specific situations only. Exceptions are Nishii (1984) who considered nested, though possibly misspecified, likelihood models for i.i.d. data, and Sin and White (1996). Consistency (weak and strong) of the BIC for data from an exponential family is obtained by Haughton (1988, 1989). Shibata (1984) studied the approximate efficiency for a small number of regression variables, while Li (1987) obtains the efficiency for Mallows's C_p, cross-validation and generalised cross-validation. Shao and Tu (1995, Section 7.4.1) show the inconsistency of leave-one-out cross-validation. Breiman and Freedman (1983) construct a different efficient criterion based on the expected prediction errors. A strongly consistent criterion based on the Fisher information matrix is constructed by Wei (1992). Strongly consistent criteria for regression are proposed by Rao and Wu (1989) and extended to include a data-determined penalty by Bai *et al.* (1999). An overview of the asymptotic behaviour of model selection methods for linear models is in Shao (1997). Zhao *et al.* (2001) construct the efficient determination criterion EDC, which allows the choice of the penalty term d_n, to be made over a wide range. In general their d_n can be taken as a sequence of positive numbers depending on n or as a sequence of positive random variables. Their main application is the determination of the order of a Markov chain with finite state space. Shen and Ye (2002) develop a data-adaptive penalty based on generalised degrees of freedom. Guyon and Yao (1999) study the probabilities of underfitting and overfitting for several model selection criteria, both in regression and autoregressive models. For nonstationary autoregressive time series models, weak consistency of order selection methods is obtained by Paulsen (1984) and Tsay (1984). For strong consistency results we refer to Pötscher (1989).

The consistency property has also been called 'the oracle property', see e.g. Fan and Li (2002). The effect of such consistency on further inference aspects has been studied by many authors. One somewhat disturbing side-effect is that the max-risk of post-selection estimators divided by the max-risk of ordinary estimators may diverge to infinity. Versions of this phenomenon have been recognised by Foster and George (1994) for the BIC in multiple regression, in Yang (2005) as mentioned above, in Leeb and Pötscher (2005, 2006, 2008) for subset selection in linear regression models, and in Hjort and Claeskens (2006) for proportional hazards models. Importantly, various other nonconsistent selection methods, like AIC and the FIC (Chapter 6), are immune to this unbounded max-risk ratio problem, as discussed in Hjort and Claeskens (2006), for example. The Leeb and Pötscher articles are also concerned with other aspects of post-model-selection inference, like the impossibility of estimating certain conditional distributions consistently. These aspects are related to the fact that it is not possible to estimate the δ parameter consistently in the local $\delta/\sqrt{n}$ framework of Chapter 5.

Exercises

4.1 *Frequency of selection:* Perform a small simulation study to investigate the frequency by which models are chosen by AIC, the BIC and the Hannan–Quinn criterion. Generate (independently for $i = 1, \ldots, n$)

$$x_{i,1} \sim \text{Uniform}(0, 1), \quad x_{i,2} \sim \text{N}(5, 1), \quad Y_i \sim \text{N}(2 + 3x_{i,1}, (1.5)^2).$$

Consider four normal regression models to fit:
$M_1: Y = \beta_0 + \varepsilon,$
$M_2: Y = \beta_0 + \beta_1 x_1 + \varepsilon,$
$M_3: Y = \beta_0 + \beta_2 x_2 + \varepsilon,$
$M_4: Y = \beta_0 + \beta_1 x_1 + \beta_2 x_2 + \varepsilon.$
For sample sizes $n = 50, 100, 200, 500$ and 1500, and 1000 simulation runs, construct a table which for each sample size shows the number of times (out of 1000 simulation runs) that each model has been chosen. Do this for each of AIC, the BIC and Hannan–Quinn. Discuss.

4.2 *Hofstedt's highway data:* Consider the data of Exercise 2.10. Use robust C_p to select variables to be used in the linear regression model. Construct a plot of the values of C_p versus the variable V_p. Compare the results to those obtained by using the original version of C_p.

4.3 *Calculating the risk for the choice between two normal models:* Let $Y_1, \ldots, Y_n$ be i.i.d. from the $\text{N}(\mu, 1)$ density, as in Section 4.2.
   (a) For the estimator (4.2), show that

$$\sqrt{n}(\widehat{\mu} - \mu) = Z(1 - A_n) - \sqrt{n}\mu A_n,$$

where $A_n = I\{|\sqrt{n}\mu + Z| \leq d_n^{1/2}\}$, and Z is a standard normal variable.
   (b) Show that the risk function $r_n(\mu)$, defined as n times mean squared error,

$$r_n(\mu) = n\,\text{E}_\mu\{(\widehat{\mu} - \mu)^2\} = \text{E}\big[\{(\sqrt{n}\mu + Z)I\{|\sqrt{n}\mu + Z| \geq d_n^{1/2}\} - \sqrt{n}\mu\}^2\big],$$

can be written as

$$r_n(\mu) = 1 - \int_{\text{low}_n}^{\text{up}_n} z^2 \phi(z)\, dz + n\mu^2 \{\Phi(\text{up}_n) - \Phi(\text{low}_n)\},$$

where

$$\text{low}_n = -d_n^{1/2} - \sqrt{n}\mu \quad \text{and} \quad \text{up}_n = d_n^{1/2} - \sqrt{n}\mu.$$

Plot the resulting risk functions corresponding to the AIC and BIC methods for different sample sizes n.

5

Bigger is not always better

Given a list of candidate models, why not select the richest, with the most parameters? When many covariates are available, for estimation, prediction or classification, is including all of them not the best one can do? The answer to this question turns out to be 'no'. The reason lies with the bias–variance trade-off. Including an extra parameter in the model means less bias but larger sampling variability; analogously, omitting a parameter means more bias but smaller variance. Two basic questions are addressed in this chapter: (i) Just how much misspecification can a model tolerate? When we have a large sample and only moderate misspecification, the answers are surprisingly simple, sharp, and general. There is effectively a 'tolerance radius' around a given model, inside of which estimation is more precise than in a bigger model. (ii) Will 'narrow model estimation' or 'wide model estimation' be most precise, for a given purpose? How can we choose 'the best model' from data?

5.1 Some concrete examples

We start with examples of model choice between two models: a simple model, and one that contains one or more extra parameters.

Example 5.1 Exponential or Weibull?

Suppose that data $Y_1, \ldots, Y_n$ come from a life-time distribution on $[0, \infty)$ and that we wish to estimate the median μ. If the density is the exponential $f(y) \doteq \theta e^{-\theta y}$, then $\mu = \log 2/\theta$, and an estimator is $\widehat{\mu}_{\mathrm{narr}} = \log 2/\widehat{\theta}_{\mathrm{narr}}$, where $\widehat{\theta}_{\mathrm{narr}} = 1/\bar{Y}$ is the maximum likelihood (ML) estimator in this narrow model. If it is suspected that the model could deviate from simple exponentiality in direction of the Weibull distribution, with

$$f(y, \theta, \gamma) = \exp\{-(\theta y)^\gamma\} \gamma(\theta y)^{\gamma-1}\theta, \quad y > 0, \tag{5.1}$$

then we should conceivably use $\widehat{\mu}_{\mathrm{wide}} = (\log 2)^{1/\widehat{\gamma}}/\widehat{\theta}$, using maximum likelihood estimators $\widehat{\theta}, \widehat{\gamma}$ in the wider Weibull model. But *if* the simple model is right, that is, when $\gamma = 1$, then $\widehat{\mu}_{\mathrm{narr}}$ is better, in terms of (for example) mean squared error. By sheer continuity it should be better also for γ close to 1. How much must γ differ from 1 in order for

$\widehat{\mu}_{\text{wide}}$ to become better? And what with similar questions for other typical parametric departures from exponentiality, like the Gamma family? ∎

Example 5.2 Linear or quadratic?

Consider a regression situation with n pairs (x_i, Y_i). The classical model takes $Y_i \sim$ $N(\alpha + \beta x_i, \sigma^2)$ for appropriate parameters α, β, σ, and encourages for example $\widehat{\mu}_{\text{narr}} = \widehat{\alpha}_{\text{narr}} + \widehat{\beta}_{\text{narr}} x$ as the estimator for the median (or mean value) of the distribution of Y for a given x value. Suppose however that the regression curve could be quadratic, $Y_i \sim N(\alpha + \beta x_i + \gamma (x_i - \bar{x})^2, \sigma^2)$ for a moderate γ, where $\bar{x} = n^{-1} \sum_{i=1}^{n} x_i$. How much must γ differ from zero in order for

$$\widehat{\mu}_{\text{wide}} = \widehat{\alpha} + \widehat{\beta} x + \widehat{\gamma}(x - \bar{x})^2,$$

with regression parameters now evaluated in the wider model, to perform better? The same questions could be asked for other parameters, like comparing $\widehat{x}_{0,\text{narr}}$ with $\widehat{x}_{0,\text{wide}}$, the narrow-model-based and the wide-model-based estimators of the point x_0 at which the regression curve crosses a certain level. Similar questions can be discussed in the framework of an omitted covariate. ∎

Example 5.3 Variance heteroscedasticity

In some situations a more interesting departure from standard regression lies in variance heterogeneity. This could for example suggest using $Y_i \sim N(\alpha + \beta x_i, \sigma_i^2)$ with $\sigma_i = \sigma \exp(\gamma x_i)$, where γ is zero under classical regression. Four-parameter inference may be carried out, leading for example to estimators of α and β that use somewhat complicated estimated weights. For what range of γ values are standard methods, all derived under the $\gamma = 0$ hypothesis, still better than four-parameter-model analysis? ∎

Example 5.4 Skewing logistic regressions

This example is meant to illustrate that one often might be interested in model extensions in more than one direction. Consider 0–1 observations $Y_1, \ldots, Y_n$ associated with say p-dimensional covariate vectors $x_1, \ldots, x_n$. The challenge is to assess how x_i influences the probability p_i that $Y_i = 1$. The classic model is the logistic regression one that takes $p_i = H(x_i^t \beta)$, with $H(u) = \exp(u)/\{1 + \exp(u)\}$. Inclusion of yet another covariate in the model, say z_i, perhaps representing an interaction term between components of x_i already inside the model, means extending the model to $H(x_i^t \beta + z_i \gamma)$. One might simultaneously wish to make the transformation itself more flexible. The logistic function treats positive and negative u in a fully symmetric manner, i.e. $H(\frac{1}{2} + u) = 1 - H(\frac{1}{2} - u)$, which means that probabilities with $x_i^t \beta$ to the right of zero dictate the values of those with $x_i^t \beta$ to the left of zero. A model with $p + 2$ parameters that extends what we started with in two directions is that of

$$p_i = H(x_i^t \beta + z_i \gamma)^\kappa = \left\{ \frac{\exp(x_i^t \beta + z_i \gamma)}{1 + \exp(x_i^t \beta + z_i \gamma)} \right\}^\kappa. \tag{5.2}$$

For what range of (γ, κ) values, around the centre point $(0, 1)$, will the simpler p-parameter model lead to more precise inference than that of the more complicated $p + 2$-parameter model? ∎

Let us summarise the common characteristics of these situations. There is a narrow and usually simple parametric model which can be fitted to the data, but there is a potential misspecification, which can be ameliorated by its encapsulation in a wider model with one or more additional parameters. Estimating a parameter assuming correctness of the narrow model involves modelling bias, but doing it in the wider model could mean larger sampling variability. Thus the choice of method becomes a statistical balancing act with perhaps deliberate bias against variance.

It is clear that the list of examples above may easily be expanded, showing a broad range of heavily used 'narrow' models along with indications of rather typical kinds of deviances from them. Many standard textbook methods for parametric inference are derived under the assumption that such narrow models are correct. Below we reach perhaps surprisingly sharp and general criteria for how much misspecification a given narrow model can tolerate in a certain direction. This is relatively easy to compute, in that it only involves the familiar Fisher information matrix computed for the wide model, but evaluated with parameter values assuming that the narrow model is correct. We shall see that the tolerance criterion does not depend upon the particular parameter estimand at all, when there is only one model extension parameter, as in Examples 5.1–5.3, but that the tolerance radius depends on the estimand under consideration in cases of two or more extension parameters, as with Example 5.4.

In addition to quantifying the degree of robustness of standard methods there are also pragmatic reasons for the present investigation. Statistical analysis will in practice still be carried out using narrow model-based methods in the majority of cases, for reasons of ignorance or simplicity; using wider-type model methods might often be more laborious, and perhaps only experts will use them. Thus it is of interest to quantify the consequences of ignorance, and it would be nice to obtain permission to go on doing analysis as if the simple model were true. Such a partial permission is in fact given here. The results can be interpreted as saying that 'ignorance is (sometimes) strength'; mild departures from the narrow model do not really matter, and more ambitious methods could perform worse.

5.2 Large-sample framework for the problem

We shall start our investigation in the i.i.d. framework, going on to regression models later in this section. Suppose $Y_1, \ldots, Y_n$ come from some common density f. The wide model, or full model, is $f(y, \theta, \gamma)$. Taking $\gamma = \gamma_0$, a known value, corresponds to the narrow model, or simplest model with density function $f(y, \theta) = f(y, \theta, \gamma_0)$. We study behaviour of estimators when γ deviates from γ_0. The parameter to be estimated is some estimand $\mu = \mu(f)$, which we write as $\mu(\theta, \gamma)$. We concentrate on maximum likelihood procedures, and write $\widehat{\theta}_{narr}$ for the estimator of θ in the narrow model and $(\widehat{\theta}, \widehat{\gamma})$ for the estimators in the wide model. This leads to

$$\widehat{\mu}_{narr} = \mu(\widehat{\theta}_{narr}, \gamma_0) \quad \text{and} \quad \widehat{\mu}_{wide} = \mu(\widehat{\theta}, \widehat{\gamma}). \tag{5.3}$$

We assume that $\theta = (\theta_1, \ldots, \theta_p)^t$ lies in some open region in Euclidean p-space, that γ lies in some open interval containing γ_0, and that the wide model is 'smooth'. The technical definition of the smoothness we have in mind is that the log-density has two continuous derivatives and that these second-order derivatives have finite means under the narrow model. The situation where parameters lie on the boundary of the parameter space requires a different approach, see Section 10.2. Regularity conditions that suffice for the following results to hold are of the type described and discussed in Hjort and Claeskens (2003, Sections 2–3).

5.2.1 A fixed true model

Suppose that the asymptotic framework is such that the Y_i come from some true fixed $f(y, \theta_0, \gamma)$, and $\gamma \neq \gamma_0$. Thus the wide model is the true model and $\mu = \mu(\theta_0, \gamma)$ is the true value. The theory of Section 2.2 tells us that $\sqrt{n}(\widehat{\mu}_{wide} - \mu)$ has a limit normal distribution, with mean zero and a certain variance. The situation is different for the narrow model estimator. Since this model is wrong, in that it does not include the γ, the narrow model's estimator will be biased. Here $\sqrt{n}(\widehat{\mu}_{narr} - \mu)$ can be represented as a sum of two terms. The first is

$$\sqrt{n}\{\mu(\widehat{\theta}_{narr}, \gamma_0) - \mu(\theta_0, \gamma_0)\},$$

which has a limiting normal distribution, with zero mean and with generally smaller variability than that of the wide model procedure. The second term is

$$-\sqrt{n}\{\mu(\theta_0, \gamma) - \mu(\theta_0, \gamma_0)\},$$

which tends to plus or minus infinity, reflecting a bias that for very large n will dominate completely (as long as the $\mu(\theta, \gamma)$ parameter depends on γ). This merely goes to show that with very large sample sizes one is penalised for any bias and one should use the wide model. This result is not particularly informative, however, and suggests that a large-sample framework which uses a local neighbourhood of γ_0 that shrinks when the sample size grows will be rather more fruitful, in the sense of reaching better descriptions of what matters in the bias versus variance balancing game. In such a framework, the wide model gets closer to the narrow model as the sample size increases.

Example 5.5 (5.1 continued) Exponential or Weibull? Asymptotic results
Let us consider again the situation of Example 5.1. The Weibull density has two parameters (θ, γ). To estimate the median, the maximum likelihood estimator is

$\widehat{\mu} = \mu(\widehat{\theta}, \widehat{\gamma}) = (\log 2)^{1/\widehat{\gamma}}/\widehat{\theta}$. From the delta method follows

$$\sqrt{n}(\widehat{\mu} - \mu) \xrightarrow{d} (\partial\mu/\partial\theta, \partial\mu/\partial\gamma)^{\mathrm{t}} \mathrm{N}_2(0, J^{-1}K J^{-1}), \tag{5.4}$$

which is a zero-mean normal with variance equal to

$$\tau^2 = \begin{pmatrix} \partial\mu/\partial\theta \\ \partial\mu/\partial\gamma \end{pmatrix}^{\mathrm{t}} J^{-1}K J^{-1} \begin{pmatrix} \partial\mu/\partial\theta \\ \partial\mu/\partial\gamma \end{pmatrix}.$$

Here J and K are as defined in (2.7). For μ equal to the median, the vector of partial derivatives with respect to (θ, γ) is equal to $(-\mu/\theta, \mu\log\log 2)$. Now we suppose that the Weibull model actual holds, so that $J = K$, cf. (2.8). Some calculations show that

$$J^{-1} = \frac{1}{\pi^2/6} \begin{pmatrix} a\theta^2/\gamma^2 & -(1 - g_e)\theta \\ -(1 - g_e)\theta & \gamma^2 \end{pmatrix} = \begin{pmatrix} 1.109\,\theta^2/\gamma^2 & -0.257\,\theta \\ -0.257\,\theta & 0.608\,\gamma^2 \end{pmatrix},$$

with $g_e = 0.577216\ldots$ the Euler–Mascheroni constant and with $a = \pi^2/6 + (1 - g_e)^2$. Inserting J^{-1} and the vector of partial derivatives in the formula of the delta method leads to

$$\sqrt{n}\left\{ \frac{(\log 2)^{1/\widehat{\gamma}}}{\widehat{\theta}} - \frac{(\log 2)^{1/\gamma}}{\theta_0} \right\} \xrightarrow{d} \mathrm{N}(0, 1.379(\log 2)^{2/\gamma}/(\theta_0\gamma)^2).$$

On the other hand, for the estimator computed in the exponential model,

$$\sqrt{n}\left\{ \frac{\log 2}{\widehat{\theta}_{\mathrm{narr}}} - \frac{(\log 2)^{1/\gamma}}{\theta_0} \right\} = \sqrt{n}\left(\frac{\log 2}{\widehat{\theta}_{\mathrm{narr}}} - \frac{\log 2}{\theta_0} \right) - \sqrt{n}\left\{ \frac{(\log 2)^{1/\gamma}}{\theta_0} - \frac{\log 2}{\theta_0} \right\}$$

has two terms; the first tends to a $\mathrm{N}(0, (\log 2)^2/\theta_0^2)$, reflecting smaller sampling variability, but the second tends to plus or minus infinity, reflecting a bias that sooner or later will dominate completely. ∎

5.2.2 Asymptotic distributions under local misspecification

Since the bias always dominates (for large n) when the true model is at a fixed distance from the smallest model, we study model selection properties in a local misspecification setting. This is defined as follows. Define model P_n, the nth model, under which

$$Y_1, \ldots, Y_n \text{ are i.i.d. from } f_n(y) = f(y, \theta_0, \gamma_0 + \delta/\sqrt{n}), \tag{5.5}$$

and where θ_0 is fixed but arbitrary. The $O(1/\sqrt{n})$ distance from the parameter $\gamma = \gamma_n$ of the nth model to γ_0 will give squared model biases of the same size as variances, namely $O(n^{-1})$. In this framework we need limit distributions for the wide model estimators $(\widehat{\theta}, \widehat{\gamma})$ and for the narrow model estimator $\widehat{\theta}_{\mathrm{narr}}$. It is also possible to work under the assumption that $f_n(y) = f(y, \theta_0 + \eta/\sqrt{n}, \gamma_0 + \delta/\sqrt{n})$ for a suitable η. This is in a sense not necessary, since the η will disappear in the end for quantities having to do with precision of $\widehat{\mu}$ estimators. The main reason for this is that θ is common to the models, with all its components included, so that there is no modelling bias for θ. Define

$$\begin{pmatrix} U(y) \\ V(y) \end{pmatrix} = \begin{pmatrix} \partial \log f(y, \theta_0, \gamma_0)/\partial\theta \\ \partial \log f(y, \theta_0, \gamma_0)/\partial\gamma \end{pmatrix},$$

the score function for the wide model, but evaluated at the null point (θ_0, γ_0). The accompanying familiar $(p + 1) \times (p + 1)$-size information matrix is

$$J_{\text{wide}} = \text{Var}_0 \begin{pmatrix} U(Y) \\ V(Y) \end{pmatrix} = \begin{pmatrix} J_{00} & J_{01} \\ J_{10} & J_{11} \end{pmatrix},$$

cf. (2.8). Note that the $p \times p$-size J_{00} is simply the information matrix of the narrow model, evaluated at θ_0, and that the scalar J_{11} is the variance of $V(Y_i)$, also computed under the narrow model. Since $\mu = \mu(\theta, \gamma)$, we first obtain the limiting distribution of $(\widehat{\theta}, \widehat{\gamma})$.

Theorem 5.1 *Under the sequence of models* P_n *of (5.5), as n tends to infinity, we have*

$$\begin{pmatrix} \sqrt{n}(\widehat{\theta} - \theta_0) \\ \sqrt{n}(\widehat{\gamma} - \gamma_0) \end{pmatrix} \xrightarrow{d} N_{p+1} \left(\begin{pmatrix} 0 \\ \delta \end{pmatrix}, J_{\text{wide}}^{-1} \right),$$

$$\sqrt{n}(\widehat{\theta}_{\text{narr}} - \theta_0) \xrightarrow{d} N_p(J_{00}^{-1}J_{01}\delta, J_{00}^{-1}).$$

Proof. Consider $\widehat{\theta}_{\text{narr}}$ first. Define $I_n(\theta) = -n^{-1}\sum_{i=1}^{n} \partial^2 \log f(Y_i, \theta, \gamma_0)/\partial\theta\partial\theta^{\text{t}}$. Then, using a Taylor series expansion,

$$0 = n^{-1} \sum_{i=1}^{n} \frac{\partial \log f(Y_i, \widehat{\theta}_{\text{narr}}, \gamma_0)}{\partial\theta} = \bar{U}_n - I_n(\widetilde{\theta}_n)(\widehat{\theta}_{\text{narr}} - \theta_0),$$

where $\bar{U}_n = n^{-1}\sum_{i=1}^{n} U(Y_i)$ and $\widetilde{\theta}_n$ lies somewhere between θ_0 and $\widehat{\theta}_{\text{narr}}$. Under the conditions stated, by using Taylor expansion arguments, $\widehat{\theta}_{\text{narr}} \to_p \theta_0$, and $I_n(\theta_0)$ as well as $I_n(\widetilde{\theta}_n)$ tend in probability to J_{00}. These statements hold under P_n. All this leads to

$$\sqrt{n}(\widehat{\theta}_{\text{narr}} - \theta_0) \doteq_d I_n(\theta_0)^{-1}\sqrt{n}\bar{U}_n \doteq_d J_{00}^{-1}\sqrt{n}\bar{U}_n, \tag{5.6}$$

where $A_n \doteq_d B_n$ means that $A_n - B_n$ tends to zero in probability. The triangular version of the Lindeberg theorem shows that $\sqrt{n}\bar{U}_n$ tends in distribution, under P_n, to $N_p(J_{01}\delta, J_{00})$. This is because

$$E_{P_n} U(Y_i) = \int f(y, \theta_0, \gamma_0 + \delta/\sqrt{n})U(y)\,dy$$

$$\doteq \int f(y, \theta_0, \gamma_0)\{1 + V(y)\delta/\sqrt{n}\}U(y)\,dy = J_{01}\delta/\sqrt{n},$$

and similar calculations show that $U(Y_i)U(Y_i)^{\text{t}}$ has expected value $J_{00} + O(\delta/\sqrt{n})$, under P_n. This proves the second part of the theorem.

Similar reasoning takes care of the first part, too, where $\bar{V}_n = n^{-1} \sum_{i=1}^{n} V(Y_i)$ is needed alongside $\bar{U}_n$. One finds

$$\begin{pmatrix} \sqrt{n}(\widehat{\theta} - \theta_0) \\ \sqrt{n}(\widehat{\gamma} - \gamma_0) \end{pmatrix} \doteq_d J_{\text{wide}}^{-1} \begin{pmatrix} \sqrt{n}\bar{U}_n \\ \sqrt{n}\bar{V}_n \end{pmatrix}$$

$$\xrightarrow{d} J_{\text{wide}}^{-1} N_{p+1} \left(\begin{pmatrix} J_{01}\delta \\ J_{11}\delta \end{pmatrix}, J_{\text{wide}} \right),$$

which is equivalent to the statement in the theorem. $\qquad\square$

This result is used to obtain the limit distribution of the estimators $\widehat{\mu}_{\text{wide}} = \mu(\widehat{\theta}, \widehat{\gamma})$ based on maximum likelihood estimation in the wide model, and $\widehat{\mu}_{\text{narr}} = \mu(\widehat{\theta}_{\text{narr}}, \gamma_0)$ in the narrow model. The true parameter is $\mu_{\text{true}} = \mu(\theta_0, \gamma_0 + \delta/\sqrt{n})$ under P_n, the nth model. First, partition the matrix J^{-1} according to the dimensions of the parameters θ and γ,

$$J_{\text{wide}}^{-1} = \begin{pmatrix} J^{00} & J^{01} \\ J^{10} & J^{11} \end{pmatrix},$$

where $\kappa^2 = J^{11} = (J_{11} - J_{10} J_{00}^{-1} J_{01})^{-1}$ will have special importance.

Corollary 5.1 *Under the sequence of models P_n of (5.5), as n tends to infinity,*

$$\sqrt{n}(\widehat{\mu}_{\text{narr}} - \mu_{\text{true}}) \xrightarrow{d} N(\omega\delta, \tau_0^2),$$
$$\sqrt{n}(\widehat{\mu}_{\text{wide}} - \mu_{\text{true}}) \xrightarrow{d} N(0, \tau_0^2 + \omega^2\kappa^2),$$

with $\omega = J_{10} J_{00}^{-1} \frac{\partial\mu}{\partial\theta} - \frac{\partial\mu}{\partial\gamma}$ and $\tau_0^2 = (\frac{\partial\mu}{\partial\theta})^{\text{t}} J_{00}^{-1} \frac{\partial\mu}{\partial\theta}$, with derivatives taken at (θ_0, γ_0).

Proof. The distribution result follows by application of the delta method, as

$$\sqrt{n}\{\mu(\widehat{\theta}, \widehat{\gamma}) - \mu(\theta_0, \gamma_0 + \delta/\sqrt{n})\}$$
$$\doteq_d (\tfrac{\partial\mu}{\partial\theta})^{\text{t}} \sqrt{n}(\widehat{\theta} - \theta_0) + \{(\tfrac{\partial\mu}{\partial\gamma}) + O(1/\sqrt{n})\} \sqrt{n}\{\widehat{\gamma} - (\gamma_0 + \delta/\sqrt{n})\}$$

tends in distribution to $N(0, \tau^2)$, where the variance is

$$\tau^2 = \begin{pmatrix} \frac{\partial\mu}{\partial\theta} \\ \frac{\partial\mu}{\partial\gamma} \end{pmatrix}^{\text{t}} J_{\text{wide}}^{-1} \begin{pmatrix} \frac{\partial\mu}{\partial\theta} \\ \frac{\partial\mu}{\partial\gamma} \end{pmatrix}.$$

The partial derivatives are computed at (θ_0, γ_0). Next, using

$$J^{01} = -J_{00}^{-1} J_{01} \kappa^2 \quad \text{and} \quad J^{00} = J_{00}^{-1} + J_{00}^{-1} J_{01} J_{10} J_{00}^{-1} \kappa^2 \qquad (5.7)$$

leads to the simplification

$$\tau^2 = (\tfrac{\partial\mu}{\partial\theta})^{\text{t}} J_{00}^{-1} (\tfrac{\partial\mu}{\partial\theta}) + (\tfrac{\partial\mu}{\partial\theta})^{\text{t}} J_{00}^{-1} J_{01} J_{10} J_{00}^{-1} (\tfrac{\partial\mu}{\partial\theta}) \kappa^2$$
$$- 2(\tfrac{\partial\mu}{\partial\theta})^{\text{t}} J_{00}^{-1} J_{01} (\tfrac{\partial\mu}{\partial\gamma}) \kappa^2 + (\tfrac{\partial\mu}{\partial\gamma})^2 \kappa^2$$
$$= \tau_0^2 + \omega^2 \kappa^2.$$

Similarly, for $\sqrt{n}\{\mu(\widehat{\theta}_{\text{narr}}, \gamma_0) - \mu(\theta_0, \gamma_0 + \delta/\sqrt{n})\}$ we have

$$\sqrt{n}\{\mu(\widehat{\theta}_{\text{narr}}, \gamma_0) - \mu(\theta_0, \gamma_0)\} - \sqrt{n}\{\mu(\theta_0, \gamma_0 + \delta/\sqrt{n}) - \mu(\theta_0, \gamma_0)\}$$

$$\doteq_d \left(\tfrac{\partial\mu}{\partial\theta}\right)^{\text{t}} \sqrt{n}(\widehat{\theta}_{\text{narr}} - \theta_0) - \sqrt{n}\tfrac{\partial\mu}{\partial\gamma}\delta/\sqrt{n} \xrightarrow{d} \text{N}(\omega\delta, \tau_0^2),$$

proving the second statement. $\square$

5.2.3 Generalisation to regression models

The above set-up assuming independent and identically distributed data can be generalised to cover regression models. The main point of departure is that independent (response) observations $Y_1, \ldots, Y_n$ are available, where Y_i comes from a density of the form $f(y \mid x_i, \theta_0, \gamma_0 + \delta/\sqrt{n})$. Here θ_0 could for example consist of regression coefficients and a scale parameter, while γ could indicate an additional shape parameter. There are analogues of Theorem 5.1 and Corollary 5.1 to this regression setting, leading to a mean squared error comparison similar to that of Theorem 5.2. To define the necessary quantities, we introduce score functions $U(y \mid x)$ and $V(y \mid x)$, the partial derivatives of $\log f(y \mid x, \theta, \gamma)$ with respect to θ and γ, evaluated at the null parameter value (θ_0, γ_0). We need the variance matrix of $U(Y \mid x)$, $V(Y \mid x)$, which is

$$J(x) = \int f(y \mid x, \theta_0, \gamma_0) \begin{pmatrix} U(y \mid x) \\ V(y \mid x) \end{pmatrix} \begin{pmatrix} U(y \mid x) \\ V(y \mid x) \end{pmatrix}^{\text{t}} \, \mathrm{d}y,$$

since $U(Y \mid x)$ and $V(Y \mid x)$ have mean zero under the narrow model. An important matrix is then

$$J_n = J_{n,\text{wide}} = n^{-1} \sum_{i=1}^{n} J(x_i) = \begin{pmatrix} J_{n,00} & J_{n,01} \\ J_{n,10} & J_{n,11} \end{pmatrix}, \tag{5.8}$$

where $J_{n,00}$ is of size $p \times p$, where $p = \text{length}(\theta)$. This matrix is assumed to converge to a positive definite matrix J_{wide} as n tends to infinity, depending on the distribution of covariates; see also the discussion of Section 2.2. Theorem 5.2 with corollaries is now valid for the regression case, with formulae for ω and τ_0 in terms of the limit J_{wide}; mild regularity conditions of the Lindeberg kind are needed, cf. Hjort and Claeskens (2003, Section 3 and Appendix). In applications we would use versions ω_n and $\tau_{0,n}$, and estimates thereof, defined in terms of J_n rather than J.

5.3 A precise tolerance limit

In this section we focus on situations where the model extension corresponds to a single extra parameter, as in Examples 5.1–5.3. From Corollary 5.1 we obtain that the limiting mean squared error of the narrow model is equal to $\omega^2\delta^2 + \tau_0^2$, while that of the wide model equals $\tau^2 = \tau_0^2 + \omega^2\kappa^2$. It is now easy to find out when the wide model estimator is better than the narrow model estimator in mean squared error sense; we simply solve the inequality $\omega^2\delta^2 + \tau_0^2 \leq \tau^2$ with respect to δ.

Theorem 5.2 *(i) Assume first* $\omega = 0$, *which typically corresponds to asymptotic independence between* $\widehat{\theta}$ *and* $\widehat{\gamma}$ *under the null model and to* μ *being functionally independent of* γ. *In such cases* $\widehat{\mu}_{\text{wide}}$ *and* $\widehat{\mu}_{\text{narr}}$ *are asymptotically equivalent, regardless of* δ. *(ii) In the case* $\omega \neq 0$, *the narrow model-based estimator is better than or as good as the wider model-based estimator*

$$\text{if and only if} \quad |\delta| \leq \kappa, \quad \text{or } |\gamma - \gamma_0| \leq \kappa/\sqrt{n}. \tag{5.9}$$

We now give some remarks to better appreciate the results of Theorem 5.2.

- The theorem holds both in the i.i.d. and regression frameworks, in view of comments made in Section 5.2.3.
- Since $\kappa^2 = J^{11} = (J_{11} - J_{10} J_{00}^{-1} J_{01})^{-1}$, the criterion $|\delta| \leq \kappa$ can be evaluated and assessed just from knowledge of J_{wide}. The J_{wide} matrix is easily computed, if not analytically then with numerical integration or simulation of score vectors at any position θ in the parameter space.
- Whether we should include γ or not depends on the size of δ in relation to the limiting standard deviation κ for $\sqrt{n}(\widehat{\gamma} - \gamma_0)$.
- Inequality (5.9) does not depend on the particularities of the specific parameter $\mu(\theta, \gamma)$ at all. Thus, in the situation of Example 5.1 calculations show that $|\gamma - 1| \leq 1.245/\sqrt{n}$ guarantees that assuming exponentiality works better than using a Weibull distribution, for *every* smooth parameter $\mu(\theta, \gamma)$. This is different in a situation with a multi-dimensional departure from the model, see Section 5.4 below.
- The κ number is dependent on the scale of the Y measurements. A scale-invariant version is $d = J_{11} J^{11} = J_{11} \kappa^2$. A large d value would signify a nonthreatening type of model departure, where sample sizes would need to be rather large before the extended model is worth using.

Remark 5.1 Bias correction
We have demonstrated that narrow estimation, which means accepting a deliberate bias to reduce variability, leads to a better estimator precision inside a certain radius around the narrow model. Can we remove the bias and do even better? About the best we can do in this direction is to use the bias-corrected $\widehat{\mu}_{\text{bc}} = \widehat{\mu}_{\text{narr}} - \widehat{\omega}(\widehat{\gamma} - \gamma_0)$, with $\widehat{\omega}$ a consistent estimate of ω. Analysis reveals that $\sqrt{n}(\widehat{\mu}_{\text{bc}} - \mu_{\text{true}})$ tends to $N(0, \tau_0^2 + \omega^2 \kappa^2)$; see Exercise 5.1. So the bias can be removed, but the price one pays amounts exactly to what was won by deliberate biasing in the first place, and the de-biased estimator is equivalent to $\widehat{\mu}_{\text{wide}}$. The reason for the extra variability is that no consistent estimator exists for δ. ∎

Remark 5.2 Connections to testing and pre-testing
We have shown that the simple θ parameter model can tolerate up to $\gamma_0 + \kappa/\sqrt{n}$ deviation from γ_0. How far is the borderline $\delta = \kappa$ from the narrow model? One way of answering this is via the probability of actually detecting that the narrow model is wrong. From Theorem 5.1,

$$D_n = \sqrt{n}(\widehat{\gamma} - \gamma_0) \xrightarrow{d} N(\delta, \kappa^2), \tag{5.10}$$

leading to the test statistic $D_n/\widehat{\kappa}$, with any consistent estimator for κ. In various cases the value of κ is even known, see examples to follow. We have $D_n/\widehat{\kappa} \to_d N(\delta/\kappa, 1)$, and testing correctness of the narrow model, against the alternative hypothesis that the additional γ parameter must be included, is done by rejecting when $|D_n/\widehat{\kappa}|$ exceeds 1.96, at significance level 0.05. The probability that this test detects that γ is not equal to γ_0, when it in fact is equal to $\gamma_0 + \delta/\sqrt{n}$, converges to

$$\text{power}(\delta) = P\big(\chi_1^2(\delta^2/\kappa^2) > 1.96^2\big), \tag{5.11}$$

in terms of a noncentral chi-squared with 1 degree of freedom and eccentricity parameter δ^2/κ^2. In particular the approximate power at the borderline case where $|\delta| = \kappa$ is equal to 17.0%. We can therefore restate the basic result as follows: provided the true model deviates so modestly from the narrow model that the probability of detecting it is 0.17 or less with the 0.05 level test, then the risky estimator is better than the safe estimator. Corresponding other values for (level, power) are (0.01, 0.057), (0.10, 0.264), (0.20, 0.400), (0.29, 0.500).

Theorem 5.2 also sheds light on the pre-testing strategy, which works as follows. Test the hypothesis $\gamma = \gamma_0$ against the alternative $\gamma \neq \gamma_0$, say at the 10% level; if accepted, then use $\widehat{\mu}_{\text{narr}}$; if rejected, then use $\widehat{\mu}_{\text{wide}}$. With the $Z_n^2 = n(\widehat{\gamma} - \gamma_0)^2/\widehat{\kappa}^2$ test, this suggestion amounts to

$$\widehat{\mu}_{\text{pretest}} = \widehat{\mu}_{\text{narr}} I\{Z_n^2 \leq 1.645^2\} + \widehat{\mu}_{\text{wide}} I\{Z_n^2 > 1.645^2\},$$

with 1.645^2 being the upper 10% point of the χ_1^2. The theory above suggests that one should rather use the much smaller value 1 as cut-off point, since $|\delta| \leq \kappa$ corresponds to $n(\gamma - \gamma_0)^2/\kappa^2 \leq 1$, and Z_n^2 estimates this ratio. Using 1 as cut-off, which appears natural in view of Theorem 5.2, corresponds to a much more relaxed significance level, indeed to 31.7%. The AIC model selection method, see Section 2.8, corresponds to using 2 as cut-off point for $D_n^2/\widehat{\kappa}^2$, with significance level 15.7%. ∎

The limit distribution of the pre-test estimator is nonstandard and in fact non-normal, and is a nonlinear mixture of two normals. Theory covering this and the more general case of compromise estimators $w(Z_n)\widehat{\mu}_{\text{narr}} + \{1 - w(Z_n)\}\widehat{\mu}_{\text{wide}}$, with $w(z)$ any weight function, is provided in Chapter 7.

5.4 Tolerance regions around parametric models

Here we shall generalise the tolerance results to the case where there are $q \geq 2$ extra parameters $\gamma_1, \ldots, \gamma_q$. The $q \geq 2$ situation is less clear-cut than the $q = 1$ case dealt with above. We work in the local misspecification setting with independent observations from the same population, with density of the form

$$f_{\text{true}}(y) = f(y, \theta_0, \gamma_0 + \delta/\sqrt{n}). \tag{5.12}$$

In the narrow model only θ is present, in the widest model we have both θ and the full vector γ. We here operate under the assumption that the widest model is correct in the sense of containing at least as many parameters as necessary. The true model is inside a distance of $O(1/\sqrt{n})$ of the narrow model. The $O(1/\sqrt{n})$ distance implies that all models are 'reasonably' close to the true model. This makes sense intuitively; we should not include a model in our list of models to choose from if we know in advance that this model is far from being adequate. The models included as candidates should all be plausible, or at least not clearly implausible.

The $(p + q) \times (p + q)$ information matrix evaluated at the narrow model is

$$J_{\text{wide}} = \begin{pmatrix} J_{00} & J_{01} \\ J_{10} & J_{11} \end{pmatrix}, \quad \text{with inverse} \quad J_{\text{wide}}^{-1} = \begin{pmatrix} J^{00} & J^{01} \\ J^{10} & J^{11} \end{pmatrix}, \tag{5.13}$$

cf. again (2.8). It is in particular assumed that J_{wide} is of full rank. An important quantity is the matrix

$$Q = J^{11} = (J_{11} - J_{10}J_{00}^{-1}J_{01})^{-1}, \tag{5.14}$$

which properly generalises the scalar κ^2 met in Sections 5.2–5.3. Also, using formulae for the inverse of a partitioned matrix (see for example Harville, 1997, Section 8.5), $J^{00} = J_{00}^{-1} + J_{00}^{-1}J_{01}QJ_{10}J_{00}^{-1}$ and $J^{01} = -J_{00}^{-1}J_{01}Q$. Generalising further from the earlier $q = 1$-related quantities, we define now

$$\omega = J_{10}J_{00}^{-1}\frac{\partial\mu}{\partial\theta} - \frac{\partial\mu}{\partial\gamma} \quad \text{and} \quad \tau_0^2 = \left(\frac{\partial\mu}{\partial\theta}\right)^{\text{t}} J_{00}^{-1}\frac{\partial\mu}{\partial\theta},$$

again with partial derivatives evaluated at (θ_0, γ_0).

To reach the appropriate generalisations about tolerance levels around narrow models in the $q \geq 2$ case, let $\mu = \mu(\theta, \gamma)$ be any parameter of interest, and consider the narrow model based $\widehat{\mu}_{\text{narr}} = \mu(\widehat{\theta}_{\text{narr}}, \gamma_0)$ and the wide model based $\widehat{\mu} = \mu(\widehat{\theta}_{\text{wide}}, \widehat{\gamma}_{\text{wide}})$, using maximum likelihood in these two models. Note that $\mu_{\text{true}} = \mu(\theta_0, \gamma_0 + \delta/\sqrt{n})$ is the appropriate parameter value under (5.12). We need at this stage to call on Theorem 6.1 of Chapter 6. As special cases of this theorem we have

$$\sqrt{n}(\widehat{\mu}_{\text{wide}} - \mu_{\text{true}}) \xrightarrow{d} \text{N}(0, \tau_0^2 + \omega^{\text{t}}Q\omega),$$

$$\sqrt{n}(\widehat{\mu}_{\text{narr}} - \mu_{\text{true}}) \xrightarrow{d} \text{N}(\omega^{\text{t}}\delta, \tau_0^2).$$

Note that this generalises Corollary 5.1. For the statements below, 'better than' refers to having smaller limiting mean squared error.

Theorem 5.3 (i) *For a given estimand μ, narrow model estimation is better than full model estimation provided δ lies in the set where $|\omega^{\text{t}}\delta| \leq (\omega^{\text{t}}Q\omega)^{1/2}$, which is an infinite band containing zero in the direction of orthogonality to ω. (ii) Narrow model estimation is better than full model estimation for all estimands provided δ lies inside the ellipsoid $\delta^{\text{t}}Q^{-1}\delta \leq 1$.*

Proof. The limiting risks involved for the narrow and wide models are respectively $\tau_0^2 + (\omega^t \delta)^2$ and $\tau_0^2 + \omega^t Q \omega$. This is equivalent to $|\omega^t \delta| \leq (\omega^t Q \omega)^{1/2}$, proving the first statement. Secondly, the narrow model being always better than the wide model corresponds to $(\omega^t \delta) \leq \omega^t Q \omega$ for all vectors ω. Writing $u = Q^{1/2} \omega$, the requirement is that $|u^t Q^{-1/2} \delta| \leq \|u\|$ for all $u \in \mathbb{R}^q$, which again is equivalent to $\|Q^{-1/2} \delta\| \leq 1$, by the Cauchy–Schwarz inequality. This proves the second statement. It is also equivalent to $Q \geq \delta \delta^t$, i.e. $Q - \delta \delta^t$ being non-negative definite. $\square$

Theorem 5.2 provides for the one-dimensional case a tolerance radius automatically valid for all estimands, but when $q \geq 2$ the tolerance region depends on the parameter under focus. There is, however, an ellipsoid of δ near zero inside which inference for all estimands will be better using narrow methods than using wide methods. The findings of Theorems 5.2–5.3 are illustrated in the next section.

Remark 5.3 Local neighbourhood models
The primary point of working with local neighbourhood models, as with (5.12), is that it leads to various precise mathematical limit distributions and therefore to useful approximations, for quantities like bias, standard deviation, mean squared error, tolerance radii, etc. The fruitfulness of the approach will also be demonstrated in later chapters, where some of the approximations lead to construction of model selection and model averaging methods. Thus we do not interpret (5.12) quite as literally as meaning that parameters in a concrete application change their values from $n = 200$ to $n = 201$; this would in fact have clashed with natural Kolmogorov coherency demands, see e.g. McCullagh (2002) and its ensuing discussion.

We also point out here that we could have taken a slightly more general version $f(y, \theta_0 + \eta/\sqrt{n}, \gamma_0 + \delta/\sqrt{n})$ of (5.12) as point of departure, i.e. including a local disturbance also for the θ. The η would however disappear in all limit expressions related to precision of estimators of μ; thus Theorems 5.1–5.3 and their consequences remain unaffected. This is perhaps intuitively clear since θ is included in both the narrow and wide models, so there is no modelling bias related to this parameter. One may of course also go through the mathematics again and verify that η drops out of the appropriate limit distributions. For a broader discussion of these and related issues, see the discussion contributions and rejoinder to Hjort and Claeskens (2003) and Claeskens and Hjort (2003). ∎

5.5 Computing tolerance thresholds and radii

We now provide answers to the questions asked in Examples 5.1–5.4.

Example 5.6 (5.1 continued) Exponential or Weibull?
In the general two-parameter Weibull model, parameterised as in (5.1), the matrix J^{-1} is as given in Example 5.5. The narrow model corresponds to $\gamma = \gamma_0 = 1$. With this value

of γ, the value $\kappa^2 = J^{11}$ equals $6/\pi^2$. We reach the following conclusion: for $|\gamma - 1| \leq \sqrt{6/\pi^2}/\sqrt{n} = 0.779/\sqrt{n}$, estimation with $\mu(1/\bar{Y}, 1)$ based on the simple exponential model performs better than the more involved $\mu(\widehat{\theta}, \widehat{\gamma})$ in the Weibull model. This is true regardless of the parameter μ to be estimated.

We find that Weibull deviance from the exponential model has scaled tolerance limit $d = J_{11}J^{11} = 1 + (1 - g_e)^2/(\pi^2/6) = 1.109$. It is instructive to compare with the corresponding value for Gamma distribution deviance from exponentiality. If $f(y) = \{\theta^\gamma / \Gamma(\gamma)\} y^{\gamma-1}e^{-\theta y}$ is the Gamma density, for which $\gamma_0 = 1$ gives back exponentiality, then $\kappa^2 = 1/(\pi^2/6 - 1)$; estimation using $\mu(1/\bar{Y}, 1)$ is more precise than $\mu(\widehat{\theta}, \widehat{\gamma})$ provided $|\gamma - 1| \leq 1.245/\sqrt{n}$. The scaled value d equals 2.551. This suggests that moderate Gamma-ness is less critical than moderate Weibull-ness for standard methods based on exponentiality. ∎

Example 5.7 (5.2 continued) Linear or quadratic?
We generalise slightly and write the wide model as $Y_i \sim N(x_i^t\beta + \gamma c(x_i), \sigma^2)$, where β and x_i are p-dimensional vectors, and $c(x)$ is some given scalar function. Exact formulae for information matrices are in practice not needed to compute the tolerance limit. We give them here for illustration purposes. By computing log-derivatives and evaluating covariances one reaches

$$J_{n,\text{wide}} = \frac{1}{\sigma^2}\begin{pmatrix} 2 & 0 & 0 \\ 0 & n^{-1}\sum_{i=1}^{n} x_i x_i^t & n^{-1}\sum_{i=1}^{n} x_i c(x_i) \\ 0 & n^{-1}\sum_{i=1}^{n} x_i^t c(x_i) & n^{-1}\sum_{i=1}^{n} c(x_i)^2 \end{pmatrix}.$$

It follows that κ^2 is equal to

$$\sigma^2 \times \text{lower right element of } \begin{pmatrix} n^{-1}\sum_{i=1}^{n} x_i x_i^t & n^{-1}\sum_{i=1}^{n} x_i c(x_i) \\ n^{-1}\sum_{i=1}^{n} x_i^t c(x_i) & n^{-1}\sum_{i=1}^{n} c(x_i)^2 \end{pmatrix}^{-1}.$$

Assume, for a concrete example, that x_i is one-dimensional and uniformly distributed over $[0, b]$, say $x_i = bi/(n + 1)$, and that the wide model has $\alpha + \beta(x_i - \bar{x}) + \gamma(x_i - \bar{x})^2$. Then $\kappa \doteq \sqrt{180}\,\sigma/b^2$. Consequently, dropping the quadratic term does not matter, and is actually advantageous, for every estimator, provided $|\gamma| \leq \sqrt{180}\,\sigma/(b^2\sqrt{n})$. In many situations with moderate n this will indicate that it is best to keep the narrow model and avoid the quadratic term.

Similar analysis can be given for the case of a wide model with an extra covariate, for example $N(x_i^t\beta + \gamma z_l, \sigma^2)$. The formulae above then hold with z_l replacing $c(x_i)$. In the case of z_i distributed independently from the x_i, the narrow x_i only model tolerates up to $|\gamma| \leq (\sigma/\sigma_z)/\sqrt{n}$, where σ_z^2 is the variance of the z_i. ∎

Example 5.8 (5.3 continued) Variance heteroscedasticity
Again we generalise slightly and write $Y_i \sim N(x_i^t\beta, \sigma_i^2)$ for the $p + 2$-parameter variance heterogeneous model, where $\sigma_i = \sigma(1 + \gamma c_i)$ for some observable $c_i = c(x_i)$; $\gamma = 0$ corresponds to the ordinary linear regression model, see e.g. Example 2.2. It is not easy

to put up simple expressions for the general information matrix, in the presence of γ, but once more it suffices for the purposes of tolerance calculations to compute $J_{n,\text{wide}}$ under the null model, that is, when $\gamma = 0$. Some calculations give

$$J_{n,\text{wide}} = n^{-1} \sum_{i=1}^{n} \text{Var} \begin{pmatrix} (\varepsilon_i^2 - 1)/\sigma \\ \varepsilon_i x_i/\sigma \\ c_i \varepsilon_i^2 \end{pmatrix} = \begin{pmatrix} 2/\sigma^2 & 0 & 2\bar{c}_n/\sigma \\ 0 & \Sigma_n/\sigma^2 & 0 \\ 2\bar{c}_n/\sigma & 0 & 2n^{-1}\sum_{i=1}^{n} c_i^2 \end{pmatrix},$$

when parameters are listed in the σ, β, γ order. Here $\varepsilon_i = (Y_i - x_i^t \beta)/\sigma_i$, and these are independent and standard normal, and $\Sigma_n = n^{-1} \sum_{i=1}^{n} x_i x_i^t$. One finds κ^2 equal to $1/(2s_{n,c}^2)$, in terms of the empirical variance $s_{n,c}^2 = n^{-1} \sum_{i=1}^{n}(c_i - \bar{c}_n)^2$ of the c_i. Thus, by Theorem 5.2, the simpler variance homogeneous standard model works better than the more ambitious variance heterogeneous model, as long as $|\gamma| \le 1/(\sqrt{2}s_{n,c}\sqrt{n})$.

For a general estimand $\mu(\sigma, \beta, \gamma)$, the formula of Corollary 5.1 leads to $\omega = \sigma\bar{c}_n \partial\mu/\partial\sigma - \partial\mu/\partial\gamma$. For estimands that are functions of the regression coefficients only, therefore, we have $\omega = 0$, and the wide model's somewhat complicated estimator

$$\widehat{\beta}_{\text{weight}} = \left\{ \sum_{i=1}^{n} \frac{x_i x_i^t}{1 + \widehat{\gamma}c(x_i)} \right\}^{-1} \sum_{i=1}^{n} \frac{x_i y_i}{1 + \widehat{\gamma}c(x_i)},$$

with estimated weights, becomes large-sample equivalent to the ordinary least squares $(\sum_{i=1}^{n} x_i x_i^t)^{-1} \sum_{i=1}^{n} x_i y_i$ estimator. For other estimands, like a probability or a quantile, the narrow or wide estimators are best depending on the size of $|\gamma|$ compared to $1/(\sqrt{2}s_{n,c}\sqrt{n})$. ∎

Example 5.9 (5.4 continued) Skewing logistic regression

For the situation described in Example 5.4, the log-likelihood contributions are of the form $Y_i \log p_i + (1 - Y_i)\log(1 - p_i)$, with p_i as in (5.2). The derivative of this log-likelihood contribution is $p_i^*(Y_i - p_i)/\{p_i(1 - p_i)\}$, where p_i^* is the derivative of p_i with respect to (β, γ, κ). Some work shows that p_i^*, evaluated at the null model where $(\gamma, \kappa) = (0, 1)$, has components $p_i^0(1 - p_i^0)x_i$, $p_i^0(1 - p_i^0)z_i$, $p_i^0 \log p_i^0$, where p_i^0 is $H(x_i^t\beta)$. The (5.8) formula therefore gives

$$J_n = n^{-1} \sum_{i=1}^{n} \frac{1}{p_i^0(1 - p_i^0)} \begin{pmatrix} p_i^0(1 - p_i^0)x_i \\ p_i^0(1 - p_i^0)z_i \\ p_i^0 \log p_i^0 \end{pmatrix} \begin{pmatrix} p_i^0(1 - p_i^0)x_i \\ p_i^0(1 - p_i^0)z_i \\ p_i^0 \log p_i^0 \end{pmatrix}^t.$$

In a situation with data one may compute the 2×2 matrix Q_n as in (5.14), with estimated values plugged in to produce an estimate of J_n. One is then in a position to apply Theorem 5.3, for specific estimands, like $P(Y = 1 \mid x_0, z_0)$ for a given (x_0, z_0). ∎

5.6 How the 5000-m time influences the 10,000-m time

We take the results of the 200 best skaters of the world, obtained from the Adelskalenderen as of April 2006; see Section 1.7. We shall actually come back to certain multivariate

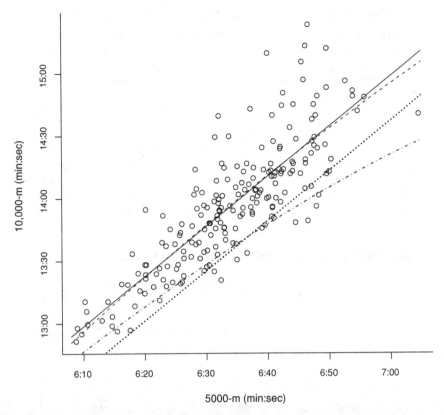

Fig. 5.1. Personal bests on the 5000-m and 10,000-m, for the world's 200 best speed-skaters, as of end-of-season 2006. There are two regression curves. The straight (solid) line is from linear regression of Y on x and the slightly curved dashed line from regression of Y on (x, z), as per the text. The other two lines are estimates of the 10% quantile for a 10,000-m result, predicted from a 5000-m result; the straight line (dotted) is based on linear regression of Y on x while the slightly curved line (dashed–dotted) comes from the five-parameter heteroscedastic model.

aspects of such data in Chapter 9, but at this moment we focus on one of the often-debated questions in speedskating circles, at all championships; how well can one predict the 10,000-m time from the 5000-m time?

Figure 5.1 shows the personal bests on the 5000-m and 10,000-m for the 200 best skaters of the world (as of end-of-season 2006), in minutes:seconds. We wish to model the 10,000-m time Y as a function of the 5000-m time x. The simplest possibility is linear regression of Y on x. The scatter plot indicates, however, both potential quadraticity and variance heterogeneity. We therefore consider models of the form

$$Y_i = a + bx_i + cz_i + \varepsilon_i, \quad \text{where } \varepsilon_i \sim \mathrm{N}(0, \sigma_i^2) \tag{5.15}$$

for $i = 1, \ldots, 200$, where z_i is a quadratic component and σ_i reflects the potentially heteroscedastic nature of the variability level. To aid identification and interpretation of

Table 5.1. *Analysis of the speedskating data, with parameter estimates, log-likelihood maxima, and values of AIC and BIC.*

	M_0	M_1	M_2	M_3
σ	16.627	16.606	15.773	15.772
a	-107.700	-102.374	-117.695	-114.424
b	2.396	2.382	2.421	2.413
c		-0.838		-0.112
ϕ			0.255	0.254
ℓ_n	-845.992	-845.745	-835.441	-835.436
AIC	-1697.983	-1699.489	-1678.883	-1680.873
BIC	-1707.878	-1712.683	-1692.076	-1717.981

the parameters involved, we choose to use standardised variables

$$z_i = (u_i - \bar{u})/s_u, \text{ with } u_i = (x_i - \bar{x})^2 \text{ and } s_u = \left\{ n^{-1} \sum_{i=1}^{n} (u_i - \bar{u})^2 \right\}^{1/2}$$

instead of simply $z_i = x_i^2$ or $z_i = (x_i - \bar{x})^2$. As usual, $\bar{x}$ and $\bar{u}$ denote averages $n^{-1} \sum_{i=1}^{n} x_i$ and $n^{-1} \sum_{i=1}^{n} u_i$. Similarly we use

$$\sigma_i = \sigma \exp(\phi v_i), \text{ with } v_i = (x_i - \bar{x})/s_x \text{ and } s_x = \left\{ n^{-1} \sum_{i=1}^{n} (x_i - \bar{x})^2 \right\}^{1/2}$$

rather than $v_i = x_i$. This implies that the v_i have average zero and that the quantity $n^{-1} \sum_{i=1}^{n} v_i^2$ is equal to one. In the final analysis one arrives at the same curves, numbers and conclusions whether one uses say $\sigma_0 \exp(\phi_0 x_i)$ or $\sigma \exp(\phi v_i)$, but parameters are rather easier to identify and interpret with these standardisations. Thus σ may be interpreted as the standard deviation of Y at position $x = \bar{x}$, with or without the variable variance parameter ϕ. To answer questions related to tolerance and influence, we write the model as $Y_i = x_i^t \beta + z_i^t \gamma + \varepsilon_i$, with $\varepsilon_i \sim N(0, \sigma^2 \exp(2\phi v_i))$. The narrow model has $p + 1 = 3$ parameters (β, σ), while the largest model has $q + 1 = 2$ more parameters γ and ϕ. Four models to consider are

M_0: the narrow model with parameters (β, σ), ordinary linear regression in x, with $\gamma = 0$ and $\phi = 0$;

M_1: linear regression in x, z, with parameters (β, γ, σ), and $\phi = 0$, including the quadratic term but having constant variance;

M_2: regression in x, but heteroscedastic variability $\sigma_i = \sigma \exp(\phi v_i)$, with parameters (β, σ, ϕ);

M_3: the fullest model, heteroscedastic regression in x, z, with parameters $(\beta, \gamma, \sigma, \phi)$.

We analyse all four models and include also their AIC and BIC values in Table 5.1. In the analysis we used the times expressed in seconds. We see that AIC and the BIC agree on model M_2 being best.

Table 5.2. *For five models aiming to explain how the 10k result depends on the 5k result, the table displays both direct and cross-validated scores for mean absolute error (mae), for root mean squared error (rmse), and for normalised log-likelihood, as well as the value of AIC.*

Model		mae	rmse	$n^{-1}\ell_{n,\max}$	AIC
M_0	direct	12.458	16.627	-4.230	-1697.983
	cv	12.580 (1)	16.790 (2)	-4.248 (3)	(3)
M_1	direct	12.475	16.606	-4.229	-1699.489
	cv	12.665 (4)	16.858 (4)	-4.253 (4)	(4)
M_2	direct	12.475	16.629	-4.177	-1678.883
	cv	12.585 (2)	16.766 (1)	-4.200 (1)	(1)
M_3	direct	12.468	16.623	-4.177	-1680.873
	cv	12.638 (3)	16.841 (3)	-4.204 (2)	(2)
M_4	direct	12.369	16.474	-4.221	-1710.278
	cv	14.952 (5)	33.974 (5)	-5.849 (5)	(5)

Example 5.10 Cross-validation for the speedskating models

Models M_0, M_1, M_2, M_3 were discussed above for explaining how the 5000-m influences the 10,000-m for top-level speedskaters. Presently, we supplement these model selection scores with results from cross-validation calculations, cf. Section 2.9. To illustrate aspects of cross-validation we shall also include a fifth model M_4, which uses a ninth-order polynomial $b_0 + b_1 w_i + \cdots + b_9 w_i^9$ as mean structure in a linear-normal model; here we use the normalised $w_i = (x_i - \bar{x}_n)/s_x$ for numerical reasons.

Table 5.2 provides direct (upper number) and cross-validated (lower number) estimates of mean absolute error, root mean squared error, and normalised log-likelihood, for each of the five models. In more detail, if $\xi(\theta, x) = E(Y \mid x)$, then the direct predictor is $\widehat{y}_i = \xi(\widehat{\theta}, x_i)$ and the cross-validated predictor is $\widehat{y}_{i,\mathrm{xv}} = \xi(\widehat{\theta}_{(i)}, x_i)$, involving for the latter a separate computation of the maximum likelihood estimates for the reduced data set of size $n - 1$. The numbers given in the table are hence $\mathrm{mae} = n^{-1} \sum_{i=1}^n |y_i - \widehat{y}_i|$ and the cross-validation version of mae $n^{-1} \sum_{i=1}^n |y_i - \widehat{y}_{i,\mathrm{xv}}|$. In the columns for root mean squared error we have $\mathrm{rmse} = (n^{-1} \sum_{i=1}^n |y_i - \widehat{y}_i|^2)^{1/2}$ and using cross-validation $(n^{-1} \sum_{i=1}^n |y_i - \widehat{y}_{i,\mathrm{xv}}|^2)^{1/2}$. The value $n^{-1}\ell_{n,\max} = n^{-1} \sum_{i=1}^n \log f(y_i, \widehat{\theta})$, while for cross-validation we use $\mathrm{xv}_n = n^{-1} \sum_{i=1}^n \log f(y_i, \widehat{\theta}_{(i)})$. Note that the direct estimates are always too optimistic, compared to the more unbiased assessment of the cross-validated estimates; this is also more notable for complex models than for simpler models, see e.g. the numbers for model M_4. We further note that the cross-validated log-likelihood numbers give the same ranking as does AIC, in agreement with the theory of Section 2.9, see e.g. (2.30). It is finally noteworthy that the simplest model M_0 is the winner when the criterion is cross-validated mean absolute error. ∎

We now continue to calculate the tolerance regions. First we obtain the necessary vectors and matrices for a general estimand $\mu(\sigma, \beta, \gamma, \phi)$, then we consider as specific examples of estimands the quantiles of the 10,000-m times and the probabilities of setting a record. For the model in (5.15) and its generalisations, we start with the log-density, which equals

$$-\frac{1}{2}\frac{1}{\sigma^2}\frac{1}{\exp(2\phi v_i)}(y_i - x_i^t\beta - z_i^t\gamma)^2 - \log\sigma - \phi v_i - \frac{1}{2}\log(2\pi). \qquad (5.16)$$

We compute its derivatives with respect to $(\sigma, \beta, \gamma, \phi)$ at the null model, and find $(\varepsilon_i^2 - 1)/\sigma$, $\varepsilon_i x_i/\sigma$, $\varepsilon_i z_i/\sigma$, $(\varepsilon_i^2 - 1)v_i$. The $\varepsilon_i = (Y_i - \xi_i)/\sigma_i$ has a standard normal distribution when $(\gamma, \phi) = (0, 0)$. This leads to

$$J_n = n^{-1}\sum_{i=1}^{n}\text{Var}\begin{pmatrix}(\varepsilon_i^2 - 1)/\sigma \\ \varepsilon_i x_i/\sigma \\ \varepsilon_i z_i/\sigma \\ (\varepsilon_i^2 - 1)v_i\end{pmatrix} = \begin{pmatrix}2/\sigma^2 & 0 & 0 & 0 \\ 0 & \Sigma_{n,00}/\sigma^2 & \Sigma_{n,01}/\sigma^2 & 0 \\ 0 & \Sigma_{n,10}/\sigma^2 & \Sigma_{n,11}/\sigma^2 & 0 \\ 0 & 0 & 0 & 2\end{pmatrix},$$

where

$$\Sigma_n = n^{-1}\sum_{i=1}^{n}\begin{pmatrix}x_i \\ z_i\end{pmatrix}\begin{pmatrix}x_i \\ z_i\end{pmatrix}^t = \begin{pmatrix}\Sigma_{n,00} & \Sigma_{n,01} \\ \Sigma_{n,10} & \Sigma_{n,11}\end{pmatrix}.$$

Matters related to the degree of influence of model deviations from the narrow model are largely determined by the $(q + 1) \times (q + 1)$-sized lower right-hand submatrix of J_n^{-1}, for which we find $Q_n = \text{diag}(\sigma^2\Sigma_n^{11}, 0)$. For a given estimand $\mu = \mu(\sigma, \beta, \gamma, \phi)$, the ω vector becomes

$$\omega = J_{n,10}J_{n,00}^{-1}\begin{pmatrix}\partial\mu/\partial\sigma \\ \partial\mu/\partial\beta\end{pmatrix} - \begin{pmatrix}\partial\mu/\partial\gamma \\ \partial\mu/\partial\phi\end{pmatrix}$$

$$= \begin{pmatrix}\Sigma_{n,10}\Sigma_{n,00}^{-1}\partial\mu/\partial\beta - \partial\mu/\partial\gamma \\ -\partial\mu/\partial\phi\end{pmatrix}, \qquad (5.17)$$

again with partial derivatives taken under the narrow M_0 model. Further,

$$\tau_0^2 = \sigma^2\begin{pmatrix}\frac{\partial\mu}{\partial\sigma} \\ \frac{\partial\mu}{\partial\beta}\end{pmatrix}^t\begin{pmatrix}\frac{1}{2} & 0 \\ 0 & \Sigma_{n,00}^{-1}\end{pmatrix}\begin{pmatrix}\frac{\partial\mu}{\partial\sigma} \\ \frac{\partial\mu}{\partial\beta}\end{pmatrix}$$

$$= \sigma^2\{\frac{1}{2}\left(\frac{\partial\mu}{\partial\sigma}\right)^2 + \left(\frac{\partial\mu}{\partial\beta}\right)^t\Sigma_{n,00}^{-1}\frac{\partial\mu}{\partial\beta}\}. \qquad (5.18)$$

Theorem 5.3 is applicable with this information. We now consider some specific examples.

Example 5.11 Quantiles of the 10,000-m times

Consider a speedskater whose 5000-m personal best is x_0. His 10,000-m result is a random variable with distribution corresponding to the log-density in (5.16). In particular, the

Table 5.3. *Speedskating data. Estimates of the
10% quantiles of the distribution of 10,000-m
times for two skaters, one with time 6:35 for the
5000-m, and the other with time 6:15 for the
5000-m. Four models are used for estimation.*

	0.10-quantile (6:35)	0.10-quantile (6:15)
M_0	13:37.25	12:49.35
M_1	13:37.89	12:48.13
M_2	13:38.05	12:57.55
M_3	13:38.12	12:57.48

qth quantile of this distribution is

$$\mu = \mu(x_0, q) = x_0^t \beta + z_0^t \gamma + d(q)\sigma \exp(\phi v_0),$$

with $d(q)$ the q-quantile of the standard normal. Also, z_0 and v_0 are defined via x_0 by the procedure described above. Following the (5.17)–(5.18) formulae, we find

$$\omega = \begin{pmatrix} \Sigma_{10}\Sigma_{00}^{-1}x_0 - z_0 \\ -d(q)\sigma v_0 \end{pmatrix}, \quad \tau_0 = \sigma\{\tfrac{1}{2}d(q)^2 + x_0^t\Sigma_{n,00}^{-1}x_0\}^{1/2}.$$

Two specific examples are as follows. First consider a medium-level skater with x_0 equal to 6:35.00, for which we estimate the 0.10 quantile of his 10,000-m time distribution; here $\widehat{\omega} = (0.733, 1.485)^t$. Then take an even better skater with x_0 equal to 6:15.00, considering now again the 0.10 quantile of his 10,000-m time; here $\widehat{\omega} = (-1.480, -39.153)^t$. Table 5.3 gives the four different estimates for the two quantile parameters in question, corresponding to the four models under discussion. Chapter 6 uses results of the present chapter to provide strategies for best selection among these four quantile estimates, in each of the two given problems. Presently our concern is with the tolerance regions. The narrow model contains parameters (β, σ) where β consists of an intercept parameter and a coefficient of the linear trend. Let $c = \delta_1/\sqrt{n}, \phi = \delta_2/\sqrt{n}$, and we are using $n = 200$ in this illustration, The vector δ has precise meaning in (c, ϕ) space. For the speedskating example, $\widehat{Q}_n = \text{diag}(284.546, 0.5)$. Figure 5.2 shows an ellipsoid inside which all narrow inference is better than full model inference. By application of Theorem 5.3, this tolerance set is that where

$$\delta_1^2/284.546 + 2\delta_2^2 = nc^2/284.546 + 2n\phi^2 \le 1.$$

Now, consider the two quantile estimands, each with their own vector ω. According to Theorem 5.3, these give rise to two infinite strips in the (c, ϕ) space inside which M_0-based estimation is better than M_3-based estimation; see Figure 5.2. For these data

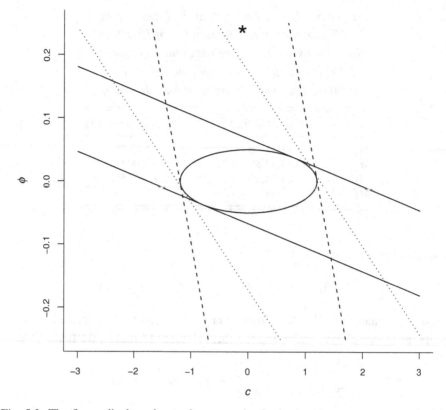

Fig. 5.2. The figure displays three tolerance strips in the (c, ϕ) parameter space, inside which the three-parameter (σ, a, b) model provides more precise inference than the correct five-parameter (σ, a, b, c, ϕ) model, for the three focus parameters in question; these are the two 10,000-m time quantile parameters corresponding to 5000-m times 6:15 (full line) and 6:35 (dashed line) and the world record 10,000-m time probability described in the text (dotted line). The figure also shows the inner ellipse of (c, ϕ) parameter values inside which narrow-based estimation is always more precise than wide-based estimation, for all estimands. The actual parameter estimate is indicated by an asterisk.

$\widehat{\delta} = (-1.586, 3.593)^{\mathrm{t}}$ and $\widehat{\delta}^t \widehat{Q}_n^{-1} \widehat{\delta} = 25.83$, hence narrow model estimation M_0 is not always better than full model estimation in M_3. For the 6:15.00 skater: $|\widehat{\omega}^t \widehat{\delta}| = 138.34 > 37.28 = (\widehat{\omega}^t \widehat{Q}_n \widehat{\omega})^{1/2}$, hence the narrow model is not sufficient. In Figure 5.2 the observed value $(\widehat{c}, \widehat{\phi}) = (-0.112, 0.254)$ is indicated by an asterisk, and indeed falls outside the tolerance band given by the solid lines. In contrast, for the 6:35.00 skater: $|\widehat{\omega}^t \widehat{\delta}| = 4.17 < 12.41 = (\widehat{\omega}^t \widehat{Q}_n \widehat{\omega})^{1/2}$. The corresponding tolerance band (dashed lines) indeed contains the pair $(\widehat{c}, \widehat{\phi})$. ∎

Example 5.12 Probabilities of setting a record
Eskil Ervik's personal best on the 5000-m is 6:10.65. What is the probability that he can set a world record on the 10,000-m? Inside our model the answer to this question is in

general terms

$$p(x_0, y_0) = \Phi\left(\frac{y_0 - x_0^t\beta - z_0^t\gamma}{\sigma \exp(\phi v_0)}\right), \tag{5.19}$$

for the fixed p-dimensional covariate vector x_0 and the fixed level y_0. In the special speedskating context (5.15) it corresponds to

$$p = \Phi\left(\frac{y_0 - a - bx_0 - cz_0}{\sigma \exp(\phi v_0)}\right),$$

where we allow a slight abuse of notation and use x_0 as the 5000-m time (corresponding to a covariate vector $(1, x_0)^t$ in the more general notation of (5.19)), where (as of end-of-2006 season) y_0 is Sven Kramer's 12:51.60, and z_0 and v_0 are defined as above as functions of x_0. The four models M_0–M_3 give rise to four estimated world record probabilities

$$0.302, \quad 0.350, \quad 0.182, \quad 0.187.$$

Chapter 6 provides methods for selecting the most appropriate among these four probability estimates. But as with the previous example, the present task is to assess and illustrate the tolerance strip in (c, ϕ) space inside which M_0-based estimation is more precise than M_3-based estimation (even when M_0 as such is 'a wrong model'); this turns out, via the ω calculation to follow, to be the strip represented by the dotted line in Figure 5.2. For (5.19) we find

$$\begin{pmatrix} \Sigma_{n,10}\Sigma_{n,00}^{-1}\partial p/\partial\beta - \partial p/\partial\gamma \\ -\partial p/\partial\phi \end{pmatrix} = \frac{f(w_0)}{\sigma \exp(\phi v_0)}\begin{pmatrix} -(\Sigma_{n,11}\Sigma_{n,00}^{-1}x_0 - z_0) \\ (y_0 - x_0^t\beta - z_0^t\gamma)v_0 \end{pmatrix},$$

writing here $w_0 = (y_0 - x_0^t\beta - z_0^t\gamma)/\{\sigma \exp(\phi v_0)\}$ and $f = \Phi'$ for the standard normal density. With partial derivatives at the narrow model this gives

$$\omega = \frac{1}{\sigma}f\left(\frac{y_0 - x_0^t\beta}{\sigma}\right)\begin{pmatrix} -(\Sigma_{10}\Sigma_{00}^{-1}x_0 - z_0) \\ (y_0 - x_0^t\beta)v_0 \end{pmatrix},$$

which leads to the estimated $\widehat{\omega} = (0.056, 0.408)^t$ for Ervik's chance of bettering the world record. Also,

$$\tau_0 = f(w_0)\{\tfrac{1}{2}(y_0 - x_0^t\beta)^2/\sigma_0^2 + x_0^t\Sigma_{n,00}^{-1}x_0\}^{1/2}$$

and leads to an estimated value of 0.870 for this situation. ∎

5.7 Large-sample calculus for AIC

In the framework of (5.12) we explore the use of the AIC method of Chapter 2 to aid selection of the 'right' set of γ parameters. The following is a generalisation of results of Section 2.8. For S a subset of $\{1, \ldots, q\}$, we work with submodel S, defined as the model with density $f(y, \theta, \gamma_S, \gamma_{0,S^c})$, where θ and γ_S are unknown parameters and γ_j

is set equal to $\gamma_{0,j}$ for $j \notin S$. We show in Section 6.3 that the estimator of the local neighbourhood parameter δ in the wide model satisfies

$$D_n = \widehat{\delta}_{\text{wide}} = \sqrt{n}(\widehat{\gamma}_{\text{wide}} - \gamma_0) \xrightarrow{d} D \sim N_q(\delta, Q), \qquad (5.20)$$

with Q defined in (5.14). To state and prove the theorem below about limits of AIC differences we need to define some matrices that are further discussed and worked with in Section 6.1. First let Q_S be the lower right-hand $|S| \times |S|$ matrix of the inverse information matrix J_S^{-1} of submodel S, where $|S|$ is the number of elements of S, and then let $Q_S^0 = \pi_S^t Q_S \pi_S$, where π_S is the $|S| \times q$ projection matrix that maps a vector $v = (v_1, \ldots, v_q)^t$ to $v_S = \pi_S v$ containing only those v_j for which $j \in S$. We start from

$$\text{AIC}_{n,S} = 2\ell_{n,S,\max} - 2(p + |S|) = 2\sum_{i=1}^{n} \log f(Y_i, \widehat{\theta}_S, \widehat{\gamma}_S, \gamma_{0,S^c}) - 2(p + |S|),$$

where $(\widehat{\theta}_S, \widehat{\gamma}_S)$ are the maximum likelihood estimators in the S submodel. A central result about AIC behaviour is then the following.

Theorem 5.4 *Under the conditions of the local large-sample framework (5.12), the AIC differences converge jointly in distribution:*

$$\text{AIC}_{n,S} - \text{AIC}_{n,\emptyset} \xrightarrow{d} \text{aic}(S, D) = D^t Q^{-1} Q_S^0 Q^{-1} D - 2|S|.$$

Proof. We start out with the likelihood-ratio statistic, expanding it to the second order. For this we need averages $\bar{U}_n = n^{-1} \sum_{i=1}^{n} U(Y_i)$ and $\bar{V}_n = n^{-1} \sum_{i=1}^{n} V(Y_i)$ of variables defined in terms of the score statistics $U(y)$ and $V(y)$, which are the derivatives of $\log f(y, \theta, \gamma)$ with respect to θ and γ, evaluated at (θ_0, γ_0). Via a variation of result (2.9) one is led to

$$\Delta_{n,S} = 2\sum_{i=1}^{n} \log \frac{f(Y_i, \widehat{\theta}_S, \widehat{\gamma}_S, \gamma_{0,S^c})}{f(Y_i, \theta_0, \gamma_0)} \doteq_d n \begin{pmatrix} \bar{U}_n \\ \bar{V}_{n,S} \end{pmatrix}^t J_S^{-1} \begin{pmatrix} \bar{U}_n \\ \bar{V}_{n,S} \end{pmatrix}.$$

For the narrow model containing only the 'protected' parameters, $\Delta_{n,\emptyset} \doteq_d n\bar{U}_n^t J_{00}^{-1} \bar{U}_n$. A convenient rearrangement of the quadratic forms here uses the algebraic fact that

$$\begin{pmatrix} u \\ v \end{pmatrix}^t J_S^{-1} \begin{pmatrix} u \\ v \end{pmatrix} - u^t J_{00}^{-1} u = \left(v - J_{10} J_{00}^{-1} u\right)^t Q_S^0 \left(v - J_{10} J_{00}^{-1} u\right).$$

This is seen to imply

$$\Delta_{n,S} - \Delta_{n,\emptyset} \doteq_d D_n^t Q^{-1} Q_S^0 Q^{-1} D_n \xrightarrow{d} D^t Q^{-1} Q_S^0 Q^{-1} D.$$

This ends the proof since the AIC difference is equal to $\Delta_{n,S} - \Delta_{n,\emptyset} - 2|S|$. $\qquad\square$

We note that the convergence of AIC differences to $\text{aic}(S, D)$ holds jointly for all subsets S, and that the theorem can be extended to the regression situation without essential difficulties. The theorem in particular says that AIC behaviour for large n is fully dictated by $D_n = \widehat{\delta}_{\text{wide}}$ of (5.20). We also learn that

$$
\begin{aligned}
\widehat{\delta}_S &= \sqrt{n}(\widehat{\gamma}_S - \gamma_{0,S}) \doteq \sqrt{n}\big(J^{10,S}\bar{U}_n + J^{11,S}\bar{V}_n\big) \\
&= \sqrt{n}Q_S\big(\bar{V}_n - J_{10}J_{00}^{-1}\bar{U}_n\big)_S \doteq Q_S(\widehat{\delta}_{\text{wide}})_S.
\end{aligned}
\tag{5.21}
$$

Thus the $\widehat{\delta}_S$ for different subsets S can be expressed as functions of $\widehat{\delta}_{\text{wide}}$, modulo terms that go to zero in probability.

There is a natural normalising transformation of D,

$$
Z = Q^{-1/2}D \sim N_q(a, I_q), \quad \text{with } a = Q^{-1/2}\delta.
\tag{5.22}
$$

The aic(S, D) can be expressed as $Z^t H_S Z - 2|S|$, where $H_S = Q^{-1/2}Q_S^0 Q^{-1/2}$. Here $Z^t H_S Z$ has a noncentral chi-squared distribution with $|S|$ degrees of freedom and excentricity parameter $\lambda_S = a^t H_S a = \delta^t Q^{-1}Q_S^0 Q^{-1}\delta$. Thus we have obtained a concise description of the simultaneous behaviour of all AIC differences, as

$$
\text{AIC}_{n,S} - \text{AIC}_{n,\emptyset} \xrightarrow{d} Z^t H_S Z - 2|S| \sim \chi^2_{|S|}(\lambda_S) - 2|S|.
$$

This in principle determines the limits of all model selection probabilities, in terms of a single multivariate normal vector Z, via

$$
\text{P(AIC selects model } S) \to p(S \mid \delta),
$$

where the right-hand limit is defined as the probability that $Z^t H_S Z - 2|S|$ is larger than all other $Z^t H_{S'} Z - 2|S'|$.

Remark 5.4 AIC as a pre-test strategy
Assume that Q of (5.14) is a diagonal matrix, with $\kappa_1^2, \ldots, \kappa_q^2$ along its diagonal. Then

$$
\text{aic}(S, D) = \sum_{j \in S}(D_j^2/\kappa_j^2 - 2),
$$

and the way to select indices for making this random quantity large is to keep the positive terms but discard the negative ones. Thus the limit version of AIC selects precisely the set of those j for which $|D_j/\kappa_j| \geq \sqrt{2}$. This is large-sample equivalent to the pre-test procedure that examines each parameter γ_j, and includes it in the model when $\sqrt{n}|\widehat{\gamma}_j - \gamma_{0,j}|$ exceeds $\sqrt{2}\widehat{\kappa}_j$. The implied significance level, per component test, is $\text{P}(|\text{N}(0, 1)| \geq \sqrt{2}) = 0.157$. ∎

Remark 5.5 When is a submodel better than the big model?
When is a candidate submodel S better than the wide model? The AIC answer to this question is to check whether

$$
2\ell_{n,S}(\widehat{\theta}_S, \widehat{\gamma}_S, \gamma_{0,S^c}) - 2(p + |S|) > 2\ell_n(\widehat{\theta}, \widehat{\gamma}) - 2(p + q).
$$

By Theorem 5.4, the limit version of this statement is

$$D^t Q^{-1} Q_S^0 Q^{-1} D - 2|S| > D^t Q^{-1} D - 2q,$$

which is equivalent to

$$T(S) = D^t(Q^{-1} - Q^{-1} Q_S^0 Q^{-1})D = Z^t(I_q - H_S)Z < 2(q - |S|).$$

The distribution of this quadratic form is found via the remarks above: $T(S) \sim \chi^2_{q-|S|}(\lambda_S)$, where $\lambda_S = a^t(I_q - H_S)a = \delta^t(Q^{-1} - Q^{-1}Q_S^0 Q^{-1})\delta$. If model S is actually correct, so that $\delta_j = 0$ for $j \notin S$, then the probability that AIC will prefer this submodel over the widest one is $P(\chi^2_{q-|S|} \leq 2(q - |S|))$. This probability is equal to 0.843, 0.865, 0.888, 0.908, 0.925, and so on, for dimensions $q - |S|$ equal to 1, 2, 3, 4, 5, and so on. The special case of testing the narrow model against the wide model is covered by these results. AIC prefers the narrow model if $T = D^t Q^{-1} D < 2q$, and the distribution of T is $\chi^2_q(\delta^t Q^{-1}\delta)$. ∎

5.8 Notes on the literature

There is a large variety of further examples of common departures from standard models that can be studied. In each case one can compute the tolerance radius and speculate about robustness against the deviation in question. Some of this chapter's results appeared in Hjort and Claeskens (2003), with origin in Hjort (1991). See also Fenstad (1992) and Kåresen (1992), who applied these methods to assess how much dependence the independence assumption typically can tolerate, and to more general problems of tolerance, respectively. The question on how much t-ness the normal model can tolerate is answered in Hjort (1994a). A likelihood treatment of misspecified models with explicit distinction between the model that generated the data and that used for fitting is White (1994). The local misspecification setting is related to contiguity and local asymptotic normality, see Le Cam and Yang (2000) or van der Vaart (1998, Chapters 6, 7). In Claeskens and Carroll (2007), contiguity is used to obtain related results in the context of a semiparametric model. Papers Bickel (1981, 1983, 1984) and from the Bayesian angle Berger (1982) all touch on the theme of 'small biases may be useful', which is related to the topic of this chapter.

Exercises

5.1 *De-biasing the narrow leads to the wide:* In the framework of Sections 5.2–5.3, which among other results led to Corollary 5.1, one may investigate the de-biased narrow-based estimator $\widehat{\mu}_{db} = \widehat{\mu}_{narr} - \omega(\widehat{\gamma} - \gamma_0)$. Show that $\sqrt{n}(\widehat{\mu}_{db} - \mu_{true})$ tends to $N(0, \tau_0^2 + \omega^2\kappa^2)$, and that in fact $\widehat{\mu}_{db}$ and $\widehat{\mu}_{wide}$ are large-sample equivalent. Show also that the conclusions remain the same when an estimator of the form $\widehat{\mu}_{narr} - \widehat{\omega}(\widehat{\gamma} - \gamma_0)$ is used, where $\widehat{\omega}$ is a consistent estimator of ω. Investigate finally behaviour and performance of estimators that remove part but not all of the bias, i.e. $\widehat{\mu}_{narr} - c\,\widehat{\omega}(\widehat{\gamma} - \gamma_0)$ for some $c \in [0, 1]$.

5.2 *The 10,000 m:* For the speedskating models discussed in Section 5.6, consider the parameter $\rho = \rho(x)$, the expected ratio Y/x for given x, where x is the 5000-m time and Y the 10,000-m time of a top skater. This is a frequently discussed parameter among speedskating fans. There are arguably clusters of 'stayers' with low ρ and 'sprinters' with big ρ in the Adelskalenderen.

 (a) Compute the matrix J, variance τ_0^2 and vector ω for this focus parameter, for some relevant models, for different types of skaters.

 (b) Why is Johann Olav Koss, who has a very impressive $\rho = 2.052$, so modestly placed in Figure 5.1?

5.3 *Drawing tolerance strips and ellipses:* Drawing tolerance figures of the type in Figure 5.2 is useful for understanding aspects of model extensions. Here we outline how to draw such strips and ellipses, using the set-up of Section 5.6 as an illustration. First one computes an estimate $\widehat{Q}$ of the Q matrix, typically via the estimate $\widehat{J}_n$ of the full information matrix J_n. The following programming steps may be used (in R language):

 (a) Specify the resolution level, for example 250 points. In this example $\delta = \sqrt{n}(c, \phi)^t$. We define a vector of c values and of ϕ values which span the range of values to be considered in the figure:

 ```
 reso = 250; cval = seq(from = −3, to = 3, length = reso)
 phival = seq(from = −0.25, to = 0.25, length = reso)
 ```

 (b) Define the values of the boundary ellipse according to Theorem 5.3. In the notation of this example $\delta^t Q^{-1} \delta = n(c, \phi)Q^{-1}(c, \phi)^t$. Hence, for the grid of (c, ϕ) values we compute `ellipse[i, j]` as $n(c_i, \phi_j)Q^{-1}(c_i, \phi_j)^t$.

 (c) The actual ellipse can be drawn as a contour plot for values of $\delta^t Q^{-1} \delta = 1$. In R language
 ```
 contour(cval, phival, ellipse, labels = " ", levels = c(1))
 ```

 (d) The tolerance strips can be obtained in the following way. Define a function which takes as input the vector ω, the combined vector (c, ϕ) and matrix Q:
 ```
 Tolerance.strip = function(omega, cphi, Q){
     arg1 = sqrt(n) * abs(sum(omega * cphi))
     arg2 = sqrt(t(omega)%*%Q%*%omega)
     return(arg1 − arg2)}
 ```

 (e) With the particular vector ω for the data, for the same grid of values of (c, ϕ), compute `tolregion[i, j]`
 ```
     = Tolerance.strip(omega, c(cval[i], phival[j]), Q)
 ```

 (f) We obtain the plotted lines by adding to the previous plot the following contours:
 ```
 contour(cval, phival, tolregion, labels = " ", levels = c(0))
 ```

5.4 *The Ladies Adelskalenderen:* Refer to the data of Exercise 2.9 and consider models as used for the male skaters. Take as focus parameters the 10% and 80% quantiles of the distribution of the 5000-m time for a female skater with time 3:59 (= 239 seconds) on the 3000-m. First compute the matrix Q_n and $\widehat{\delta}$. Then estimate ω for the two different focus parameters. With this information, construct a figure similar to that of Figure 5.2, showing the ellipse where narrow model estimation is better than wide model estimation for all estimands, and the two tolerance bands for the specific focus parameters. Is the narrow model best for both focus parameters?

5.5 *Beating the records:* Go to internet websites that deal with the current Adelskalenderen (e.g. for the International Skating Union, or Wikipedia) and re-do Example 5.12 for other speedskaters and with the most recently updated 10k world record.

5.6 *Narrow or wide for Weibull:* Consider the situation of the Weibull model as in Examples 5.1 and 5.5.

(a) Show that the r-quantile of the Weibull distribution is equal to $\mu = A^{1/\gamma}/\theta$, where $A = -\log(1-r)$. For the maximum likelihood estimator $\widehat{\mu} = A^{1/\widehat{\gamma}}/\widehat{\theta}$, show that $\sqrt{n}(\widehat{\mu} - \mu) \to_d N(0, \tau^2)$, with

$$\tau^2 = (\mu^2/\gamma^2)[1 + (6/\pi^2)\{\log A - (1-g_e)\}^2].$$

For the median, this yields limit distribution $N(0, 1.1742^2 \mu^2/\gamma^2)$.

(b) If data really come from the Weibull, how much better is the median estimator $\widehat{\mu} = (\log 2)^{1/\widehat{\gamma}}/\widehat{\theta}$ than M_n, the standard (nonparametric) sample median? Show that the parametric estimator's standard deviation divided by that of the nonparametric estimator converges to

$$\rho(r) = \frac{A\exp(-A)}{\{r(1-r)\}^{1/2}}[1 + (6/\pi^2)\{\log A - (1-g_e)\}^2]^{1/2},$$

again with $A = -\log(1-r)$. Use the fact that the standard deviation of the limiting distribution for $\sqrt{n}(M_n - \mu)$ is $\frac{1}{2}/f(\mu)$ with $f(\mu)$ the population density computed at the median μ, cf. Serfling (1980, Section 2.3).

(c) Verify the claims made in Section 5.2 about narrow and wide estimation of quantiles for the Weibull family.

5.7 *The tolerance threshold:* Theorem 5.2 may be used to explicitly find the tolerance limit in a range of situations involving models being extended with one extra parameter, as seen in Section 5.5. Here are some further illustrations.

(a) Consider the normal linear regression model where

$$Y_i = \beta_0 + \beta_1 x_{i,1} + \beta_2 x_{i,2} + \beta_3 x_{i,3} + \varepsilon_i \quad \text{for } i = 1, \ldots, n,$$

and where the ε_i are $N(0, \sigma^2)$. Assume for concreteness that the covariates $x_{i,1}, x_{i,2}, x_{i,3}$ behave as independent standard normals. How much interaction can be tolerated for this model, in terms of an added term $\gamma_{1,2} x_{i,1} x_{i,2}$ to the regression structure? Answer also the corresponding question when the term to be potentially added takes the form $\gamma(x_{i,1} x_{i,2} + x_{i,1} x_{i,3} + x_{i,2} x_{i,3})$.

(b) The Gamma model has many important and convenient properties, and is in frequent use in many application areas. Consider the three-parameter extension $f(y, a, b, c) = k(a, b, c) y^{a-1} \exp(-by^c)$, with $k(a, b, c)$ the appropriate normalisation constant. Here $c = 1$ corresponds to the Gamma model, with $k(a, b, 1) = b^a/\Gamma(a)$. How much non-Gamma-ness can the Gamma model tolerate, in terms of c around 1?

(c) There are many extensions of the normal model in the literature, and for each one can ask about the tolerance threshold. For one such example, consider

$f(y, \xi, \sigma, \gamma) = k(\sigma, \gamma) \exp\{-\frac{1}{2}|(y - \xi)/\sigma|^{\gamma}\}$. How much must γ differ from $\gamma_0 = 2$ in order for the more complicated estimators $\mu(\widehat{\xi}_{\text{wide}}, \widehat{\sigma}_{\text{wide}}, \widehat{\gamma}_{\text{wide}})$ to give better precision than the ordinary normality-based estimators?

5.8 *Should the Poisson model be stretched?* Sometimes the Poisson approximation to a given distribution $g(y)$ is too crude, for example when the dispersion $\delta = \text{Var}_g Y / \text{E}_g Y$ differs noticeably from 1. Let this be motivation for studying an extended model.

(a) Let $Y_1, \ldots, Y_n$ be a sample from the stretched Poisson model

$$f(y, \theta, \gamma) = \frac{1}{k(\theta, \gamma)} \frac{\theta^y}{(y!)^{\gamma}} \quad \text{for } y = 0, 1, 2, \ldots$$

How far away must γ be from $\gamma_0 = 1$ in order for the stretched Poisson to provide more precise estimators of $\mu(\theta, \gamma)$ parameters than under the simple Poisson model?

(b) More generally, suppose regression data (x_i, Y_i) are available for $i = 1, \ldots, n$, where $Y_i \mid x_i$ has the distribution $k(\xi_i, \gamma)^{-1} \xi_i^y / (y!)^{\gamma}$ for $y = 0, 1, 2, \ldots$, where $\xi_i = \exp(x_i^t \beta)$. How different from 1 must γ be in order for this more involved analysis to be advantageous?

5.9 *Quadraticity and interaction in regression:* Consider a Poisson regression model with two covariates x_1 and x_2 that influence the intensity parameter λ. The simple narrow model takes $Y_i \sim \text{Pois}(\lambda_i)$, with $\lambda_i = \exp(\beta_0 + \beta_1 x_{i,1} + \beta_2 x_{i,2})$. We shall investigate this model's tolerance against departures of quadraticity and interaction; more specifically, the widest model has six parameters, with

$$\lambda_i = \exp(\beta_0 + \beta_1 x_{i,1} + \beta_2 x_{i,2} + \gamma_1 z_{i,1} + \gamma_2 z_{i,2} + \gamma_3 z_{i,3}),$$

where $z_{i,1} = x_{i,1}^2, z_{i,2} = x_{i,2}^2, z_{i,3} = x_{i,1} x_{i,2}$.

(a) Show that the general method of (5.8) leads to the 6×6 information matrix

$$J_n = n^{-1} \sum_{i=1}^{n} \lambda_i \begin{pmatrix} 1 \\ x_i \\ z_i \end{pmatrix} \begin{pmatrix} 1 \\ x_i \\ z_i \end{pmatrix}^t,$$

computed at the null model.

(b) Use Theorem 5.3 to find expressions for the tolerance levels, against specific departures or against all departures, involving the matrix Q_n, the lower right-hand 3×3 submatrix of J_n^{-1}.

(c) Carry out some experiments for this situation, where the ranges of $x_{i,1}$ and $x_{i,2}$ are varied. Show that if the $x_{i,1}$ and $x_{i,2}$ span short ranges, then confidence intervals for $\gamma_1, \gamma_2, \gamma_3$ will be quite broad, and inference based on the narrow model typically outperforms that of using the bigger model. If the covariates span broader ranges, however, then precision becomes better for γ estimation, and the narrow model has a more narrow tolerance.

5.10 *Tolerance with an L_1 view:* Theorem 5.2 reached a clear result for the tolerance threshold, under the L_2 viewpoint corresponding to mean squared error, via limit versions of

$\mathrm{E}\,n(\widehat{\mu} - \mu_{\text{true}})^2$. Similar questions may be asked for other loss functions. For the L_1 loss function, show from Corollary 5.1 that the risk limits to compare are

$$\text{risk}_{\text{narr}}(\delta) = \tau_0 A(\omega\delta/\tau_0) \quad \text{and} \quad \text{risk}_{\text{wide}}(\delta) = \left(\tau_0^2 + \omega^2\kappa^2\right)^{1/2} A(0),$$

where $A(a)$ is the function $\mathrm{E}|a + \mathrm{N}(0, 1)|$. Show that $A(a) = a\{2\Phi(a) - 1\} + 2\phi(a)$, so in particular $A(0) = \sqrt{2/\pi}$. Characterise the region of δ for which the narrow limit risk is smaller than the wider limit risk. Discuss implications for the range of situations studied in Sections 5.1 and 5.5.

6

The focussed information criterion

The model selection methods presented earlier (such as AIC and the BIC) have one thing in common: they select one single 'best model', which should then be used to explain all aspects of the mechanisms underlying the data and predict all future data points. The tolerance discussion in Chapter 5 showed that sometimes one model is best for estimating one type of estimand, whereas another model is best for another estimand. The point of view expressed via the focussed information criterion (FIC) is that a 'best model' should depend on the parameter under focus, such as the mean, or the variance, or the particular covariate values, etc. Thus the FIC allows and encourages different models to be selected for different parameters of interest.

6.1 Estimators and notation in submodels

In model selection applications there is a list of models to consider. We shall assume here that there is a 'smallest' and a 'biggest' model among these, and that the others lie between these two extremes. More concretely, there is a *narrow model*, which is the simplest model that we possibly might use for the data, having an unknown parameter vector θ of length p. Secondly, in the *wide model*, the largest model that we consider, there are an additional q parameters $\gamma = (\gamma_1, \ldots, \gamma_q)$. We assume that the narrow model is a special case of the wide model, which means that there is a value γ_0 such that with $\gamma = \gamma_0$ in the wide model, we get precisely the narrow model. Further *submodels* correspond to including some, but excluding others, among the γ_j parameters. We index the models by subsets S of $\{1, \ldots, q\}$. A model S, or more precisely, a model indexed by S, contains those parameters γ_j for which j belongs to S; cf. Section 5.7. The empty set $\emptyset$ corresponds to no additional γ_j, hence identifying the narrow model.

The submodel S corresponds to working with the density $f(y, \theta, \gamma_S, \gamma_{0,S^c})$, with S^c denoting the complementary set with indices not belonging to S. This slight abuse of notation indicates that for model S the values of γ_j for which j does not belong to S are set to their null values $\gamma_{0,j}$. The maximum likelihood estimators for the parameters of this submodel are denoted $(\widehat{\theta}_S, \widehat{\gamma}_S)$. These lead to maximum likelihood estimator

$\widehat{\mu}_S = \mu(\widehat{\theta}_S, \widehat{\gamma}_S, \gamma_{0,S^c})$ for $\mu = \mu(\theta, \gamma)$, where μ is some parameter under consideration. Note that the vector $\widehat{\theta}_S$ has always p components, for all index sets S. Because the value of the estimator can change from one submodel to another (depending on which components γ_j are included), we explicitly include the subscript S in the notation of the estimator.

There are up to 2^q such submodel estimators, ranging from the narrow $S = \emptyset$ model to the wide $S = \{1, \ldots, q\}$. Sometimes the range of candidate models is restricted on a priori grounds, as with nested models that have some natural order of complexity. In the sequel, therefore, the range of the sets S is not specified, allowing the user the freedom to specify only some of the possible subsets, or to be conservative and allow all subset models. Below we need various quantities that partly depend on J_{wide}, the $(p + q) \times (p + q)$ information matrix in the fullest model, evaluated at (θ_0, γ_0), assumed to be an inner point in its parameter space.

For a vector v we use v_S to denote the subset of components v_j with $j \in S$. This may also be written $v_S = \pi_S v$ with π_S the appropriate $|S| \times q$ projection matrix of zeros and ones; again $|S|$ denotes the cardinality of S. We let J_S be the $(p + |S|) \times (p + |S|)$-sized submatrix of J_{wide} that corresponds to keeping all first p rows and columns (corresponding to θ) and taking of the last q rows and columns only those with numbers j belonging to S. Hence J_S is the Fisher information matrix for submodel S. In block notation, let

$$J_S = \begin{pmatrix} J_{00} & J_{01,S} \\ J_{10,S} & J_{11,S} \end{pmatrix} \quad \text{and} \quad J_S^{-1} = \begin{pmatrix} J^{00,S} & J^{01,S} \\ J^{10,S} & J^{11,S} \end{pmatrix}.$$

We have worked with $Q = J^{11}$ in Chapter 5 and shall also need $Q_S = J^{11,S}$ for submodel S; in fact, $Q_S = (\pi_S Q^{-1} \pi_S^{\text{t}})^{-1}$ in terms of the projection matrix π_S introduced above. Let next $Q_S^0 = \pi_S^{\text{t}} Q_S \pi_S$; it is a $q \times q$ matrix with elements equal to those of Q_S apart from those where rows and columns are indexed by S^c, and where elements of Q_S^0 are zero. Define finally the $q \times q$ matrix $G_S = Q_S^0 Q^{-1} = \pi_S^{\text{t}} Q_S \pi_S Q^{-1}$. Since G_S multiplies Q by part of its inverse, it is seen that $\text{Tr}(G_S) = |S|$; see Exercise 6.1. It is helpful to note that when Q is diagonal, then Q_S is the appropriate $|S| \times |S|$ subdiagonal matrix of Q, and G_S is the $q \times q$ matrix with diagonal elements 1 for $j \in S$ and 0 for $j \notin S$.

6.2 The focussed information criterion, FIC

Consider a *focus parameter* $\mu = \mu(\theta, \gamma)$, i.e. a parameter of direct interest and that we wish to estimate with good precision. For the estimation we want to include all components of θ in the model, but are not sure about which components of γ to include when forming a final estimate. Perhaps all γ_j shall be included, perhaps none. This leads to considering estimators of the form $\widehat{\mu}_S = \mu(\widehat{\theta}_S, \widehat{\gamma}_S, \gamma_{0,S^c})$, see Section 6.1. The 'best' model for estimation of the focus parameter μ is this model for which the mean squared error of $\sqrt{n}(\widehat{\mu}_S - \mu_{\text{true}})$ is the smallest. The focussed information criterion (FIC) is based on an estimator of these mean squared errors. The model with the lowest value of the FIC is selected.

In order not to use the 'wide' subscript too excessively, we adopt the convention that $(\widehat{\theta}, \widehat{\gamma})$ signals maximum likelihood estimation in the full $p + q$-parameter model. Let $D_n = \sqrt{n}(\widehat{\gamma} - \gamma_0)$, as in (5.20), and let ω be as in Section 5.4. We shall now define the FIC score, for each of the submodels indexed by S; its proper derivation is given in Section 6.3. Several equivalent formulae may be used, including

$$
\begin{aligned}
\text{FIC}(S) &= \widehat{\omega}^{\text{t}}(I_q - \widehat{G}_S)D_n D_n^{\text{t}}(I_q - \widehat{G}_S)^{\text{t}}\widehat{\omega} + 2\widehat{\omega}^{\text{t}}\widehat{Q}_S^0\widehat{\omega}, \\
&= n\widehat{\omega}^{\text{t}}(I_q - \widehat{G}_S)(\widehat{\gamma} - \gamma_0)(\widehat{\gamma} - \gamma_0)^{\text{t}}(I_q - \widehat{G}_S)^{\text{t}}\widehat{\omega} + 2\widehat{\omega}^{\text{t}}\widehat{Q}_S^0\widehat{\omega} \\
&= (\widehat{\psi}_{\text{wide}} - \widehat{\psi}_S)^2 + 2\widehat{\omega}_S^{\text{t}}\widehat{Q}_S\widehat{\omega}_S.
\end{aligned}
\tag{6.1}
$$

In the latter expression, $\widehat{\psi} = \widehat{\omega}^{\text{t}}D_n$ and $\widehat{\psi}_S = \widehat{\omega}^{\text{t}}G_S D_n$ may be seen as the wide model based and the S-model based estimates of $\psi = \omega^{\text{t}}\delta$. Note that $\widehat{\omega}^{\text{t}}\widehat{Q}_S^0\widehat{\omega} = \widehat{\omega}_S^{\text{t}}\widehat{Q}_S\widehat{\omega}_S$, see Exercise 6.1. Further information related to the computation of the FIC is in Section 6.5. This FIC is the criterion that we use to select the best model, with smaller values of FIC(S) favoured over larger ones. A bigger S makes the first term small and the second big; correspondingly, a smaller S makes the first term big and the second small. The two extremes are $2\widehat{\omega}^{\text{t}}\widehat{Q}\widehat{\omega}$ for the wide model and $(\widehat{\omega}^{\text{t}}D_n)^2$ for the narrow model. In practice we compute the FIC value for each of the models that are deemed plausible a priori, i.e. not always for the full list of 2^q submodels.

A revealing simplification occurs when $\widehat{Q}$ is a diagonal matrix $\text{diag}(\widehat{\kappa}_1^2, \ldots, \widehat{\kappa}_q^2)$. In this case $\widehat{G}_S$ is a diagonal matrix containing a 1 on position j if variable γ_j is in model S, and 0 otherwise. The FIC expression simplifies to

$$
\text{FIC}(S) = \left(\sum_{j \notin S} \widehat{\omega}_j D_{n,j}\right)^2 + 2\sum_{j \in S} \widehat{\omega}_j^2 \widehat{\kappa}_j^2.
\tag{6.2}
$$

The first term is a squared bias component for those parameters not in the set S, while the second term is twice the variance for those parameter estimators of which the index belongs to the index set S.

Remark 6.1 Asymptotically AIC and the FIC agree for $q = 1$
Note that for the narrow model ($S = \emptyset$) and for the wide model

$$
\text{FIC}(\emptyset) = (\widehat{\omega}^{\text{t}}D_n)^2 = n\{\widehat{\omega}^{\text{t}}(\widehat{\gamma}_{\text{wide}} - \gamma_0)\}^2 \quad \text{and} \quad \text{FIC}(\text{wide}) = 2\widehat{\omega}^{\text{t}}\widehat{Q}\widehat{\omega}.
$$

In the simplified model selection problem where only these two extremes are considered, the largest model is preferred when $2\widehat{\omega}^{\text{t}}\widehat{Q}\widehat{\omega} \leq (\widehat{\omega}^{\text{t}}D_n)^2$. When q is one-dimensional, the $\widehat{\omega}$ term cancels (as long as it is nonzero), and the criterion for preferring the full model over the narrow is $2\widehat{Q} \leq D_n^2$, or $D_n^2/\widehat{\kappa}^2 \geq 2$, writing as in Chapter 5 $Q = \kappa^2$ for the J^{11} number. But this is the same, to the first order of approximation, as for AIC (see Section 2.8). Thus the FIC and AIC are first-order equivalent for $q = 1$, but not for $q \geq 2$. ∎

6.3 Limit distributions and mean squared errors in submodels

The derivation of the FIC of (6.1) is based on analysis and estimation of the mean squared error (mse) of the estimators $\widehat{\mu}_S$. We continue to work in the local misspecification setting used in Chapter 5, see for example Sections 5.2, 5.4 and 5.7. We assume in the i.i.d. case that the true model P_n, the nth model, has

$$Y_1, \ldots, Y_n \text{ i.i.d. from } f_n(y) = f(y, \theta_0, \gamma_0 + \delta/\sqrt{n}). \tag{6.3}$$

The results that follow are also valid in general regression models for independent data, where the framework is that

$$Y_i \text{ has a density } f(y \mid x_i, \theta_0, \gamma_0 + \delta/\sqrt{n}) \quad \text{for } i = 1, \ldots, n,$$

cf. Section 5.2.3. The reasons for studying mean squared errors of estimators inside this local misspecification setting are as in the mentioned sections, and are summarised in Remark 5.3.

Under certain natural and mild conditions, Hjort and Claeskens (2003) prove the following theorem about the distribution of the maximum likelihood estimator $\widehat{\mu}_S$ in the submodel S, under the sequence of models P_n. To present it, recall first the quantities

$$\omega = J_{10} J_{00}^{-1} \frac{\partial \mu}{\partial \theta} - \frac{\partial \mu}{\partial \gamma} \quad \text{and} \quad \tau_0^2 = \left(\frac{\partial \mu}{\partial \theta}\right)^t J_{00}^{-1} \frac{\partial \mu}{\partial \theta}$$

from Section 5.4, with partial derivatives evaluated at the null point (θ_0, γ_0), and introduce independent normal variables $D \sim N_q(\delta, Q)$ and $\Lambda_0 \sim N(0, \tau_0^2)$.

Theorem 6.1 *First, for the maximum estimator of δ in the wide model,*

$$D_n = \widehat{\delta}_{\text{wide}} = \sqrt{n}(\widehat{\gamma}_{\text{wide}} - \gamma_0) \xrightarrow{d} D \sim N_q(\delta, Q). \tag{6.4}$$

Secondly, for the maximum likelihood estimator $\widehat{\mu}_S$ of $\mu = \mu(\theta, \gamma)$,

$$\sqrt{n}(\widehat{\mu}_S - \mu_{\text{true}}) \xrightarrow{d} \Lambda_S = \Lambda_0 + \omega^t(\delta - G_S D). \tag{6.5}$$

Proof. We shall be content here to give the main ideas and steps in the proof, leaving details and precise regularity conditions to Hjort and Claeskens (2003, Section 3 and appendix). For the i.i.d. situation, maximum likelihood estimators are to the first asymptotic order linear functions of score function averages $\bar{U}_n = n^{-1} \sum_{i=1}^{n} U(Y_i)$ and $\bar{V}_n = n^{-1} \sum_{i=1}^{n} V(Y_i)$, where $U(y)$ and $V(y)$ are the log-derivatives of the density $f(y, \theta, \gamma)$ evaluated at the null point (θ_0, γ_0). Taylor expansion arguments combined with central limit theorems of the Lindeberg variety lead to

$$\begin{pmatrix} \sqrt{n}(\widehat{\theta}_S - \theta_0) \\ \sqrt{n}(\widehat{\gamma}_S - \gamma_{0,S}) \end{pmatrix} \doteq_d J_S^{-1} \begin{pmatrix} \sqrt{n}\bar{U}_n \\ \sqrt{n}\bar{V}_{n,S} \end{pmatrix} \xrightarrow{d} \begin{pmatrix} C_S \\ D_S \end{pmatrix} = J_S^{-1} \begin{pmatrix} J_{01}\delta + U' \\ \pi_S J_{11}\delta + V_S' \end{pmatrix},$$

say, where (U', V') has the $N_{p+q}(0, J)$ distribution; the '$\doteq_d$' means that the difference tends to zero in probability.

This result, which already characterises the joint limit behaviour of all subset estimators of the model parameters, then leads with additional Taylor expansion arguments to

$$\sqrt{n}(\widehat{\mu}_S - \mu_{\text{true}}) \xrightarrow{d} \Lambda_S = (\tfrac{\partial\mu}{\partial\theta})^t C_S + (\tfrac{\partial\mu}{\partial\gamma_S})^t D_S - (\tfrac{\partial\mu}{\partial\gamma})^t \delta.$$

The rest of the proof consists in re-expressing this limit in the intended orthogonal fashion. The two independent normals featuring in the statement of the theorem are in fact

$$\Lambda_0 = (\tfrac{\partial\mu}{\partial\theta})^t J_{00}^{-1} U' \quad \text{and} \quad D = \delta + Q(V' - J_{10} J_{00}^{-1} U'),$$

for which one checks independence and that they are respectively a $N(0, \tau_0^2)$ and a $N_Q(\delta, Q)$. With algebraic efforts one finally verifies that the Λ_S above is identical to the intended $\Lambda_0 + \omega^t(\delta - G_S D)$.

The generalisation to regression models follows via parallel arguments and some extra work, involving among other quantities and arguments the averages of score functions $U(Y_i \mid x_i)$ and $V(Y_i \mid x_i)$, as per Section 5.2.3. $\qquad \square$

We note that the theorem and its proof are related to Theorem 5.4 and that Theorem 5.1 with Corollary 5.1 may be seen to be special cases. Observe next that the limit variables Λ_S in the theorem are normal, with means $\omega^t(I_q - G_S)\delta$ and variances

$$\tau_S^2 = \text{Var}\,\Lambda_S = \tau_0^2 + \omega^t G_S Q G_S^t \omega = \tau_0^2 + \omega^t Q_S^0 \omega = \tau_0^2 + \omega_S^t Q_S \omega_S,$$

see Exercise 6.1 for some of the algebra involved here. The wide model has the largest variance, $\tau_0^2 + \omega^t Q\omega$, and the smallest bias, zero; on the other side of the spectrum is the narrow model with smallest variance, τ_0^2, and largest bias, $\omega^t\delta$.

The theorem easily leads to expressions for the limiting mean squared errors for the different subset estimators. Adding squared bias and variance we find that $\sqrt{n}(\widehat{\mu}_S - \mu_{\text{true}})$ has limiting mean squared error

$$\text{mse}(S, \delta) - \tau_0^2 + \omega^t Q_S^0 \omega + \omega^t(I_q - G_S)\delta\delta^t(I_q - G_S)^t\omega. \tag{6.6}$$

The idea is to estimate this quantity for each of the candidate models S and choose that model which gives the smallest estimated mean squared error $\widehat{\text{mse}}(S)$. This will yield the focussed model choice criterion (6.1).

When attempting to estimate the limiting mean squared error (6.6) it is important to note the crucial difference between parameters τ_0, ω, G_S, Q_S on the one hand and δ on the other. The first parameters can be estimated without problem with methods that are consistent and converge at the usual rate, i.e. estimators $\widehat{\omega}$ are available for which $\sqrt{n}(\widehat{\omega} - \omega)$ would have a limiting distribution, etc. This is rather different for the more elusive parameter δ, since in fact no consistent estimator exists, and D_n of (6.4) is about the best one can do.

In (6.6) what is required is not to estimate δ per se but rather $\delta\delta^t$. Since DD^t has mean $\delta\delta^t + Q$ (see Exercise 6.2), we use the estimator $D_n D_n^t - \widehat{Q}$ for $\delta\delta^t$. An asymptotically unbiased estimator of the limiting mean squared error is accordingly

$$
\begin{aligned}
\widehat{\mathrm{mse}}(S) &= \widehat{\tau}_0^2 + \widehat{\omega}^t \widehat{Q}_S^0 \widehat{\omega} + \widehat{\omega}^t (I_q - \widehat{G}_S)(D_n D_n^t - \widehat{Q})(I_q - \widehat{G}_S)^t \widehat{\omega} \\
&= \widehat{\omega}^t (I_q - \widehat{G}_S) D_n D_n^t (I_q - \widehat{G}_S)^t \widehat{\omega} + 2\widehat{\omega}^t Q_S^0 \widehat{\omega} + \widehat{\tau}_0^2 - \widehat{\omega}^t \widehat{Q} \widehat{\omega}.
\end{aligned}
\tag{6.7}
$$

In the last equation we have used that $Q_S^0 Q^{-1} Q_S^0 = Q_S^0$ (again, see Exercise 6.1). Since the last two terms do not depend on the model, we may leave them out, and this leads precisely to the (6.1) formula.

Since $\omega = J_{10} J_{00}^{-1} \frac{\partial \mu}{\partial \theta} - \frac{\partial \mu}{\partial \gamma}$ depends on the focus parameter $\mu(\theta, \gamma)$ through its partial derivatives, it is clear that different models might be selected for different foci μ: three different parameters $\mu_1(\theta, \gamma)$, $\mu_2(\theta, \gamma)$, $\mu_3(\theta, \gamma)$ might have three different optimally selected S subsets. This is because the lists of estimated limiting mean squared errors, $\widehat{\mathrm{mse}}_1(S)$, $\widehat{\mathrm{mse}}_2(S)$, $\widehat{\mathrm{mse}}_3(S)$, may have different rankings from smallest to largest. The illustrations exemplify this.

Remark 6.2 Correlations between estimators
All submodel estimators are aiming for the same parameter quantity, and one expects them to exhibit positive correlations. This may actually be read off from Theorem 6.1, the key being that the convergence described there holds simultaneously across submodels, as may be seen from the proof. The correlation between $\widehat{\mu}_S$ and $\widehat{\mu}_{S'}$ will, with mild conditions, converge to that of their limit variables in the limit experiment, i.e.

$$
\mathrm{corr}(\Lambda_S, \Lambda_{S'}) = \frac{\tau_0^2 + \omega^t G_S Q G_{S'}^t \omega}{(\tau_0^2 + \omega^t G_S Q G_S^t \omega)^{1/2}(\tau_0^2 + \omega^t G_{S'} Q G_{S'}^t \omega)^{1/2}}.
$$

These are, for example, all rather high in situations where null model standard deviation τ_0 dominates $(\omega^t Q \omega)^{1/2}$, but may otherwise even be small. The limit correlation between estimators from the narrow and the wide model is $\tau_0 / (\tau_0^2 + \omega^t Q \omega)^{1/2}$. ∎

6.4 A bias-modified FIC

The FIC is based on an estimator of squared bias plus variance, but it might happen that the squared bias estimator is negative. To avoid such cases, we define the bias-modified FIC value

$$
\mathrm{FIC}^*(S) = \begin{cases} \mathrm{FIC}(S) & \text{if } N_n(S) \text{ does not take place}, \\ \widehat{\omega}^t (I_q + \widehat{G}_S) \widehat{Q} \widehat{\omega} & \text{if } N_n(S) \text{ takes place}, \end{cases}
\tag{6.8}
$$

where $N_n(S)$ is the event of negligible bias,

$$\{\widehat{\omega}^{\mathrm{t}}(I_q - \widehat{G}_S)\widehat{\delta}_{\mathrm{wide}}\}^2 = n\{\widehat{\omega}^{\mathrm{t}}(I_q - \widehat{G}_S)\widehat{\gamma}_{\mathrm{wide}}\}^2 < \widehat{\omega}^{\mathrm{t}}(\widehat{Q} - \widehat{Q}_S^0)\widehat{\omega}.$$

To understand the nature of this modification it is useful to first consider the following simple problem: assume $X \sim \mathrm{N}(\xi, \sigma^2)$ with unknown ξ and known σ, and suppose that it is required to estimate $\kappa = \xi^2$. We consider three candidate estimators: (i) The direct $\widehat{\kappa}_1 = X^2$, which is also the maximum likelihood estimator, because X is the maximum likelihood estimator for ξ. The mean of X^2 is $\xi^2 + \sigma^2$, so it overshoots. (ii) The unbiased version $\widehat{\kappa}_2 = X^2 - \sigma^2$. (iii) The modified version that restricts the latter estimate to be non-negative,

$$\widehat{\kappa}_3 = (X^2 - \sigma^2)I\{|X| \geq \sigma\} = \begin{cases} X^2 - \sigma^2 & \text{if } |X| \geq \sigma, \\ 0 & \text{if } |X| \leq \sigma. \end{cases} \tag{6.9}$$

Figure 6.1 displays the square root of the three risk functions $\mathrm{E}_\xi(\widehat{\kappa}_j - \xi^2)^2$ as functions of ξ; cf. Exercise 6.7. The implication is that the modified estimator $\widehat{\kappa}_3$ provides uniform improvement over the unbiased estimator $\widehat{\kappa}_2$.

This has implications for the FIC construction, since one of its components is precisely an unbiased estimator of a squared bias component. The limit risk $\mathrm{mse}(S, \delta)$ is a sum of

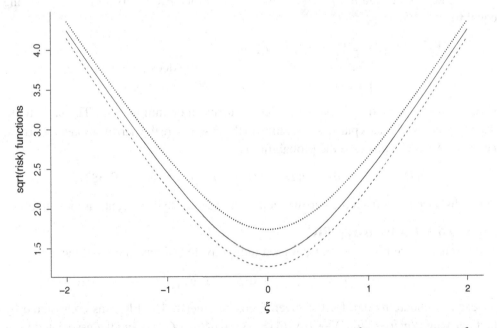

Fig. 6.1. Square root of risk functions for three estimators for ξ^2 in the normal (ξ, σ^2) model, with $\sigma = 1$. The modified $(X^2 - \sigma^2)I\{|X| \geq \sigma\}$ (dashed) is best, followed by the unbiased $X^2 - \sigma^2$ (solid), and then by the maximum likelihood estimator X^2 (dotted).

variance and squared bias, say

$$V(S) = \tau_0^2 + \omega^t Q_S^0 \omega \quad \text{and} \quad \text{sqb}(S) = \{\omega^t(I_q - G_S)\delta\}^2.$$

For the squared bias term there are (at least) three options, in terms of the limit variable D of (6.4),

$$\widehat{\text{sqb}}_1(S) = \{\omega^t(I_q - G_S)D\}^2,$$
$$\widehat{\text{sqb}}_2(S) = \omega^t(I_q - G_S)(DD^t - Q)(I_q - G_S^t)\omega,$$
$$\widehat{\text{sqb}}_3(S) = \max\{\omega^t(I_q - G_S)(DD^t - Q)(I_q - G_S^t)\omega, 0\}.$$

The squared bias estimator used in the construction of the FIC score (6.1) is the second version, coming from $\widehat{r}_2(S) = V(S) + \widehat{\text{sqb}}_2(S)$. In view of the finding above, that $\widehat{\kappa}_3$ is better than $\widehat{\kappa}_2$, we define the modified limiting mean squared error

$$\begin{aligned} \text{mse}^*(S) &= V(S) + \widehat{\text{sqb}}_3(S) \\ &= \begin{cases} \text{mse}(S) & \text{if } N(S) \text{ does not take place,} \\ V(S) & \text{if } N(S) \text{ takes place,} \end{cases} \end{aligned} \tag{6.10}$$

where $N(S)$ is the event of negligible bias, that is,

$$\{\omega^t(I_q - G_S)D\}^2 < \omega^t(I_q - G_S)Q(I_q - G_S)^t\omega = \omega^t(Q - Q_S^0)\omega.$$

In (6.6) we obtained the FIC by subtracting the term $\widehat{\tau}_0^2 - \widehat{\omega}^t\widehat{Q}\widehat{\omega}$ that was independent of S. In the present situation we end up with the modified FIC (in its limit version) being equal to

$$\begin{aligned} \text{FIC}^*(S) &= \text{mse}^*(S) - (\tau_0^2 - \omega^t Q\omega) \\ &= \begin{cases} \text{FIC}(S) & \text{if } N(S) \text{ does not take place,} \\ V(S) - (\tau_0^2 - \omega^t Q\omega) & \text{if } N(S) \text{ takes place.} \end{cases} \end{aligned}$$

Estimators are inserted for unknown values to arrive at definition (6.8). The probability that the event $N(S)$ takes place is quite high in the case where the narrow model is correct, i.e. when $\delta = 0$. In this case the probability is

$$P(DD^t - Q < 0) = P(DQ^{-1}D^t < 1) = P(\chi_1^2 \le 1) = 0.6827.$$

Some distance away from the narrow model, the negligible bias events are less likely.

Remark 6.3 The bootstrap-FIC *
The essence of the FIC method is, at least indirectly, to estimate the risk function

$$r_n(S, \theta, \gamma) = n \, \text{E}\{\widehat{\mu}_S - \mu(\theta, \gamma)\}^2$$

for each candidate model, for the given focus parameter. The FIC was constructed by first deriving $\sqrt{n}(\widehat{\mu}_S - \mu_{\text{true}}) \to_d \Lambda_S(\delta) = \Lambda_0 + \omega^t(\delta - G_S D)$ and the associated limit risk $r(S, \delta)$, and then constructing an asymptotically unbiased estimator for the risk. A method that, in some sense, is simpler is to plug in D for δ, in effect estimating limit risk $r(S, \delta)$ with $\widehat{r}(S) = r(S, D)$. This yields a valid model selection scheme as such, choosing the model with smallest $\widehat{r}(S)$.

One may think of this as a bootstrap method: $\Lambda_S(\delta)$ has an unknown mean square, and one estimates this quantity by plugging in the wide model's $\widehat{\delta} = D$. These risk estimates may also be computed via simulations: at the estimated position $\widehat{\delta}$, simulate a large number of (Λ_0, D) and thence $\Lambda_S(\delta)$, and take the mean of their squares to be the risk estimate. This is not really required, since we already know a formula for $\mathrm{E}\,\Lambda_S(\delta)^2$, but we point out the connection since it matches the bootstrap paradigm and since it opens the door to other and less analytically tractable loss functions, for example. The method just described amounts to 'bootstrapping in the limit experiment'. It is clear that we also may bootstrap in the finite-sample experiment, computing the plug-in estimated risk $r_n(S, \widehat{\theta}_S, \widehat{\gamma}_S)$ by simulating a large number of $\sqrt{n}\{\widehat{\mu}_S^* - \mu(\widehat{\theta}_S, \widehat{\gamma}_S, \gamma_{0,S^c})\}$, where the $\widehat{\mu}_S^*$ are bootstrap estimates using the $\widehat{\mu}_S$ procedure on data sets generated from the estimated biggest model. The mean of their squares will be close to the simpler $r(S, \widehat{\delta})$, computable in the limit experiment. It is important to realise that these bootstrap schemes just amount to alternative and also imperfect risk estimates, plugging in a parameter instead of estimating the risks unbiasedly. Bootstrapping alone cannot 'solve' the model selection problems better than say the FIC; see some further discussion in Hjort and Claeskens (2003, Section 10). ∎

6.5 Calculation of the FIC

Let us summarise the main ingredients for carrying out FIC analysis. They are

(i) specifying the focus parameter of interest, and expressing it as a function $\mu(\theta, \gamma)$ of the model parameters;
(ii) deciding on the list of candidate models to consider;
(iii) estimating J_{wide}, from which estimates of the matrices Q, Q_S, G_S can be obtained;
(iv) estimating γ in the largest model, which leads to the estimator $D_n = \sqrt{n}(\widehat{\gamma} - \gamma_0)$ for δ;
(v) estimating ω.

The first step is an important one in the model selection process, as it requires us to think about why we wish to select a model in the first place. Step (ii) is a reminder that models that appear a priori unlikely ought to be taken out of the candidate list.

There are various strategies for estimating the information matrix $J = J(\theta_0, \gamma_0)$. For some of the examples of Section 6.6, we find explicit formulae. If an explicit formula for J is not available, an empirical Hessian matrix can be used; we may use the narrow-based $J_{n,\text{wide}}(\widehat{\theta}_{\text{narr}}, \gamma_0)$ of the wide-based $J_{n,\text{wide}}(\widehat{\theta}, \widehat{\gamma})$, where

$$J_{n,\text{wide}}(\theta, \gamma) = -n^{-1} \sum_{i=1}^{n} \frac{\partial^2}{\partial \zeta \partial \zeta^{\text{t}}} \log f(y_i, \theta, \gamma), \tag{6.11}$$

writing here $\zeta = (\theta, \gamma)$. Often this comes as a by-product of the maximisation procedure (such as Newton–Raphson for example) to obtain the estimators of θ and γ. For example, the R function nlm provides the Hessian as a result of the optimisation routine. A simple alternative if no formula or Hessian matrix is available is to calculate the variance matrix of say 10,000 simulated score vectors at the estimated null model, which is sometimes much easier than from the estimated full model.

From the theoretical derivation of the FIC in Section 6.3 it is clear that consistency of $\widehat{J}$ for J under the P_n sequence of models is sufficient, so both plug-ins $(\widehat{\theta}_{\text{narr}}, \gamma_0)$ and $(\widehat{\theta}, \widehat{\gamma})$ may be used (or in fact anything in between, from any submodel S). The wide model estimator has the advantage of being consistent even when the narrow model assumptions do not hold. This builds in some model robustness by not having to rely on γ being close to γ_0. In our applications we shall mostly use the fitted widest model to estimate τ_0, ω, J and further implied quantities, for these reasons of model robustness. In certain cases the formulae are simpler to express and more attractive to interpret when J is estimated from the narrow model, however, and this is particularly true when the Q population matrix is diagonal under the narrow model. In such cases we shall often use this narrow estimator to construct the FIC.

Estimators for $\partial\mu/\partial\theta$ and $\partial\mu/\partial\gamma$ can be constructed by plugging in an estimator of θ in explicit formulae, if available, or via numerical approximations. See the functions D and deriv in the software package R, which provide such derivatives. Alternatively, compute $\{\mu(\widehat{\theta} + \eta e_i, \gamma_0) - \mu(\widehat{\theta}, \gamma_0)\}/\eta$ for the components of $\partial\mu/\partial\theta$ and $\{\mu(\widehat{\theta}, \gamma_0 + \eta e_j) - \mu(\widehat{\theta}, \gamma_0)\}/\eta$ for the components of $\partial\mu/\partial\gamma$, for a small η value. Here e_i is the ith unit vector, consisting of a one at index i and zeros elsewhere.

6.6 Illustrations and applications

We provide a list of examples and data illustrations of the construction of the FIC and its use for model selection.

6.6.1 FIC in logistic regression models

The focussed information criterion can be used for discrete or categorical outcome data. For a logistic regression model, there is a binary outcome variable Y_i which is either one or zero. The most widely used model for relating probabilities to the covariates takes

$$P(Y_i = 1 \mid x_i, z_i) = p_i = \frac{\exp(x_i^t\beta + z_i^t\gamma)}{1 + \exp(x_i^t\beta + z_i^t\gamma)},$$

or equivalently $\text{logit}\{P(Y_i = 1 \mid x_i, z_i)\} = x_i^t\beta + z_i^t\gamma$, where $\text{logit}(v) = \log\{v/(1 - v)\}$. The vector of covariates is split into two parts; $x_i = (x_{i,1}, \ldots, x_{i,p})^t$ has the protected covariates, meant to be present in all of the models we consider, while $z_i = (z_{i,1}, \ldots, z_{i,q})^t$ are the open or nonprotected variables from which we select the most adequate or important ones. Likewise, the coefficients are vectors $\beta = \theta = (\beta_1, \ldots, \beta_p)^t$ and $\gamma = (\gamma_1, \ldots, \gamma_q)^t$. In order to obtain the matrix J_{wide}, or its empirical version $J_{n,\text{wide}}$, we compute the second-order partial derivatives of the log-likelihood function

$$\ell_n(\beta, \gamma) = \sum_{i=1}^{n} [y_i \log p(x_i, z_i) + (1 - y_i) \log\{1 - p(x_i, z_i)\}].$$

Note that the inverse logit function $H(u) = \exp(u)/\{1 + \exp(u)\}$ has $H'(u) = H(u)\{1 - H(u)\}$, so $p_i = H(x_i^t\beta + z_i^t\gamma)$ has $\partial p_i/\partial \beta = p_i(1 - p_i)x_i$ and $\partial p_i/\partial \gamma = p_i(1 - p_i)z_i$, which leads to

$$\begin{pmatrix} \partial \ell_n/\partial \beta \\ \partial \ell_n/\partial \gamma \end{pmatrix} = \sum_{i=1}^{n} \{y_i - p(x_i, z_i)\} \begin{pmatrix} x_i \\ z_i \end{pmatrix}.$$

It readily follows that for a logistic regression model

$$J_{n,\text{wide}} = n^{-1} \sum_{i=1}^{n} p_i(1 - p_i) \begin{pmatrix} x_i x_i^t & x_i z_i^t \\ z_i x_i^t & z_i z_i^t \end{pmatrix} = \begin{pmatrix} J_{n,00} & J_{n,01} \\ J_{n,10} & J_{n,11} \end{pmatrix}.$$

We estimate this matrix and the matrices Q, Q_S and G_S using the estimates for the β and γ parameters in the full model; it is also possible to estimate $J_{n,\text{wide}}$ using estimators from the narrow model where $\gamma_0 = 0$. The vector $\delta/\sqrt{n}$ measures the departure distance between the smallest and largest model and is estimated by $\widehat{\gamma} - \gamma_0$. The narrow model in this application corresponds to $\gamma_0 = 0_{q \times 1}$. Thus we construct $\widehat{\delta} = \sqrt{n}\widehat{\gamma}$.

The odds of an event: As a focus parameter which we wish to estimate we take the odds of the event of interest taking place, at a given position in the covariate space:

$$\mu(\theta, \gamma) = \frac{p(x, z)}{1 - p(x, z)} = \exp(x^t\beta + z^t\gamma).$$

An essential ingredient for computing the FIC is the vector ω, which here is given by

$$\omega(x, z) = \frac{p(x, z)}{1 - p(x, z)}(J_{n,10} J_{n,00}^{-1}x - z).$$

With this information, we compute the FIC for each of the models of interest, and select the model with the lowest value of the FIC.

The probability that an event occurs: The second focus parameter we consider is the probability that the event of interest takes place. For a specified set of covariates x, z, select variables in a logistic regression model to estimate $\mu(\beta, \gamma; x, z) = P(Y = 1 \mid x, z)$ in the best possible way, measured by the mean squared error. By construction, the FIC will serve this purpose. The required FIC component ω is for this focus parameter equal to $p(x, z)\{1 - p(x, z)\}(J_{n,10} J_{n,00}^{-1}x - z)$.

When ω has been estimated for the given estimand, we compute the values of the bias-modified focussed information criterion (6.8) for all models in our search path as

$$\text{FIC}_S^* = \widehat{\omega}^t(I_q - \widehat{G}_S)D_n D_n^t(I_q - \widehat{G}_S)^t\widehat{\omega} + 2\widehat{\omega}_S^t\widehat{Q}_S\widehat{\omega}_S$$

when $\widehat{\omega}_S^t(I_q - \widehat{G}_S)(\widehat{\delta\delta}^t - \widehat{Q})(I_q - \widehat{G}_S)^t\widehat{\omega}$ is positive, and $\text{FIC}_S^* = \widehat{\omega}^t(\widehat{Q} + \widehat{Q}_S^0)\widehat{\omega}$ otherwise. This is the method described in Section 6.4.

Example 6.1 Low birthweight data: FIC plots and FIC variable selection

As an illustration we apply this to the low birthweight data described in Section 1.5. Here Y is an indicator variable for low birthweight, and a list of potentially influential covariates is defined and discussed in the earlier section. For the present illustration, we include in every candidate model the intercept $x_1 = 1$ and the weight x_2 (in kg) of the mother prior to pregnancy. Other covariates, from which we wish to select a relevant subset, are

- z_1 (age, in years);
- z_2 (indicator for smoking);
- z_3 (history of hypertension);
- z_4 (uterine irritability);
- the interaction term $z_5 = z_1 z_2$ between smoking and age; and
- the interaction term $z_6 = z_2 z_4$ between smoking and uterine irritability.

We shall build and evaluate logistic regression models for the different subsets of $z_1, \ldots, z_6$. We do not allow all $2^6 = 64$ possible combinations of these, however, as it is natural to request that models containing an interaction term should also contain both corresponding main effects: thus z_5 may only be included if z_1 and z_2 are both present, and similarly z_6 may only be considered if z_2 and z_4 are both present. A counting exercise shows that 26 subsets of $z_1, \ldots, z_6$ define proper candidate models. We fit each of the 26 candidate logistic regression models, and first let AIC decide on the best model. It selects the model '0 1 1 1 0 0', in the notation of Table 6.1, i.e. z_2, z_3, z_4 are included (in addition to the protected x_1, x_2), the others not, with best AIC value -222.826. The BIC, on the other hand, chooses the much more parsimonious model '0 0 1 0 0 0', with only z_3 present, and best BIC value -236.867. Note again that AIC and the BIC work in 'overall modus' and are, for example, not concerned with specific subgroups of mothers.

Now let us perform a focussed model search for the probability of low birthweight in two subgroups: smokers and non-smokers. We let these two groups be represented by average values of covariates weight and age, and find $(59.50, 23.43)$ for non-smokers and $(58.24, 22.95)$ for smokers; for our evaluation we furthermore focus on mothers without hypertension or uterine irritation (i.e. z_3 and z_4 are set equal to zero). Figure 6.2 and Table 6.1 give the result of this analysis. The 'FIC plots' summarise the relevant information in a convenient fashion, displaying all 26 estimates, plotted against the FIC scores of the associated 26 candidate models. Thus estimates to the left of the diagram are

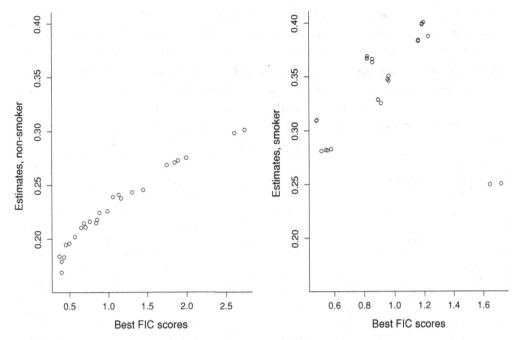

Fig. 6.2. FIC plots for the low birthweight probabilities, for the smoking and non-smoking strata of mothers. For each stratum, all 26 candidate estimates are plotted, as a function of the FIC score. Estimates to the left, with lower FIC values, have higher precision.

more reliable than those on the right. The table gives for each of the two strata the five best models, as ranked by FIC*, along with the estimated probability of low birthweight for each of these models, then the estimated standard deviation, bias and root mean squared error, and finally the FIC* score, as per formulae developed in Section 6.4. In some more detail, these are given by

$$\widehat{sd}_S = \widehat{V}(S)^{1/2}/\sqrt{n}, \quad \widehat{bias}_S = \widehat{\omega}^{t}(I_q - \widehat{G}_S)\widehat{\delta}/\sqrt{n}, \quad \widehat{mse}(S)^{1/2}/\sqrt{n},$$

with the appropriate $\widehat{V}(S) = \widehat{\tau}_0^2 + \widehat{\omega}^{t}\widehat{Q}_S^0\widehat{\omega}$ and

$$\widehat{mse}(S) = \widehat{V}(S) + \max\{\widehat{\omega}^{t}(I_q - \widehat{G}_S)(\widehat{\delta\delta}^{t} - \widehat{Q})(I_q - \widehat{G}_S)^{t}\widehat{\omega}, 0\}.$$

The primary aspect of the FIC analysis for these two strata is that the low birthweight probabilities are so different; the best estimates are around 0.18 for non-smokers and 0.30 for smokers, as seen in both Figure 6.2 and Table 6.1. We also gain insights from studying and comparing the best models themselves; for predicting low birthweight chances for non-smokers, one uses most of the $z_1, \ldots, z_6$ covariates, while for the smokers group, the best models use very few covariates. ■

Table 6.1. *FIC model selection for the low birthweight data: the table gives for the smoking and non-smoking strata the five best models, as ranked by the FIC, with '1 0 1 1 0 0' indicating inclusion of z_1, z_3, z_4 but exclusion of z_2, z_5, z_6, etc. Given further, for each model, are the estimated probability, the estimated standard deviation, bias and root mean squared error, each in percent, and the FIC score. For comparison also the AIC and BIC scores are given.*

Variables	Estimate	sd	Bias	$\sqrt{\text{mse}}$	FIC*	AIC	BIC
non-smoker:							
1 1 1 1 1 0	18.39	3.80	1.53	3.860	0.367	−223.550	−246.242
1 1 1 1 0 1	17.91	3.94	1.30	4.042	0.394	−224.366	−247.059
1 1 1 1 1 1	16.90	4.04	0.00	4.042	0.394	−224.315	−250.249
0 1 1 1 0 1	18.31	3.84	2.20	4.249	0.426	−223.761	−243.211
1 1 1 1 0 0	19.46	3.67	2.91	4.385	0.449	−223.778	−243.228
smoker:							
0 0 0 0 0 0	30.94	3.85	−1.57	3.849	0.480	−232.691	−239.174
1 0 0 0 0 0	30.98	3.86	−1.81	3.863	0.482	−233.123	−242.849
0 0 1 0 0 0	28.10	4.07	−4.82	4.070	0.513	−227.142	−236.867
0 0 0 1 0 0	28.20	4.25	−4.66	4.249	0.542	−231.096	−240.821
1 0 1 0 0 0	28.17	4.08	−5.06	4.322	0.553	−227.883	−240.850

Example 6.2 Onset of menarche

The onset of menarche varies between different segments of societies, nations and epochs. The potentially best way to study the onset distribution, for a given segment of a population, would be to ask a random sample of women precisely when their very first menstrual period took place, but such data, even if available, could not automatically be trusted for accuracy. For this illustration we instead use a more indirect data set pertaining to 3918 Warsaw girls, each of whom in addition to providing their age answered 'yes' or 'no' to the question of whether they had reached menarche or not; see Morgan (1992). Thus the number y_j of the m_j girls in age group j who had started menstruation is recorded, for each of $n = 25$ age groups. We shall see that the FIC prefers different models for making inference about different quantiles of the onset distribution.

We take y_j to be the result of a Bin(m_j, p_j) experiment, where $p_j = p(x_j)$, writing

$$p(x) = \text{P(onset} \le x) = H(b_0 + b_1 z + b_2 z^2 + \cdots + b_k z^k),$$

with $z = x - 13$, and where $H(u) = \exp(u)/\{1 + \exp(u)\}$ is the inverse logistic transform. We choose the narrow model to be ordinary logistic regression in z alone, corresponding to $k = 1$, and the wide model to be the sixth-order model with seven parameters $(b_0, b_1, \ldots, b_6)$. In other words, $p = 2$ and $q = 5$ in our usual notation. Unlike various earlier examples, this situation is naturally ordered, so the task is to select one of the six models, of order 1, 2, 3, 4, 5, 6. See Figure 6.3.

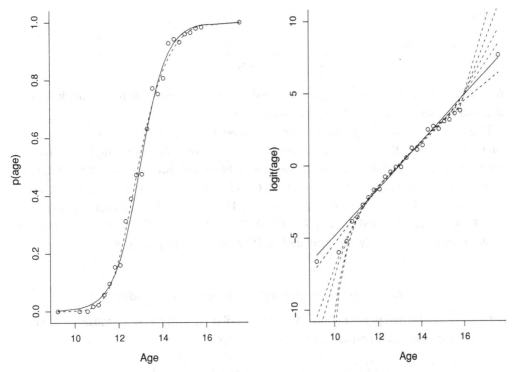

Fig. 6.3. Age at onset of menarche: the left panel shows relative frequencies y_j/m_j along with fitted logistic regression curves of order 1 and 3. The right panel displays logits of the frequencies along with fitted logit(x) curves for models of order 1, 2, 3, 4, 5, 6.

The AIC values for models of order 1, 2, 3, 4, 5, 6 are easy to compute, via log-likelihoods $\ell_n(b)$ of the form

$$\sum_{j=1}^{n}\{y_j \log p_j + (m_j - y_j)\log(1 - p_j)\} = \sum_{j=1}^{n}[y_j u_j^t b - m_j \log\{1 + \exp(u_j^t b)\}],$$

where we let u_j denote the vector $(1, z, \ldots, z^k)^t$ at position x_j. For the data at hand, AIC prefers the third-order model, $k = 3$. We also note that the normalised information matrix J_n is $n^{-1}\sum_{j=1}^{n} m_j p_j(1 - p_j)u_j u_j^t$. For the FIC application below we estimate J_n by inserting $\widehat{p}_j$ estimates based on the full sixth-order model.

Let now $\mu = \mu(r)$ be equal to the rth quantile of the onset distribution, in other words the solution $z = z(b)$ to the equation

$$G(b, z) = b_0 + b_1 z + \cdots + b_6 z^6 = H^{-1}(r) = \log \frac{r}{1 - r}.$$

Taking derivatives of the identity $G(b, z(b)) = $ constant gives

$$\frac{\partial G}{\partial b_j}(b, z(b)) + \frac{\partial G}{\partial z}(b, z(b))\frac{\partial z(b)}{\partial b_j} = 0,$$

leading to

$$\frac{\partial \mu}{\partial b_j} = -\frac{\mu^j}{b_1 + 2b_2\mu + \cdots + 6b_6\mu^5} \quad \text{for } j = 0, 1, \ldots, 6.$$

We estimate $\omega = J_{10}J_{00}^{-1}\frac{\partial \mu}{\partial \theta} - \frac{\partial \mu}{\partial \gamma}$, of dimension 5×1, for example for the median onset age, and let the FIC determine which model is best.

It turns out that the FIC prefers the simplest model, of degree $k = 1$, for r in the middle range (specifically, this is so for the median), but that it prefers more complicated models for r closer to 0, i.e. the very start of the menarche onset distribution. For $r = 0.01$, for example, ω is estimated at $(1.139, -0.942, 6.299, -4.788, 33.244)^t$, and the best model is of order $k = 4$ (with estimated 0.01-quantile 10.76 years). For another example, take $r = 0.30$, where one finds $\widehat{\omega} = (-0.163, 0.881, -0.191, 4.102, 2.234)^t$, and the best model is of order $k = 2$ (with estimated 0.30-quantile 12.46 years). ∎

6.6.2 FIC in the normal linear regression model

Let $Y_1, \ldots, Y_n$ be independent and normally distributed with the same variance. To each response variable Y_i corresponds a covariate vector

$$(x_i^t, z_i^t) = (x_{i,1}, \ldots, x_{i,p}, z_{i,1}, \ldots, z_{i,q})$$

where the x_i part is to be included in each candidate model. The x_i may consist of an intercept only, for example, in which case $x_i = 1$, or there may be some covariates included that are known beforehand to have an effect on the response variable. Components of the q-vector of covariates z_i may or may not be included in the finally selected model.

We fit a normal linear regression model

$$Y_i = x_i^t \beta + z_i^t \gamma + \sigma \varepsilon_i, \tag{6.12}$$

where the error terms $\varepsilon_1, \ldots, \varepsilon_n$ are independent and standard normal. The unknown coefficient vectors are $\beta = (\beta_1, \ldots, \beta_p)^t$ and $\gamma = (\gamma_1, \ldots, \gamma_q)^t$. Denote by X the $n \times p$ design matrix with ith row equal to x_i, and Z the design matrix of dimension $n \times q$ with ith row equal to z_i. The full model corresponds to including all covariates, that is, $Y_i \sim N(x_i^t \beta + z_i^t \gamma, \sigma^2)$, and has $p + q + 1$ parameters. In the narrow model $Y_i \sim N(x_i^t \beta, \sigma^2)$, with $p + 1$ parameters. For the normal linear regression model, the vector of parameters common to all models is $\theta = (\sigma, \beta)$ and the q-vector of extra coefficients γ is as defined above. For the narrow model $\gamma_0 = 0_{q \times 1} = (0, \ldots, 0)^t$.

Note here that σ is used as somewhat generic notation for a model parameter that may change its interpretation and value depending on which of the γ_js are included or excluded in the model. If many $z_{i,j}$s are included, then variability of Y_i minus the predicting part becomes smaller than if few $z_{i,j}$s are present; the more γ_js put into the model, the smaller the σ.

To derive formulae for the FIC, we start with the log-likelihood function

$$\ell_n = \sum_{i=1}^{n} \log f(Y_i; \theta, \gamma) = \sum_{i=1}^{n} \{-\log \sigma - \tfrac{1}{2}(Y_i - x_i^t \beta - z_i^t \gamma)^2 / \sigma^2 - \tfrac{1}{2} \log(2\pi)\}.$$

From this we compute the empirical $J_{n,\text{wide}}$ as the expected value of minus the second partial derivatives of the log-likelihood function with respect to $\theta = (\sigma, \beta)$ and γ, divided by the sample size n. We keep the order of the parameters as above, that is, (σ, β, γ), and find the formula for the $(p+q+1) \times (p+q+1)$ information matrix

$$J_{n,\text{wide}} = n^{-1} \sum_{i=1}^{n} \frac{1}{\sigma^2} \begin{pmatrix} 2 & 0 \\ 0 & \Sigma_n \end{pmatrix},$$

where we write

$$\Sigma_n = \begin{pmatrix} \Sigma_{n,00} & \Sigma_{n,01} \\ \Sigma_{n,10} & \Sigma_{n,11} \end{pmatrix} = n^{-1} \begin{pmatrix} X \\ Z \end{pmatrix}^t \begin{pmatrix} X \\ Z \end{pmatrix} = n^{-1} \begin{pmatrix} X^t X & X^t Z \\ Z^t X & Z^t Z \end{pmatrix}. \tag{6.13}$$

As indicated above, we partition the information matrix into four blocks, as in (5.13), with $J_{n,00}$ of size $(p+1) \times (p+1)$ and so on. It is now straightforward to compute the matrix

$$Q_n = (J_{n,11} - J_{n,10} J_{n,00}^{-1} J_{n,01})^{-1}$$
$$= \sigma^2 (\Sigma_{n,11} - \Sigma_{n,10} \Sigma_{n,00}^{-1} \Sigma_{n,01})^{-1} = \sigma^2 \{n^{-1} Z^t (I_n - H) Z\}^{-1},$$

where H denotes the familiar 'hat matrix' $H = X(X^t X)^{-1} X^t$. For each considered subset S of $\{1, \ldots, q\}$, corresponding to allowing a subset of the variables $z_1, \ldots, z_q$ in the final model, define, as before, $Q_{n,S} = J_n^{11,S}$ and $\widehat{G}_{n,S} = Q_{n,S}^0 Q_n^{-1}$. The $Q_{n,S}$ matrix is proportional to σ^2, and is estimated by inserting $\widehat{\sigma}$ for σ, where we use $\widehat{\sigma} = \widehat{\sigma}_{\text{wide}}$, for example. The $\widehat{G}_{n,S}$ matrix is scale-free and can be computed directly from the X and Z matrices.

Together with $D_n = \sqrt{n}\widehat{\gamma}_{\text{wide}}$, these are the quantities we can compute without specifying a focus parameter $\mu(\theta, \gamma)$. Next, we give several examples of focus parameters and give in each case a formula for

$$\omega = J_{n,10} J_{n,00}^{-1} \frac{\partial \mu}{\partial \theta} - \frac{\partial \mu}{\partial \gamma}$$
$$= (0, \Sigma_{n,10}) \begin{pmatrix} \frac{1}{2} & 0 \\ 0 & \Sigma_{n,00}^{-1} \end{pmatrix} \begin{pmatrix} \frac{\partial \mu}{\partial \sigma} \\ \frac{\partial \mu}{\partial \beta} \end{pmatrix} - \frac{\partial \mu}{\partial \gamma} = \Sigma_{n,10} \Sigma_{n,00}^{-1} \frac{\partial \mu}{\partial \beta} - \frac{\partial \mu}{\partial \gamma} \tag{6.14}$$

and the corresponding FIC expressions. Note that $\frac{\partial \mu}{\partial \sigma}$ drops out here, having to do with the fact that σ is estimated equally well in all submodels, asymptotically, inside the large-sample framework, and which again is related to the asymptotic independence between σ estimators and (β, γ) estimators.

Example 6.3 Response at fixed position
Let the focus parameter be the mean of Y at covariate position (x, z), that is,

$$\mu(\sigma, \beta, \gamma; x, z) = x^t \beta + z^t \gamma.$$

Since the partial derivative of μ with respect to σ is equal to zero, this gives

$$\omega = Z^t X (X^t X)^{-1} x - z = \Sigma_{n,10} \Sigma_{n,00}^{-1} x - z.$$

Next, construct the estimator $\widehat{\psi}_{\text{wide}} = \omega^t \widehat{\delta}$ together with the submodel estimators $\widehat{\psi}_S = \omega^t G_{n,S} \widehat{\delta}$. This leads for each model S to

$$\text{FIC}(S) = \omega^t (I_q - \widehat{G}_S) \widehat{\delta}_{\text{wide}} \widehat{\delta}_{\text{wide}}^t (I_q - \widehat{G}_S)^t \omega + 2 \omega^t \widehat{Q}_S^0 \omega,$$

where $\widehat{Q}_S^0$ contains the scale factor $\widehat{\sigma}^2$, with estimation carried out in the wide model. The bias-modified FIC, when different from FIC(S), is equal to $\omega^t (I_q + \widehat{G}_S) \widehat{Q} \omega$, specifically in cases where $\omega^t (I_q - \widehat{G}_S)(\widehat{\delta}\widehat{\delta}^t - \widehat{Q})(I_q - \widehat{G}_S)^t \omega$ is negative. Thus the FIC depends in this case on the covariate position (x, z) via the vector $\omega = \omega(x, z)$, indicating that there could be different suggested covariate models in different covariate regions. This is not a paradox, and stems from our wish to estimate the expected value $\text{E}(Y \mid x, z)$ with optimal precision, for each given (x, z). See Section 6.9 for FIC-averaging strategies. ∎

Example 6.4 Focussing on a single covariate
Sometimes interest focusses on the impact of a particular covariate on the mean structure. For the purpose of investigating the influence of the kth covariate z_k, we define

$$\mu(\sigma, \beta, \gamma) = \text{E}(Y \mid x + e_k, z) - \text{E}(Y \mid x, z) = \beta_k,$$

using the notation e_k for the kth unit vector, having a one on the kth entry and zero elsewhere. The FIC can then be set to work, with $\omega = Z^t X (X^t X)^{-1} e_k$. ∎

Example 6.5 Linear inside quadratic and cubic
Let us illustrate the variable selection method in a situation where one considers augmenting a normal linear regression trend with a quadratic and/or cubic term. This fits the above with a model $Y_i = x_i^t \beta + z_i^t \gamma + \sigma \varepsilon_i$, where $x_i = (1, x_{i,2})^t$ and $z_i = (x_{i,2}^2, x_{i,2}^3)^t$. The focus parameter is

$$\mu(\theta, \gamma; x) = \text{E}(Y \mid x) = \beta_1 + \beta_2 x + \gamma_1 x^2 + \gamma_2 x^3,$$

for which we find

$$\omega = Z^t X (X^t X)^{-1} \begin{pmatrix} 1 \\ x \end{pmatrix} - \begin{pmatrix} x^2 \\ x^3 \end{pmatrix}.$$

This can be written in terms of the first four sample moments of the covariate x_i. In the full model we estimate $\widehat{\delta} = \sqrt{n}(\widehat{\gamma}_1, \widehat{\gamma}_2)^t$. The four FIC values can now readily be computed. In this simple example they can be calculated explicitly (see Exercise 6.6), though in most cases it will be more convenient to work numerically with the matrices, using, for example, R. ∎

Example 6.6 Quantiles and probabilities
Another type of interest parameter is the 0.90-quantile of the response distribution at a given position (x, z),

$$\mu = G^{-1}(0.90 \mid x, z) = x^t\beta + z^t\gamma + \sigma\Phi^{-1}(0.90).$$

But by (6.14) the $\frac{\partial\mu}{\partial\sigma}$ part drops out, and $\omega = \Sigma_{n,10}\Sigma_{n,00}^{-1}x - z$ again, leading to the same FIC values as for the simpler mean parameter $E(Y \mid x, z)$. Consider finally the cumulative distribution function $\mu = \Phi((y - x^t\beta - z^t\gamma)/\sigma)$, for fixed position (x, z), at value y. By (6.14) again, one finds

$$\omega = -(1/\sigma)\phi((y - x^t\beta - z^t\gamma)/\sigma)(\Sigma_{n,10}\Sigma_{n,00}^{-1}x - z).$$

This is different from, but only a constant factor away from, the ω formula for $E(Y \mid x, z)$, and it is an invariance property of the FIC method that the same ranking of models is being generated for two estimands if the two ωs in question are proportional to each other. Hence the FIC-ranking of candidate models for $E(Y \mid x, z)$ is valid also for inference about the full cumulative response distribution at (x, z). ∎

In Section 6.7 it is shown that the FIC, which has been derived under large-sample approximations, happens to be exact in this linear-normal model, as long as the focus parameter is linear in the mean parameters.

6.6.3 FIC in a skewed regression model

Data are observed in a regression context with response variables Y_i and covariates x_i for individuals $i = 1, \ldots, n$. For the following illustrations four possible models are considered. (1) The simplest model, or narrow model, is a constant mean, constant variance model where $Y_i \sim N(\beta_0, \sigma^2)$. In the model selection process we consider model departures in two directions, namely in the mean and in skewness. Model (2) includes a linear regression curve, but no skewness: $Y_i = \beta_0 + \beta_1 x + \sigma\varepsilon_{0,i}$ where $\varepsilon_{0,i} \sim N(0, 1)$; whereas model (3) has a constant regression curve ($\beta_1 = 0$), but skewness: $Y_i = \beta_0 + \sigma\varepsilon_i$, where ε_i comes from the skewed distribution with density $\lambda\Phi(u)^{\lambda-1}\phi(u)$. Here ϕ and Φ are the density and cumulative distribution function for the standard normal distribution, and λ is the parameter dictating the degree of skewness of the distribution, with $\lambda = 1$ corresponding to the ordinary symmetric normal situation. Models (2) and (3) are in between the narrow model and the full model (4), the latter allowing for both a linear regression relationship and for skewness. The full model is

$$Y_i = \beta_0 + \beta_1 x_i + \sigma\varepsilon_i \quad \text{with } \varepsilon_i \sim \lambda\Phi(u)^{\lambda-1}\phi(u).$$

Note as in Section 6.6.2 that σ changes interpretation and value with β_1 and λ.

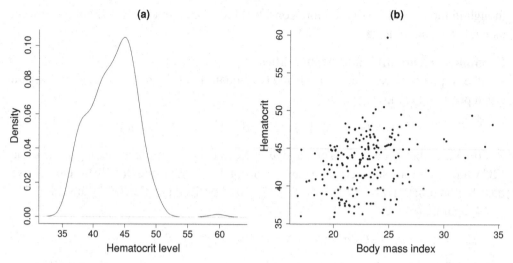

Fig. 6.4. (a) Estimated density function of the hematocrit level in the Australian Institute of Sports data set, and (b) scatterplot of the body mass index versus hematocrit level. The sample size equals 202.

For illustration we use the Australian Institute of Sports data (Cook and Weisberg, 1994), available as data set `ais` in the R library `sn`. For our example we use as outcome variable y the hematocrit level and x the body mass index (defined as the weight, in kg, divided by the square of the height, in m). In Figure 6.4 we plot a kernel smoother of the density of the body mass index (using the default smoothing parameters for the function `density` in R). The data are somewhat skewed to the right. In the scatterplot of body mass index versus hematocrit level there is arguably an indication of a linear trend.

Each of these models contains the parameter $\theta = (\beta_0, \sigma)$, while components of the parameter $\gamma = (\beta_1, \lambda)$ are included in only some of the models. The narrow model has $\gamma = \gamma_0 = (0, 1)$. A local misspecification situation translates here to $\mu_{\text{true}} = \mu(\beta_0, \sigma, \beta_1, \lambda) = \mu(\beta_0, \sigma, \delta_1/\sqrt{n}, 1 + \delta_2/\sqrt{n})$. There are four estimators, one for each of the submodels. In the narrow model we insert the ordinary sample mean and standard deviation, resulting in the estimator $\widehat{\mu}_{\text{narr}} = \mu(\bar{y}, s, 0, 1)$. Maximum likelihood estimation is used also inside each of the larger alternative models. For the full model this gives the estimator $\widehat{\mu}_{\text{wide}} = \mu(\widehat{\beta_0}, \widehat{\sigma}, \widehat{\beta_1}, \widehat{\lambda})$. We wish to select one of these four models by the FIC. The main FIC ingredients are as specified at the start of Section 6.5. With the knowledge of μ and $\widehat{J}$ we construct estimators for ω and the matrices Q and G_S. We obtain the FIC first with a general μ, inserting estimators for all required quantities, and then make the criterion more specific for several choices of μ.

First we write down the log-density function for Y_i in the full model. Since ε_i has distribution function $\Phi(\cdot)^\lambda$, the log-density equals

$$\log \lambda + (\lambda - 1) \log \Phi\left(\frac{y_i - \beta_0 - \beta_1 x_i}{\sigma}\right) - \log \sigma - \frac{1}{2}\left(\frac{y_i - \beta_0 - \beta_1 x_i}{\sigma}\right)^2$$

plus the constant $-\frac{1}{2}\log(2\pi)$. After some algebra one finds the score vector of partial derivatives, evaluated at the narrow model where $\beta_1 = 0$ and $\lambda = 1$, and then the information matrix

$$
J_{n,\text{wide}} = \left(\begin{array}{cc|cc} 1/\sigma^2 & 0 & \bar{x}/\sigma^2 & c/\sigma \\ 0 & 2/\sigma^2 & 0 & d/\sigma \\ \hline \bar{x}/\sigma^2 & 0 & (v_n^2 + \bar{x}^2)/\sigma^2 & c\bar{x}/\sigma \\ c/\sigma & d/\sigma & c\bar{x}/\sigma & 1 \end{array}\right) = \left(\begin{array}{cc} J_{n,00} & J_{n,01} \\ J_{n,10} & J_{n,11} \end{array}\right),
$$

with 2×2 blocks as per the sizes of the parameter vectors θ and γ. Here $\bar{x}$ and v_n^2 are the empirical mean and variance of the x_is, while

$$
c = \text{cov}\{\varepsilon_i, \log \Phi(\varepsilon_i)\} = 0.9032 \quad \text{and} \quad d = \text{cov}\{\varepsilon_i^2, \log \Phi(\varepsilon_i)\} = -0.5956,
$$

with values arrived at via numerical integration. For the vector $\omega = J_{n,10}J_{n,00}^{-1}\frac{\partial\mu}{\partial\theta} - \frac{\partial\mu}{\partial\gamma}$ one finds

$$
\omega_1 = \bar{x}\frac{\partial\mu}{\partial\beta_0} - \frac{\partial\mu}{\partial\beta_1} \quad \text{and} \quad \omega_2 = c\sigma\frac{\partial\mu}{\partial\beta_0} + \tfrac{1}{2}d\sigma\frac{\partial\mu}{\partial\sigma} - \frac{\partial\mu}{\partial\lambda}.
$$

Their estimators are denoted $\widehat{\omega}_1$ and $\widehat{\omega}_2$, respectively. Next, we obtain the matrix $Q_n = J_n^{11} = (J_{n,11} - J_{n,10}J_{n,00}^{-1}J_{n,01})^{-1}$. After straightforward calculation, it follows that Q_n is a diagonal matrix $\text{diag}(\kappa_{n,1}^2, \kappa_{n,2}^2)$ with $\kappa_{n,1}^2 = \sigma^2/v_n^2$ and $\kappa_{n,2}^2 = 1/(1 - c^2 - \tfrac{1}{2}d^2) = 12.0879^2$. This allows us to use the simpler formulation for FIC from (6.2). The focussed information criterion in this setting reads

$$
\text{FIC} = \left(\sum_{j \notin S} \widehat{\omega}_j\widehat{\delta}_j\right)^2 + 2\sum_{j \in S} \widehat{\omega}_j^2\widehat{\kappa}_{n,j}^2, \tag{6.15}
$$

where $\widehat{\delta}_1 = \sqrt{n}\widehat{\beta}_{1,\text{wide}}$ and $\widehat{\delta}_2 = \sqrt{n}(\widehat{\lambda}_{\text{wide}} - 1)$. The partial derivatives are evaluated (and estimated, when necessary) in the narrow model where $(\beta_1, \lambda) = (0, 1)$.

Using the R function $\texttt{nlm}$ to maximise the log-likelihood, we get for the Australian Institute of Sports data that the estimate in the full model equals $(\widehat{\beta}_0, \widehat{\sigma}, \widehat{\beta}_1, \widehat{\lambda}) = (31.542, 4.168, 0.404, 1.948)$, with estimated standard deviations $(3.637, 0.984, 0.085, 1.701)$ (obtained from inverting the Hessian matrix in the output). From this it follows that $\widehat{\delta}_{\text{wide}} = (5.749, 45.031)^t$. The estimate for $\kappa_{1,n}$ is 1.455, found by plugging in the estimator for σ in the full model.

The submodel with the smallest value of the FIC is chosen:

$$
\text{FIC} = \begin{cases} (\widehat{\omega}_1\widehat{\delta}_1 + \widehat{\omega}_2\widehat{\delta}_2)^2 & \text{for the narrow model,} \\ \widehat{\omega}_2^2\widehat{\delta}_2^2 + 2\widehat{\omega}_1^2\kappa_1^2 & \text{for including } \beta_1, \text{ not } \lambda, \\ \widehat{\omega}_1^2\widehat{\delta}_1^2 + 2\widehat{\omega}_2^2\kappa_2^2 & \text{for including } \lambda, \text{ not } \beta_1, \\ 2(\widehat{\omega}_1^2\kappa_1^2 + \widehat{\omega}_2^2\kappa_2^2) & \text{for the full model.} \end{cases} \tag{6.16}
$$

Now we specify the focus parameter in four different applications.

Example 6.7 The mean

Let μ be the mean of $Y = \beta_0 + \beta_1 x + \sigma \varepsilon$ for some given covariate value x, that is, $\mu(\beta_0, \sigma, \beta_1, \lambda) = \beta_0 + \beta_1 x + \sigma e(\lambda)$, where $e(\lambda) = \int u \lambda \Phi(u)^{\lambda-1} \phi(u) \, du$. Here one finds $\omega_1 = \bar{x} - x$ and $\omega_2 = 0$, using the result $e'(1) = c$. For estimating the mean there is no award in involving the λ aspects of the data, as the added complexity does not alter the large-sample performance of estimators. We learn this from the fact that $\omega_2 = 0$. Only two of the four FIC cases need to be considered: the FIC value for the full model is the same as the value for the model including a linear term but no skewness, and the value for the model including skewness but only a constant mean is the same as the value for the narrow model. Hence the question is reduced to choosing between the narrow model with parameters (β_0, σ) or the broader model with three parameters $(\beta_0, \sigma, \beta_1)$.

$$\text{FIC} = \begin{cases} (\widehat{\omega}_1 \widehat{\delta}_1)^2 & \text{for the narrow model,} \\ 2\widehat{\omega}_1^2 \widehat{\kappa}_1^2 & \text{for including } \beta_1, \text{ not } \lambda. \end{cases}$$

For the Australian Institute of Sports data, the focus points at which we choose to estimate the hematocrit level are (i) at a body mass index of 23.56, which corresponds to the median value for male athletes, and (ii) at value 21.82, which is the corresponding median value for female athletes. For this data set we get $\widehat{\omega}_1 = -0.604$ for males and $\widehat{\omega}_1 = 1.141$ for females. This leads, when focus is on the males, to FIC values 12.061 for the narrow model and 1.546 for the model including a slope (though no skewness parameter). The FIC decides here that a linear slope is needed. For females the FIC values are, respectively, 43.016 and 5.515, and we come to the same conclusion.

Since $\widehat{\omega}_1^2$ is common to both FIC values, we can simplify the model comparison situation. FIC chooses the narrow model when $\widehat{\delta}_1^2 < 2\widehat{\kappa}_1^2$, or equivalently, when $\sqrt{n} |\widehat{\beta}_{1,\text{wide}} v_n / \widehat{\sigma}_{\text{wide}} < \sqrt{2}$. For the $\mathtt{ais}$ data, we get the value 5.558, which indeed confirms the choice of the model including a slope parameter. Notice that since ω_1 drops out of the equation, this conclusion is valid for all choices of x. ∎

Example 6.8 The median

Let μ be the median at covariate value x, that is, $\mu(\beta_0, \sigma, \beta_1, \lambda) = \beta_0 + \beta_1 x + \sigma \Phi^{-1}((\frac{1}{2})^{1/\lambda})$. Here $\omega_1 = \bar{x} - x$ and $\omega_2 = \sigma\{c - \frac{1}{2}(\log 2)/\phi(0)\} = 0.1313\,\sigma$. Neither of the ω values is equal to zero, and all four models are potential candidates. We insert an estimator for σ (computed in the full model) in ω_2 and construct the FIC values according to (6.16).

Now let us turn to the Australian Institute of Sports data, where the point of interest is to estimate the hematocrit level. We use the same covariate values as above. This leads to $\widehat{\omega}_2 = 0.144$ for both males and females, and the following four FIC values. For males we find 8.979, 43.399, 18.092, 7.578. The FIC picks out the last model, which includes both the skewness parameter λ and a slope parameter. For the median body mass index at the median female body fat percentage, we come to the same conclusion, based on the following FIC values: 169.730, 47.368, 49.048, 11.546.

Notice that here there is no easy testing criterion on the β_1 and σ separately, as there is in Example 6.7 where the focus parameter is the mean. ∎

Example 6.9 The third central moment
Consider the third central moment $\mu(\beta_0, \sigma, \beta_1, \lambda) = E(Y - EY)^3 = \sigma^3 E\{\varepsilon - e(\lambda)\}^3$, which is a measure of skewness of the distribution. Here $\omega_1 = 0$, implying that the inference is not influenced by inclusion or exclusion of the β_1 parameter. Some work yields $\omega_2 = -0.2203 \sigma^3$. The FIC values for the two models without skewness are the same. Likewise, the FIC values for the two models with skewness are the same. Model choice is between the narrow model and the model with skewness but constant mean.

$$\text{FIC} = \begin{cases} (\widehat{\omega}_2 \widehat{\delta}_2)^2 & \text{for the narrow model,} \\ 2\widehat{\omega}_2^2 \kappa_2^2 & \text{for including } \lambda, \text{ not } \beta_1. \end{cases}$$

The narrow model will be preferred by FIC when $\widehat{\omega}_2^2 \widehat{\delta}_2^2 < 2\widehat{\omega}_2^2 \kappa_2^2$. If $\sqrt{n}|\widehat{\lambda}_{\text{wide}} - 1|/\kappa_{n,2} < \sqrt{2}$ the narrow model is sufficient; otherwise we include the skewness parameter λ. There is a similar conclusion when the skewness $E(Y - EY)^3/\{E(Y - EY)^2\}^{3/2}$ is the focus parameter. ∎

Example 6.10 The cumulative distribution function
Consider the cumulative distribution function at y associated with a given x value $\mu(\beta_0, \sigma, \beta_1, \lambda) = P(Y \le y) = \Phi((y - \beta_0 - \beta_1 x)/\sigma)^\lambda$. Computing the derivatives with respect to $\beta_0, \sigma, \beta_1, \lambda$ one finds

$$\widehat{\omega}_1 = \frac{x - \bar{x}}{s} \phi\left(\frac{y - \bar{y}}{s}\right),$$

$$\widehat{\omega}_2 = -\left(\tfrac{1}{2}d\frac{y - \bar{y}}{s} + c\right)\phi\left(\frac{y - \bar{y}}{s}\right) - \Phi\left(\frac{y - \bar{y}}{s}\right)\log\Phi\left(\frac{y - \bar{y}}{s}\right),$$

in terms of sample average $\bar{y}$ and standard deviation s. Again, we need to consider all four models. The FIC values are obtained by inserting the estimators above in the formula (6.16).

The situation considered here can be generalised to $Y_i = \beta_0 + x_i^t \beta_1 + z_i^t \beta_2 + \sigma \varepsilon_i$, where the x_is are always to be included in the model whereas the z_is are extra candidates, along with the extra λ parameter for skewness of the error terms. ∎

6.6.4 FIC for football prediction

In Example 2.8 we considered four different Poisson rate models M_0, M_1, M_2, M_3 for explaining football match results in terms of the national teams' official FIFA ranking scores one month prior to the tournaments in question. These were special cases of the general form

$$\lambda(x) = \begin{cases} \exp\{a + c(x - x_0)\} & \text{if } x \le x_0, \\ \exp\{a + b(x - x_0)\} & \text{if } x \ge x_0, \end{cases}$$

with $x = \log(\text{fifa}/\text{fifa}')$ and $x_0 = -0.21$. In that example we described the four associated estimators of

$$\mu = P(\text{team 1 wins against team 2})$$
$$= P\big(\text{Pois}(\lambda(\text{fifa}, \text{fifa}')) > \text{Pois}(\lambda(\text{fifa}', \text{fifa}))\big).$$

Determining which estimator is best is a job for the FIC, where we may anticipate different optimal models for different matches, that is, for different values of fifa/fifa'. We illustrate this for two matches below, Norway–Belgium (where model M_1 is best for estimating μ) and Italy–Northern Ireland (where model M_2 is best for estimating μ). We estimated the 3×3 information matrix J_{wide}, with consequent estimates of Q and so on, and furthermore τ_0 and $\omega = J_{10} J_{00}^{-1} \frac{\partial \mu}{\partial a} - (\frac{\partial \mu}{\partial b}, \frac{\partial \mu}{\partial c})^t$, the latter requiring numerical approximations to partial derivatives of $\mu(a, b, c)$ at the estimated position $(\widehat{a}, \widehat{b}, \widehat{c})$ equal to $(-0.235, 1.893, -1.486)$ (as per Example 2.8).

The table below gives first of all the estimated probabilities of win, draw and loss, for the match in question, based on models M_0, M_1, M_2, M_3 respectively. This is followed by estimated standard deviation, estimated bias and estimated root mean squared error, for the estimate $\widehat{\mu} = \widehat{\mu}_{\text{win}}$. These numbers are given on the original scale of probabilities, but expressed in percent. In other words, rather than giving estimates of τ_S and $\omega^t(I - G_S)\delta$, which are quantities on the $\sqrt{n}(\widehat{\mu} - \mu)$ scale, we divide by $\sqrt{n}$ to get back to the original and most directly interpretable scale. Thus the columns of standard errors and estimated biases below are in percent. Note finally that the $\sqrt{\text{mse}}$ scale of the last column is a monotone transformation of the FIC score.

First, for Norway versus Belgium, with March 2006 FIFA rankings 643 and 605 (ratio 1.063): ω is estimated at $(-3.232, -0.046)^t/100$, and model M_1 is best:

| Model | Probability of | | | se | bias | $\sqrt{\text{mse}}$ |
	win	draw	loss			
M_0	0.364	0.272	0.364	0.484	5.958	5.978
M_1	0.423	0.273	0.304	0.496	0.482	0.691
M_2	0.426	0.274	0.299	1.048	0.000	1.048
M_3	0.428	0.276	0.296	1.073	0.000	1.073

Secondly, for Italy versus Northern Ireland, with FIFA rankings 738 and 472 (ratio 1.564): ω is estimated at $(-24.204, -5.993)^t/100$, and model M_2 is best:

| Model | Probability of | | | se | bias | $\sqrt{\text{mse}}$ |
	win	draw	loss			
M_0	0.364	0.272	0.364	0.761	32.552	32.561
M_1	0.804	0.135	0.061	3.736	12.412	12.962
M_2	0.767	0.144	0.089	7.001	0.000	7.001
M_3	0.693	0.161	0.146	12.226	0.000	12.226

Application of the FIC to this example actually requires a slight extension of our methodology. The problem lies with model M_1. The widest model M_3 uses three parameters (a, b, c); the hockey-stick model M_2 takes $c = 0$; model M_1 takes $b = c$ and corresponds to a log-linear Poisson rate model; and the narrow model M_0 takes $b = 0$ and $c = 0$. Setting $b = c$ does not fit immediately with the previous results. We have therefore uncovered yet more natural submodels than the $2^2 = 4$ associated with setting one or both of the two parameters b and c to zero.

This problem can be solved by introducing an invertible $q \times q$ contrast matrix A, such that

$$\Upsilon = A(\gamma - \gamma_0) = A \begin{pmatrix} b \\ c \end{pmatrix} = \begin{pmatrix} 1 & 0 \\ 1 & -1 \end{pmatrix} \begin{pmatrix} b \\ c \end{pmatrix} = \begin{pmatrix} b \\ b - c \end{pmatrix}.$$

Model M_1 now corresponds to $b - c = 0$. The reparametrised model density is $f(y, \theta, A^{-1}\Upsilon)$ and the focus parameter $\mu = \mu(\theta, \gamma)$ can be represented as

$$v = v(\theta, \Upsilon) = \mu(\theta, \gamma_0 + A^{-1}\Upsilon).$$

In the submodel that uses parameters Υ_j for $j \in S$ whereas $\Upsilon_j = 0$ for $j \notin S$, the estimator of μ takes the form $\widehat{v}_S = v(\widehat{\theta}_S, \widehat{\Upsilon}_S, 0_{S^c})$, with maximum likelihood estimators in the (θ, Υ_S) model. The $\widehat{v}_S$ is also identical to

$$\widehat{\mu}_{A,S} = \mu(\widehat{\theta}_S, \gamma_0 + A^{-1}\widetilde{\Upsilon}_S),$$

where $\widetilde{\Upsilon}_S$ has zeros for $j \notin S$ and otherwise components agreeing with $\widehat{\Upsilon}_S$; this is the maximum likelihood estimator of the focus parameter in the (θ, Υ_S) model that employs those parameters Υ_j among $\Upsilon = A(\gamma - \gamma_0)$ for which $j \in S$. It can be shown that, with $\widetilde{Q} = AQA^t$, the limit distribution of $\sqrt{n}(\widehat{\mu}_{A,S} - \mu_{\text{true}})$ has mean $\omega^t(I_q - A^{-1}\widetilde{G}_S A)\delta$ and variance $\tau_0^2 + \omega^t A^{-1}\widetilde{Q}_S^0(A^{-1})^t\omega$; see Exercise 6.8, where also the required matrix $\widetilde{G}_S$ is identified. The limiting mean squared error is therefore

$$\tau_0^2 + \omega^t(I_q - A^{-1}\widetilde{G}_S A)\delta\delta^t(I_q - A^t\widetilde{G}_S^t(A^{-1})^t)\omega + \omega^t A^{-1}\widetilde{Q}_S^0(A^{-1})^t\omega.$$

As with our earlier FIC construction, which corresponds to $A = I_q$, an appropriately extended FIC formula is now obtained, by again estimating the $\delta\delta^t$ term with $D_n D_n^t - \widehat{Q}$.

6.6.5 FIC for speedskating prediction

We continue the model selection procedure started in Section 5.6, where both AIC and the BIC decided on a linear model with heteroscedasticity for prediction of the time on the 10,000-m using the time on the 5000-m. The four models considered were either linear or quadratic regression with homoscedastic or heteroscedastic normal errors. Three focus parameters are of interest for this illustration. Consider Example 5.11, concerned with the estimation of the 10% quantile of the 10,000-m distribution for a skater whose personal best time on the 5000-m distance is x_0; $\mu(x_0, q) = x_0^t\beta + z_0^t\gamma + d(q)\sigma \exp(\phi v_0)$. In

Table 6.2. *Focussed model selection for the 10% quantiles of the 10,000-m distribution for a skater with 5000-m time 6:35 (μ_1) and another skater with 6:15 time (μ_2), and for the probability of a skater with 5000-m time equal to 6:10.65 setting a world record on the 10,000-m (μ_3).*

		FIC scale:			Real scale:		
	Estimate	std.dev	Bias	$\sqrt{\text{FIC}}$	std.dev	Bias	$\sqrt{\text{mse}}$
For focus $\mu_1 = \mu(6{:}35, 0.1)$:							
M_0	13:37.25	0.0000	0.000	0.000	1.589	0.000	1.589 (1)
M_1	13.37.89	12.361	5.233	13.423	1.813	0.370	1.851 (4)
M_2	13.38.05	1.050	0.000	1.050	1.590	0.000	1.590 (2)
M_3	13.38.12	12.405	0.000	12.405	1.815	0.000	1.815 (3)
For focus $\mu_2 = \mu(6{:}15, 0.1)$:							
M_0	12:49.34	0.000	133.227	133.227	2.685	9.421	9.796 (3)
M_1	12:48.13	24.964	137.940	140.181	3.213	9.754	10.269 (4)
M_2	12:57.55	27.685	0.000	27.685	3.323	0.000	3.323 (1)
M_3	12.57.48	37.278	0.000	37.278	3.762	0.000	3.762 (2)
For focus $\mu_3 = p(6{:}10.65, 12{:}51.60)$:							
M_0	0.302	0.000	0.952	0.952	0.062	0.067	0.091 (2)
M_1	0.350	0.950	1.436	1.722	0.091	0.1025	0.136 (4)
M_2	0.182	0.288	0.000	0.288	0.065	0.000	0.065 (1)
M_3	0.187	0.993	0.000	0.993	0.093	0.000	0.093 (3)

particular, we take a medium-level skater with 5000-m time x_0 equal to 6:35.00, and a top-level skater with x_0 equal to 6:15.00. This defines two focus parameters, $\mu_1 = \mu(6{:}35, 0.1)$ and $\mu_2 = \mu(6{:}15, 0.1)$. The third focus parameter is as in Example 5.12, the probability that Eskil Ervik, with a personal best on the 5000-m equal to 6:10.65, sets a world record on the 10,000-m, that is,

$$\mu_3 = p(x_0, y_0) = \Phi\left(\frac{y_0 - a - bx_0 - cz_0}{\sigma \exp(\phi v_0)}\right),$$

with $x_0 = 6{:}10.65$ and $y_0 = 12{:}51.60$ (as of the end of the 2005–2006 season).

Table 6.2 gives the results of the FIC analysis. The table presents the estimated value of the focus parameters in each of the four models, together with the estimated bias and standard deviation of $\sqrt{n}\widehat{\mu}_j$, as well as the square root of the FIC values, all expressed in the unit seconds. The last three columns give the more familiar values for bias and standard deviation for $\widehat{\mu}_j$ (not multiplied by $\sqrt{n}$), and the square root of the mse. For μ_1 the FIC-selected model is the narrow model M_0, linear and homoscedastic. This agrees with the discussion based on tolerance regions in Example 5.12. For the other two parameters the model with linear trend, though including a heteroscedasticity parameter, is considered the best model.

6.6.6 FIC in generalised linear models

Assume that there are independent observations Y_i that conditional on covariate information (x_i, z_i) follow densities of the form

$$f(y_i, \theta_i, \phi) = \exp\left\{ \frac{y_i\theta_i - b(\theta_i)}{a(\phi)} + c(y_i, \phi) \right\} \quad \text{for } i = 1, \ldots, n,$$

for suitable a, b, c functions, where θ_i is a smooth transformation of the linear predictor $\eta_i = x_i^t\beta + z_i^t\gamma$. There is a link function g such that

$$g(\xi_i) = x_i^t\beta + z_i^t\gamma, \quad \text{where} \quad \xi_i = \mathrm{E}(Y_i \mid x_i, z_i) = b'(\theta_i).$$

The notation is again chosen so that x denotes protected covariates, common to all models, while components of z are open for ex- or inclusion.

To compute the FIC, we need the Fisher information matrix J_n, computed via the second-order partial derivatives of the log-likelihood function with respect to (β, γ). Define first the diagonal weight matrix $V = \mathrm{diag}\{v_1, \ldots, v_n\}$ with

$$v_i = b''(\theta_i)r(x_i^t\beta + z_i^t\gamma)^2 = \left\{ b''(\theta_i)g'(\xi_i)^2 \right\}^{-1}, \tag{6.17}$$

which is a function of $\eta_i = x_i^t\beta + z_i^t\gamma$, and $r(x_i^t\beta + z_i^t\gamma) = \partial\theta_i/\partial\eta_i$. We then arrive at

$$J_n = a(\phi)^{-1}n^{-1}\sum_{i=1}^{n} v_i \begin{pmatrix} x_i \\ z_i \end{pmatrix}\begin{pmatrix} x_i \\ z_i \end{pmatrix}^t = a(\phi)^{-1}n^{-1}\begin{pmatrix} X^tVX & X^tVZ \\ Z^tVX & Z^tVZ \end{pmatrix},$$

and the matrix Q_n takes the form

$$\begin{aligned} Q_n &= (J_{n,11} - J_{n,10}J_{n,00}^{-1}J_{n,01})^{-1} \\ &= a(\phi)\left\{ n^{-1}Z^tV(I - X(X^tVX)^{-1}X^tV)Z \right\}^{-1}. \end{aligned} \tag{6.18}$$

Here X is the $n \times p$ design matrix of x_i variables while Z is the $n \times q$ design matrix of additional z_i variables. It follows from the parameter orthogonality property of generalised linear models that the mixed second-order derivative with respect to ϕ and (β, γ) has mean zero, and hence that the Q_n formula is valid whether the scale parameter ϕ is known or not. In detail, if ϕ is present and unknown, then

$$J_{n,\text{wide}} = \begin{pmatrix} J_{n,\text{scale}} & 0 & 0 \\ 0 & n^{-1}a(\phi)^{-1}X^tVX & n^{-1}a(\phi)^{-1}X^tVZ \\ 0 & n^{-1}a(\phi)^{-1}Z^tVX & n^{-1}a(\phi)^{-1}Z^tVZ \end{pmatrix},$$

where $J_{n,\text{scale}}$ is the required $-n^{-1}\sum_{i=1}^{n} \mathrm{E}\{\partial^2 \log f(Y_i; \theta_i, \phi)/\partial\phi^2\}$. For the normal distribution with $\phi = \sigma$ and $a(\sigma) = \sigma^2$, for example, one finds $J_{n,\text{scale}} = 2/\sigma^2$. The block-diagonal form of $J_{n,\text{wide}}$ implies that the $q \times q$ lower right-hand corner of the inverse information matrix remains as in (6.18).

The next required FIC ingredient is the ω vector, for a given parameter of interest $\mu = \mu(\phi, \beta, \gamma)$. For generalised linear models

$$\omega = (0, n^{-1}a(\phi)^{-1}Z^{t}VX) \begin{pmatrix} J_{n,\text{scale}} & 0 \\ 0 & n^{-1}a(\phi)^{-1}X^{t}VX \end{pmatrix}^{-1} \begin{pmatrix} \frac{\partial\mu}{\partial\phi} \\ \frac{\partial\mu}{\partial\beta} \end{pmatrix} - \frac{\partial\mu}{\partial\gamma}$$

$$= Z^{t}VX(X^{t}VX)^{-1}\frac{\partial\mu}{\partial\beta} - \frac{\partial\mu}{\partial\gamma},$$

again taking different forms for different foci. It is noteworthy that it does not depend on the scale parameter ϕ, though, even in cases where it is explicitly involved in the interest parameter. With $G_{n,S} = Q^{0}_{n,S}Q^{-1}_{n}$ and $\widehat{\gamma} = \widehat{\gamma}_{\text{wide}}$, we have arrived at a FIC formula for generalised linear models:

$$\text{FIC}(S) = n\widehat{\omega}^{t}(I_{q} - \widehat{G}_{n,S})\widehat{\gamma}\widehat{\gamma}^{t}(I_{q} - \widehat{G}_{n,S})^{t}\widehat{\omega} + 2\widehat{\omega}^{t}\widehat{Q}^{0}_{n,S}\widehat{\omega}. \qquad (6.19)$$

The bias-modified version takes

$$\text{FIC}^{*}(S) = \widehat{\omega}^{t}(I_{q} + \widehat{G}_{n,S})\widehat{Q}\widehat{\omega}$$

if $\{\widehat{\omega}^{t}(I_{q} - \widehat{G}_{n,S})\widehat{\gamma}\}^{2} < \widehat{\omega}^{t}(\widehat{Q} - \widehat{Q}^{0}_{S})\widehat{\omega}$, and takes $\text{FIC}^{*}(S) = \text{FIC}(S)$ otherwise.

When μ is the linear predictor $\mu(x, z) = x^{t}\beta + z^{t}\gamma$, for example, we have $\omega = Z^{t}VX(X^{t}VX)^{-1}x - z$, with an appropriate specialisation of (6.19). The same model ranking will be found for any smooth function $\mu = m(x^{t}\beta + z^{t}\gamma)$, like $\xi = \text{E}(Y \mid x, z) = g^{-1}(x^{t}\beta + z^{t}\gamma)$, since the ω for this problem is merely a common constant times the one just given.

Example 6.11 Logistic regression as a GLM
For a logistic regression model, each Y_{i} follows a Bernoulli distribution

$$p_{i}^{y_{i}}(1 - p_{i})^{1-y_{i}} = \exp\left\{y_{i}\log\frac{p_{i}}{1 - p_{i}} + \log(1 - p_{i})\right\}.$$

It follows that $\theta_{i} = \log\{p_{i}/(1 - p_{i})\}$, which indeed is a function of the mean p_{i}, $b(\theta_{i}) = \log(1 + \theta_{i})$ and $c(y_{i}, \phi) = 0$. In this model there is no scale parameter ϕ (or one may set $a(\phi) = 1$ to force it into the generalised linear models formulae). The canonical link function is the logit function $g(p_{i}) = \log\{p_{i}/(1 - p_{i})\} = x_{i}^{t}\beta + z_{i}^{t}\gamma$. With this input, we find the previous FIC calculations for logistic regression. ∎

6.7 Exact mean squared error calculations for linear regression *

We show here that for a normal linear regression model, the exact mean squared error calculations and the FIC large-sample risk approximations exactly coincide, as long as the focus parameter is linear in the mean parameters. We start with a linear model as in (6.12), or in matrix notation

$$Y = X\beta + Z\gamma + \varepsilon.$$

with X the $n \times p$ design matrix with rows being the p-vectors $x_i^t = (x_{i,1}, \ldots, x_{i,p})$ and with Z the $n \times q$ design matrix with rows the q-vectors $z_i^t = (z_{i,1}, \ldots, z_{i,q})$. A homoscedastic model assumes that $\text{Var } \varepsilon = \sigma^2 I_n$. Combining both matrices into a single design matrix $B = (X, Z)$ leads to the familiar least squares formula for the estimated coefficients

$$\begin{pmatrix} \widehat{\beta} \\ \widehat{\gamma} \end{pmatrix} = \begin{pmatrix} X^t X & X^t Z \\ Z^t X & Z^t Z \end{pmatrix}^{-1} \begin{pmatrix} X^t \\ Z^t \end{pmatrix} Y = (B^t B)^{-1} B^t Y.$$

Here

$$\text{Var} \begin{pmatrix} \widehat{\beta} \\ \widehat{\gamma} \end{pmatrix} = (\sigma^2/n) \Sigma_n = \sigma^2 (B^t B)^{-1},$$

partitioned as in (6.13). We let its inverse matrix be partitioned in a similar way,

$$\Sigma_n^{-1} = \begin{pmatrix} \Sigma^{00} & \Sigma^{01} \\ \Sigma^{10} & \Sigma^{11} \end{pmatrix}.$$

When not all q variables z are included, the design matrix Z is replaced by its submatrix Z_S, only including those columns corresponding to j in S. This leads to similarly defined matrices $\Sigma_{n,S}$ and their inverse matrices with submatrices $\Sigma^{00,S}$, $\Sigma^{01,S}$, $\Sigma^{10,S}$ and $\Sigma^{11,S}$.

The FIC is an estimator for the mean squared error of the estimator for the focus parameter $\mu = x^t\beta + z^t\gamma$. Since the mean squared error decomposes into a variance part and a squared bias part, we study first the variance of the least squares estimator and compare it to the variance used in the expression for the FIC. Next, we repeat this for the bias.

The *exact* variance of the least squares estimator for $\mu = x^t\beta + z^t\gamma$, in the model including extra variables z_S, is equal to

$$\sigma^2 \begin{pmatrix} x \\ z_S \end{pmatrix}^t \begin{pmatrix} X^t X & X^t Z_S \\ Z_S^t X & Z_S^t Z_S \end{pmatrix}^{-1} \begin{pmatrix} x \\ z_S \end{pmatrix}.$$

Using the notation defined above, this can be rewritten as

$$n^{-1}\sigma^2(x^t\Sigma^{00,S}x + z_S^t\Sigma^{10,S}x + x^t\Sigma^{01,S}z_S + z_S\Sigma^{11,S}z_S). \tag{6.20}$$

Assuming normality, the FIC uses in its approximation to the mean squared error the limiting variance term

$$\tau_0^2 + \omega^t Q_S^0 \omega. \tag{6.21}$$

For normal linear models, $J_n = \sigma^{-2}\Sigma_n$, which implies that $Q = \sigma^2\Sigma^{11}$ and further that $\omega = \Sigma_{10}\Sigma_{00}^{-1}x - z$ and that $\tau_0^2 = \sigma^2 x^t\Sigma_{00}^{-1}x$. The variance (6.21) used in the limit experiment can now be rewritten exactly as in (6.20), modulo the appropriate n^{-1} factor.

For the bias, we compute the *exact* expected value of the least squares estimator in the model indexed by S, and find

$$
\begin{aligned}
\mathrm{E}\begin{pmatrix} \widehat{\beta}_S \\ \widehat{\gamma}_S \end{pmatrix} &= \Sigma_{n,S}^{-1} \begin{pmatrix} X^t X & X^t Z \\ Z_S^t X & Z_S^t Z \end{pmatrix} \begin{pmatrix} \beta \\ \gamma \end{pmatrix} \\
&= \Sigma_{n,S}^{-1} \begin{pmatrix} \Sigma_{00}\beta + \Sigma_{01}\gamma \\ \Sigma_{10,S}\beta + (\Sigma_{11}\gamma)_S \end{pmatrix} \\
&= \begin{pmatrix} (\Sigma^{00,S}\Sigma_{00} + \Sigma^{01,S}\Sigma_{10,S})\beta + \Sigma^{00,S}\Sigma_{01}\gamma + \Sigma^{01,S}(\Sigma_{11}\gamma)_S \\ (\Sigma^{10,S}\Sigma_{00} + \Sigma^{11,S}\Sigma_{10,S})\beta + \Sigma^{10,S}\Sigma_{01}\gamma + \Sigma^{11,S}(\Sigma_{11}\gamma)_S \end{pmatrix} \\
&= \begin{pmatrix} \beta + \Sigma_{00}^{-1}\Sigma_{01}(I_q - (\Sigma^{11,S})^0(\Sigma^{11})^{-1})\gamma \\ 0 + \Sigma^{11,S}((\Sigma^{11})^{-1}\gamma)_S \end{pmatrix}.
\end{aligned}
$$

The simplification in the last step is obtained by using the formulae for the blocks of inverse matrices, cf. (5.7). After working out the matrix multiplication, the exact bias expression is equal to

$$
\begin{aligned}
\mathrm{E}(x^t\widehat{\beta}_S + z_S^t\widehat{\gamma}_S) &- (x^t\beta + z^t\gamma) \\
&= (x^t\Sigma_{00}^{-1}\Sigma_{01} - z^t)(I_q - (\Sigma^{11,S})^0(\Sigma^{11})^{-1})\gamma = \omega^t(I_q - G_S)\gamma.
\end{aligned}
$$

This exactly matches the bias used to obtain the mean squared error in the large-sample experiment.

The main message is that although the FIC expression is obtained by estimating a quantity in the limit experiment, there are some situations, such as when estimating linear functions of the mean parameters in the linear regression model, where the obtained FIC expression is exact (and not an approximation).

6.8 The FIC for Cox proportional hazard regression models

We first make the comment that the method above, leading via Theorem 6.1 and estimation of risk functions under quadratic loss to the FIC, may be generalised without serious obstacles to the comparison of *parametric* models for hazard rates, in survival and event history analysis. Thus we may construct FIC methods for the Gamma process threshold models used in Example 3.10, for example, with an appropriate definition for the J matrix and its submatrices. This remark applies also to parametric proportional hazard models.

The situation is different and more challenging for the Cox proportional hazard model in that it is *semiparametric*, with an unspecified baseline hazard $h_0(t)$. It specifies that individual i has hazard rate $h_0(t)\exp(x_i^t\beta + z_i^t\gamma)$, see Section 3.4. This set-up requires an extension of the construction and definition of the FIC since now the focus parameter might depend on both vector parameters β and γ and the cumulative baseline hazard function H_0, that is, $\mu = \mu(\beta, \gamma, H_0(t))$. We continue to use the notation introduced in

Section 3.4, and shall work with the local neighbourhood model framework formalised by taking .

$$h_i(t) = h_0(t) \exp(x_i^t \beta + z_i^t \delta / \sqrt{n}) \quad \text{for } i = 1, \ldots, n. \tag{6.22}$$

We need some further definitions. Let first

$$G_n^{(0)}(u, \beta, \gamma) = n^{-1} \sum_{i=1}^n Y_i(u) \exp(x_i^t \beta + z_i^t \gamma),$$

$$G_n^{(1)}(u, \beta, \gamma) = n^{-1} \sum_{i=1}^n Y_i(u) \exp(x_i^t \beta + z_i^t \gamma) \begin{pmatrix} x_i \\ z_i \end{pmatrix},$$

where sufficient regularity is assumed to secure convergence in probability of these average functions to appopriate limit functions; see Hjort and Claeskens (2006) for relevant discussion. We denote the limit in probability of $G_n^{(0)}$ by $g^{(0)}(u, \beta, \gamma)$, and also write $e(u, \beta, 0)$ for the limit of $G_n^{(1)}(u, \beta, \delta / \sqrt{n}) / G_n^{(0)}(u, \beta, \delta / \sqrt{n})$. Define next the $(p + q)$-vector function

$$F(t) = \int_0^t e(u, \beta, 0) \, dH_0(u) = \begin{pmatrix} F_0(t) \\ F_1(t) \end{pmatrix},$$

where $F_0(t)$ and $F_1(t)$ have respectively p and q components. The semiparametric information matrix J is the limit in probability of $-n^{-1} I_n(\beta, 0)$, where $I_n(\beta, \gamma)$ is the matrix of second-order derivatives of the log-partial-likelihood function $\ell_n^p(\beta, \gamma)$, as in Section 3.4. Parallelling notation of Section 6.1, we need quantites defined via J and J^{-1}, such as $Q = J^{11}$, and for each submodel S we have use for the $(p + |S|) \times (p + |S|)$ submatrix J_S, with $Q_S = J_S^{11}$ and $G_S = Q_S^0 Q^{-1}$.

For a focus parameter of the type $\mu = \mu(\beta, \gamma, H_0(t))$, subset estimators of the form $\widehat{\mu}_S = \widehat{\mu}(\widehat{\beta}_S, \widehat{\gamma}_S, 0_{S^c}, \widehat{H}_{0,S}(t))$ may be formed, for each submodel S identified by inclusion of those γ_j for which $j \in S$. Here $(\widehat{\beta}_S, \widehat{\gamma}_S, \widehat{H}_{0,S})$ are the Cox estimators and Aalen–Breslow estimator in the S submodel, see e.g. Andersen *et al.* (1993, Chapter IV). Hjort and Claeskens (2006) obtained the following result. Define

$$\omega = J_{10} J_{00}^{-1} \frac{\partial \mu}{\partial \beta} - \frac{\partial \mu}{\partial \gamma}, \quad \kappa(t) = \{ J_{10} J_{00}^{-1} F_0(t) - F_1(t) \} \frac{\partial \mu}{\partial H_0},$$

and

$$\tau_0(t)^2 = \left(\frac{\partial \mu}{\partial H_0} \right)^2 \int_0^t \frac{dH_0(u)}{g^{(0)}(u, \beta, 0)} + \{ \frac{\partial \mu}{\partial \beta} - \frac{\partial \mu}{\partial H_0} F_0(t) \}^t J_{00}^{-1} \{ \frac{\partial \mu}{\partial \beta} - \frac{\partial \mu}{\partial H_0} F_0(t) \},$$

with all derivatives evaluated at $(\beta, 0, H_0(t))$. Under (6.22), the true value μ_{true} of the focus parameter is $\mu(\beta, \delta / \sqrt{n}, H_0(t))$.

Theorem 6.2 *Under the sequence of models identified in equation (6.22), $D_n = \widehat{\delta} = \sqrt{n}\widehat{\gamma}$ tends in distribution to $D \sim N_q(\delta, Q)$. Furthermore,*

$$\sqrt{n}(\widehat{\mu}_S - \mu_{\text{true}}) \xrightarrow{d} \Lambda_0 + \{\omega - \kappa(t)\}^{\text{t}}(\delta - G_S D),$$

where $\Lambda_0 \sim N(0, \tau_0(t)^2)$ is independent of D.

The similarity with Theorem 6.1 is striking. Differences are the dependence on time t of τ_0^2 and the introduction of $\kappa(t)$. The reason for the $\kappa(t)$ is that the focus parameter $\mu(\beta, \gamma, H_0(t))$ is allowed to depend on $H_0(t)$; for parameters depending only on β and γ, the $\kappa(t)$ term disappears. As in Section 6.3 we may use the new theorem to obtain expressions for limiting mean squared error $\text{mse}(S, \delta)$ associated with estimators $\widehat{\mu}_S$, for each submodel,

$$\tau_0^2(t) + \{\omega - \kappa(t)\}^{\text{t}}\{(I_q - G_S)\delta\delta^{\text{t}}(I_q - G_S^{\text{t}}) + G_S Q G_S^{\text{t}}\}\{\omega - \kappa(t)\}.$$

When the full $(p + q)$-parameter model is used, $G_S = I_q$ and the mse is $\tau_0^2 + (\omega - \kappa)^{\text{t}}Q(\omega - \kappa)$, constant in δ. The other extreme is to select the narrow p-parameter model, for which $S = \emptyset$, and $G_S = 0$. This leads to $\text{mse}(\emptyset, \delta) = \tau_0^2 + \{(\omega - \kappa)^{\text{t}}\delta\}^2$.

In these risk expressions, quantities τ_0, ω, κ, G_S, and Q can all be estimated consistently; for details see the above-mentioned article. As in Section 6.3, the difficulty lies with the $\delta\delta^{\text{t}}$ parameter, which we as there estimate using $D_n D_n^{\text{t}} - \widehat{Q} = n\widehat{\gamma}\widehat{\gamma}^{\text{t}} - \widehat{Q}$. Thus we have for each S an asymptotically unbiased risk estimator $\widehat{\text{mse}}(S)$, equal to

$$\widehat{\tau}_0^2 + \{\widehat{\omega} - \widehat{\kappa}(t)\}^{\text{t}}\{(I_q - \widehat{G}_S)(D_n D_n^{\text{t}} - \widehat{Q})(I_q - \widehat{G}_S)^{\text{t}} + \widehat{G}_S \widehat{Q} \widehat{G}_S^{\text{t}}\}\{\widehat{\omega} - \widehat{\kappa}(t)\}.$$

The focussed information criterion consists in selecting the model with smallest estimated $\text{mse}(S)$. Since the constant $\widehat{\tau}_0^2$ does not affect the model comparison, and similarly $(\widehat{\omega} - \widehat{\kappa}(t))^{\text{t}}\widehat{Q}(\widehat{\omega} - \widehat{\kappa}(t))$ is common to each risk estimate, both quantities can be subtracted. These rearrangements lead to defining

$$\text{FIC}(S) = (\widehat{\psi} - \widehat{\psi}_S)^2 + 2(\widehat{\omega} - \widehat{\kappa})^{\text{t}}\widehat{Q}_S^0(\widehat{\omega} - \widehat{\kappa}),$$

where $\widehat{\psi} = (\widehat{\omega} - \widehat{\kappa})^{\text{t}}\widehat{\delta}$ and $\widehat{\psi}_S = (\widehat{\omega} - \widehat{\kappa})^{\text{t}}\widehat{G}_S\widehat{\delta}$, and where finally $\widehat{\kappa}$ is short-hand for $\widehat{\kappa}(t)$. The FIC is oriented towards selecting an optimal model, in the mean squared error sense, for the particular task at hand; different estimands $\mu(\beta, \gamma, H_0(t))$ correspond to different $\omega - \kappa(t)$ and different ψ.

We refer to Hjort and Claeskens (2006) for an extension of the FIC that is able to deal with focus parameters that are medians, or quantiles. This is outside the scope of Theorem 6.2, since a quantile does not depend on the baseline cumulative hazard function H_0 at only a single value. The structure of the FIC for such a case is similar to that above, though a bit more involved.

Some applications of the methods above, to specific parameters of interest, are as follows.

Relative risk: The relative risk at position (x, z) in the covariate space is often the quantity of interest to estimate, i.e.

$$\mu = \exp(x^t\beta + z^t\gamma).$$

Here $\omega = \exp(x^t\beta)(J_{10}J_{00}^{-1}x - z)$ and $\kappa = 0_q$. This is the relative risk in comparison with an individual with covariates $(x, z) = (0_p, 0_q)$. If the covariates have been centered to have mean zero, the comparison is to 'the average individual'. Similarly, if x and z represent risk factors, scaled such that zero level corresponds to normal healthy conditions and positive values correspond to increased risk, then $\mu = \mu(x, z)$ is relative risk increase at level (x, z) in comparison with normal health level. In other situations it would be more natural to compare individuals with an existing or hypothesised individual with suitable given covariates (x_0, z_0). This corresponds to focussing on the relative risk

$$\mu = \exp\{(x - x_0)^t\beta + (z - z_0)^t\gamma\}.$$

Here, $\kappa = 0_q$ and $\omega = \exp\{(x - x_0)^t\beta\}\{J_{10}J_{00}^{-1}(x - x_0) - (z - z_0)\}$. Note in particular that different covariate levels give different ω vectors. In view of constructing a FIC criterion based on the mean squared error this implies that there might well be different optimal S submodels for different covariate regions.

The cumulative baseline hazard: While the relative risk is independent of the cumulative baseline hazard, estimating $H_0(t)$ is of separate interest. Since there is no dependence on (x, z), $\omega = 0_p$ while $\kappa = J_{10}J_{00}^{-1}F_0(t) - F_1(t)$.

A survival probability: Estimating a survival probability for a given individual translates to focussing on

$$\mu = S(t \mid x, z) = \exp\{-\exp(x^t\beta + z^t\gamma)H_0(t)\},$$

for which one finds

$$\omega = -S(t \mid x, z)\exp(x^t\beta + z^t\gamma)H_0(t)(J_{10}J_{00}^{-1}x - z),$$
$$\kappa = -S(t \mid x, z)\exp(x^t\beta + z^t\gamma)\{J_{10}J_{00}^{-1}F_0(t) - F_1(t)\}.$$

Here both the covariate position (x, z) and the time value t play a role.

Example 6.12 Danish melanoma study: FIC plots and FIC variable selection
For an illustration of the methods developed above we return to the survival data set about 205 melanoma patients treated in Example 3.9. The current purpose is to identify those among seven covariates $x_1, z_1, \ldots, z_6$ that are most useful for estimating different focus parameters.

Three focus parameters are singled out for this illustration, corresponding to versions of those discussed above. FIC_{μ_1} takes the relative risk $\mu_1 = \exp\{(x - x_0)^t\beta + (z - z_0)^t\gamma\}$, where (x_0, z_0) corresponds to average tumour thickness amongst all women participating in the study, infection infiltration level $z_2 = 4$, epithelioid cells present ($z_3 = 1$),

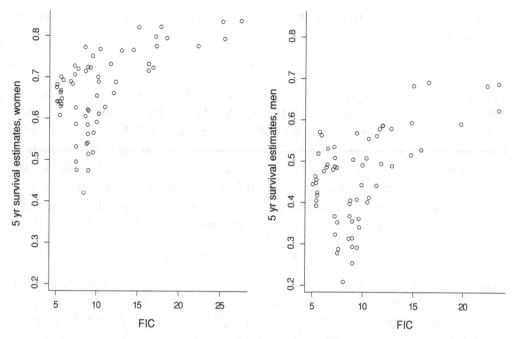

Fig. 6.5. FIC plots for five-year survival probabilities, for women (left panel) and men (right). For each group, all 64 estimates of the survival probability $\mathcal{S}(t \mid x, z)$ are shown, plotted against the FIC score. Estimates associated with smaller FIC scores have higher precision.

ulceration present ($z_4 = 1$), invasion depth $z_5 = 2$, and average women's age in the study. The variables (x, z) take average tumour thickness for men and average men's age, all other variables remain the same. For FIC_{μ_2} the focus parameter is $\mu_2 = H_0(t)$ at time $t = 3$ years after operation. The third focus parameter is the five-year survival probability $\mu_3 = \mathcal{S}(5 \mid x, z)$ for respectively a woman and a man, both with $z_2 = 4$, $z_3 = 1$, $z_4 = 1$, $z_5 = 2$, and with average values for z_1 and z_6 among respectively women and men. See Figure 6.5.

Table 6.3 shows the five highest ranked values for the three versions of FIC, allowing all $2^6 = 64$ candidate models. The table also shows the selected variables, and can be compared to Table 3.6, where AIC and the BIC were used for variable selection. Note that the values are sorted in importance per criterion.

With the FIC there is not one single answer for the 'best model', as with AIC and the BIC, since the model chosen depends on the focus. The relative risk as a focus parameter lets the FIC point to the narrow model. To estimate the cumulative hazard $H_0(t)$ only variable z_1 (tumour thickness) is selected. For the survival probability, focus μ_3, z_3 and z_5 are the most important. The fact that different models are selected for different purposes should be seen as a way of strengthening the biostatistician's ability to produce more precise estimates or predictions for a specific patient. ∎

Table 6.3. *Danish melanoma study. The five best FIC models are indicated, among all*
$2^6 = 64$ *candidate models for inclusion or exclusion of* $z_1, \ldots, z_6$, *for each of four*
situations. The first corresponds to μ_1, *a relative risk; the second to* μ_2, *the cumulative*
baseline hazard at three years after operation; and the third and fourth correspond to
five-year survival probabilities for women and men, respectively; see the text for
more details.

	μ_1			μ_2			μ_3			μ_4	
var's	est	FIC	var's	est	FIC	var's	est	FIC	var's	est	FIC
none	1.939	5.070	none	0.063	0.691	4	0.675	5.057	3 4	0.444	5.037
5	1.978	5.123	1	0.065	0.691	4 6	0.681	5.096	3 4 6	0.444	5.039
6	1.891	5.261	5	0.029	0.715	3 4	0.641	5.098	4 5	0.463	5.289
5 6	1.986	5.374	1 5	0.040	0.721	3 4 6	0.640	5.124	3 5	0.447	5.376
1 5	2.053	5.386	3 5	0.075	0.773	4 5	0.641	5.392	3 4 5	0.492	5.387

As in Section 6.4 there is a modification of the FIC in cases where the estimated
squared bias happens to be negative. This happens when $N_n(s)$ takes place, that is when

$$n\{(\widehat{\omega} - \widehat{\kappa}(t))^{\mathrm{t}}(I_q - \widehat{\Omega}_S)\widehat{\gamma}\}^2 < (\widehat{\omega} - \widehat{\kappa}(t))^{\mathrm{t}}(\widehat{Q} - \widehat{Q}_S^0)(\widehat{\omega} - \widehat{\kappa}(t)).$$

In such cases the modified FIC, similar to (6.8), is defined as

$$\mathrm{FIC}^*(S) = \begin{cases} \mathrm{FIC}(S) & \text{if } N_n(S) \text{ does not take place,} \\ (\widehat{\omega} - \widehat{\kappa})^{\mathrm{t}}(\widehat{Q}_S^0 + \widehat{Q})(\widehat{\omega} - \widehat{\kappa}) & \text{if } N_n(S) \text{ takes place.} \end{cases} \tag{6.23}$$

6.9 Average-FIC

The FIC methodology allows and encourages 'sharpened questions'. This has been
demonstrated in illustrations above where we could, for example, specify the covariate vector for an individual and proceed to find the best model for that single-focus
individual. In other words, we could perform a subject-wise model search. There are
however often situations where we wish a good model valid for a range of individuals
or situations simultaneously, say for a subgroup of the population. Here we develop an
'average-focussed information criterion' for dealing with such questions, to be termed
the AFIC.

Suppose in general terms that a parameter of interest $\mu(u)$ depends on some quantity
u that varies in the population being studied. For each u, Theorem 6.1 applies to the
subset model-based estimators $\widehat{\mu}_S(u)$, for which we have

$$\sqrt{n}\{\widehat{\mu}_S(u) - \mu_{\mathrm{true}}(u)\} \overset{d}{\to} \Lambda_S(u) = (\tfrac{\partial \mu(u)}{\partial \beta})^{\mathrm{t}} J_{00}^{-1} U' + \omega(u)^{\mathrm{t}}(\delta - G_S D),$$

where $\omega(u) = J_{10}J_{00}^{-1}\partial\mu(u)/\partial\theta - \partial\mu(u)/\partial\gamma$, $U' \sim N_p(0, J_{00})$, $D \sim N_q(\delta, Q)$, and G_S is the matrix $Q_S^0 Q^{-1}$. Now consider the loss function

$$L_n(S) = n\int\{\widehat{\mu}_S(u) - \mu_{\text{true}}(u)\}^2\,\mathrm{d}W_n(u), \qquad (6.24)$$

where W_n represents some relevant distribution of u values. Some examples include

- W_n represents the empirical distribution over one or more covariates;
- W_n gives equal weight to the deciles $u = 0.1, 0.2, \ldots, 0.9$ for estimation of the quantile distribution;
- W_n represents the real distribution of covariates in the population;
- W_n may be a distribution that focusses on some segment of the population, like when wishing a good logistic regression model aimed specifically at good predictive performance for white non-smoking mothers;
- somewhat more generally, W_n does not have to represent a distribution, but could be a weight function across the covariate space, with more weight associated with 'important' regions.

The AFIC that we shall derive can be applied when one wishes to consider average risk across both covariates and quantiles, for example.

Assume that W_n converges to a weight distribution W, or that it simply stays fixed, independent of sample size. It then follows that, under mild conditions,

$$L_n(S) \overset{d}{\to} L(S) = \int \Lambda_S(u)^2\,\mathrm{d}W(u).$$

The total averaged risk is obtained as the expected value of L_n, which converges to

$$\text{a-risk}(S, \delta) = \mathrm{E}\,L(S) = \int \mathrm{E}\Lambda_S(u)^2\,\mathrm{d}W(u).$$

Using earlier results, discussed in connection with (6.6), we may write

$$\begin{aligned}
\mathrm{E}\Lambda_S(u)^2 &= \tau_0(u)^2 + \omega(u)^{\mathrm{t}}\{(I_q - G_S)\delta\delta^{\mathrm{t}}(I_q - G_S)^{\mathrm{t}} + Q_S^0\}\omega(u) \\
&= \tau_0(u)^2 + \mathrm{Tr}\{(I_q - G_S)\delta\delta^{\mathrm{t}}(I_q - G_S)^{\mathrm{t}}\omega(u)\omega(u)^{\mathrm{t}}\} \\
&\quad + \mathrm{Tr}\{Q_S^0\omega(u)\omega(u)^{\mathrm{t}}\}.
\end{aligned}$$

For the limit risk, after subtracting $\int \tau_0(u)^2\,\mathrm{d}W(u)$ which does not depend on S, we find the expression

$$\text{a-risk}(S, \delta) = \mathrm{Tr}\{(I_q - G_S)\delta\delta^{\mathrm{t}}(I_q - G_S)^{\mathrm{t}}A\} + \mathrm{Tr}(Q_S^0 A), \qquad (6.25)$$

writing A for the matrix $\int \omega(u)\omega(u)^{\mathrm{t}}\,\mathrm{d}W(u)$. The limit version of the average-focussed information criterion, which generalises the bias-corrected FIC, is therefore defined as

$$\text{AFIC}_{\text{lim}}(S) = \max\{\mathrm{I}(S), 0\} + \mathrm{II}(S), \qquad (6.26)$$

where

$$I(S) = \text{Tr}\{(I_q - G_S)(DD^t - Q)(I_q - G_S)^t A\}$$
$$= D^t(I_q - G_S)^t A(I_q - G_S)D - \text{Tr}\{(Q - Q_S^0)A\},$$
$$II(S) = \text{Tr}(Q_S^0 A).$$

The model with lowest AFIC(S) score is selected. Note that the AFIC score is not simply the weighted average of FIC scores. The I(S) is an unbiased estimator of the first term in (6.25), and we truncate it to zero lest an estimate of a squared bias quantity should turn negative. In cases of non-truncation,

$$\text{AFIC}_{\lim}(S) = I(S) + II(S)$$
$$= D^t(I_q - G_S)^t A(I_q - G_S)D + 2\text{Tr}(Q_S^0 A) - \text{Tr}(QA),$$

where the last term is independent of S.

For real data situations, estimates are inserted for unknown parameters and one uses AFIC$(S) = \max\{\widehat{I}(S), 0\} + \widehat{II}(S)$, with

$$\widehat{I}(S) = \text{Tr}\{(I_q - \widehat{G}_{n,S})(\widehat{\delta\delta^t} - \widehat{Q}_n)(I_q - \widehat{G}_{n,S})^t \widehat{A}\},$$
$$\widehat{II}(S) = \text{Tr}(\widehat{Q}_{n,S}^0 \widehat{A}), \tag{6.27}$$

where $\widehat{A}$ is a sample-based estimate of the A matrix.

It is clear that the limit risk (6.25) and its estimator (6.27) depend crucially on the $q \times q$ matrix A, which again is dictated by the user-specified weight distribution W of positions u at which $\mu(u)$ is to be estimated. Writing

$$B = \int \begin{pmatrix} \partial\mu(u)/\partial\theta \\ \partial\mu(u)/\partial\gamma \end{pmatrix} \begin{pmatrix} \partial\mu(u)/\partial\theta \\ \partial\mu(u)/\partial\gamma \end{pmatrix}^t dW(u) = \begin{pmatrix} B_{00} & B_{01} \\ B_{10} & B_{11} \end{pmatrix},$$

we find

$$A = \int \left(J_{10}J_{00}^{-1}\tfrac{\partial\mu(u)}{\partial\theta} - \tfrac{\partial\mu(u)}{\partial\gamma}\right)\left(J_{10}J_{00}^{-1}\tfrac{\partial\mu(u)}{\partial\theta} - \tfrac{\partial\mu(u)}{\partial\gamma}\right)^t dW(u) \tag{6.28}$$
$$= J_{10}J_{00}^{-1}B_{00}J_{00}^{-1}J_{01} - J_{10}J_{00}^{-1}B_{01} - B_{10}J_{00}^{-1}J_{01} + B_{11}.$$

Similar expressions hold when an estimator $\widehat{A}$ is required.

Remark 6.4 AFIC versus AIC *
Different distributions of u values lead to different $q \times q$ matrices A, by the formulae above, and hence to different AFIC model selection criteria. The particular special case that W is concentrated in a single point corresponds to the (pointwise) FIC, for example. Suppose at the other side of the spectrum that $\mu(u)$ and the W distribution of u values are such that the B matrix above is equal to $J = J_{\text{wide}}$. Then A simplifies to $J_{11} - J_{10}J_{00}^{-1}J_{01}$,

which is the same as Q^{-1}, cf. (5.14), and (6.25) becomes

$$\text{a-risk}(S, \delta) = \delta^t(I_q - G_S)^t Q^{-1}(I_q - G_S)\delta + \text{Tr}(Q_S^0 Q^{-1})$$
$$= |S| + \delta^t(Q^{-1} - Q^{-1}Q_S^0 Q^{-1})\delta$$

(where we used that $\text{Tr}(G_S) = |S|$). It can be shown that this is equivalent to the limit risk function associated with AIC. Also, for this situation, the AFIC(S) scheme (disregarding the squared bias truncation modification) becomes equivalent to that of the AIC method itself; see Claeskens and Hjort (2008) for discussion. These considerations also suggest that the weighted FIC may be considered a useful generalisation of AIC, with context-driven weight functions different from the implied default of $B = J_{\text{wide}}$. ∎

Example 6.13 AFIC for generalised linear models

Consider the generalised linear model set-up of Section 6.6.6. We take the linear predictor $\mu(x, z) = x^t\beta + z^t\gamma$ as the focus parameter, and assume that this needs to be estimated across the observed population, with weights $w(x_i, z_i)$ dictating the degree of importance in different regions of the (x, z) space. The weighted average quadratic loss, on the scale of the linear predictor, is then of the form

$$L_n(S) = \sum_{i=1}^{n} w(x_i, z_i)\{\widehat{\mu}_S(x_i, z_i) - \mu_{\text{true}}(x_i, z_i)\}^2$$
$$= \sum_{i=1}^{n} w(x_i, z_i)(x_i^t\widehat{\beta}_S + z_{i,S}^t\widehat{\gamma}_S - x_i^t\beta_0 - z_i^t\delta/\sqrt{n})^2.$$

The weighted average-FIC can be computed as in (6.27) with the A matrix associated with the weighting scheme used here. Define the matrix

$$\Omega_n = n^{-1}\sum_{i=1}^{n} w(x_i, z_i)\begin{pmatrix} x_i \\ z_i \end{pmatrix}\begin{pmatrix} x_i \\ z_i \end{pmatrix}^t = n^{-1}\begin{pmatrix} X^t W X & X^t W Z \\ Z^t W X & Z^t W Z \end{pmatrix},$$

writing $W = \text{diag}\{w(x_1, z_1), \ldots, w(x_n, z_n)\}$. Then, by (6.28),

$$\widehat{A} = \widehat{J}_{10}\widehat{J}_{00}^{-1}\Omega_{n,00}\widehat{J}_{00}^{-1}\widehat{J}_{01} - \widehat{J}_{10}\widehat{J}_{00}^{-1}\Omega_{n,01} - \Omega_{n,10}\widehat{J}_{00}^{-1}\widehat{J}_{01} + \Omega_{n,11},$$

and the weighted FIC method proceeds by computing the AFIC(S) value as

$$\max[\text{Tr}\{(I_q - \widehat{G}_{n,S})(n\widehat{\gamma}\widehat{\gamma}^t - \widehat{Q}_n)(I_q - \widehat{G}_{n,S})^t\widehat{A}\}, 0] + \text{Tr}(\widehat{Q}_{n,S}^0\widehat{A}).$$

This expression can be simplified when the weights happen to be chosen such that $W = V$, that is,

$$w(x_i, z_i) = v_i = b''(\theta_i)r(x_i^t\beta + z_i^t\gamma)^2,$$

cf. (6.17). In that case $\Omega_n = J_n$ and $\widehat{A}$ becomes $\widehat{Q}_n^{-1}$. Hence, for this particular choice of weights, and assuming a positive estimated squared bias,

$$\text{AFIC}(S) = n\widehat{\gamma}^t(Q_n^{-1} - Q_n^{-1}Q_{n,S}^0 Q_n^{-1})\widehat{\gamma} + 2|S| - q.$$

This again points to the link with AIC, as per Remark 6.4. The AFIC method, when using the GLM variances as weights, is essentially large-sample equivalent to the AIC method. The truncation of a negative estimated squared bias can lead to different results, however. ■

Example 6.14 AFIC for CH$_4$ concentrations

As an application of a weighted FIC model search, we consider the CH$_4$ data, which are atmospheric CH$_4$ concentrations derived from flask samples collected at the Shetland Islands of Scotland. The response variables are monthly values expressed in parts per billion by volume (ppbv), the covariate is time. Among the purposes of studies related to these data are the identification of any time trends, and to construct CH$_4$ predictions for the coming years. In total there are 110 monthly measurements, starting in December 1992 and ending in December 2001. The regression variable x = time is rescaled to the (0, 1) interval. We use an orthogonal polynomial estimator in the normal regression model $Y_i = \mu(x_i) + \varepsilon_i$ with

$$\mu(x) = \mathrm{E}(Y \mid X = x) = \gamma_0 + \sum_{j=1}^{m} \gamma_j c_j(x),$$

where the $c_j(\cdot)$ represent orthogonalised polynomials of degree j, to facilitate computations. When m is varied, this defines a sequence of nested models. This setting fits into the regression context of the previous sections when defining $z_j = c_j(x)$. We wish to select the best order m. In our modelling efforts, we let m be any number between 1 and 15 (the wide model). The narrow model corresponds to

$$\mu(x) = \gamma_0 + \gamma_1 c_1(x).$$

A scatterplot of the data is shown in Figure 6.6(a). We applied the AFIC(S) method with equal weights $w_i = 1$, for the nested sequence of $S = \{1, \ldots, m\}$ with $m = 1, \ldots, 15$, and found that the best model is for $m = 2$; see Figure 6.6(b).

Another set of weights which makes sense for variables measured in time, is that which gives more weight to more recent measurements. As an example we used the weighting scheme i/n (for $i = 1, \ldots, n = 110$), and found in this particular case the same FIC-selected model, namely the model with truncation point $m = 2$. ■

6.10 A Bayesian focussed information criterion *

Our aim in this section is to put a Bayesian twist on the FIC story, without losing the F of the focus. Thus consider the framework of submodels and candidate estimators $\widehat{\mu}_S$ for a given focus parameter $\mu = \mu(\theta, \gamma)$. Which submodel S would a Bayesian select as best? When the question is focussed in this fashion the answer is not necessarily 'the one selected by BIC'.

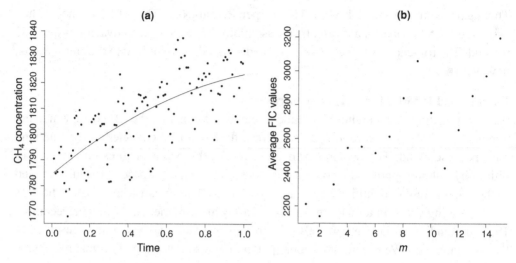

Fig. 6.6. (a) Scatterplot of the CH_4 data, along with the estimated mean curve for the $m = 2$ model. (b) The AFIC values, for the nested sequence of models, with equal weights $w_i = 1$.

The FIC was derived using the squared error loss function, which we now write as

$$L_n(S) = L_n(S, \theta_0, \gamma_0 + \delta/\sqrt{n}) = n(\widehat{\mu}_S - \mu_{\text{true}})^2.$$

This is the random loss incurred if the statistician selects submodel S, and uses its asso ciated estimator of μ, when the true state of affairs is that of the local model alternatives (6.3). Then, as sample size increases, the random loss converges in distribution, by Theorem 6.1,

$$L_n(S, \theta_0, \gamma_0 + \delta/\sqrt{n}) \xrightarrow{d} L(S, \delta) = \{\Lambda_0 + \omega^{\text{t}}(\delta - G_S D)\}^2.$$

The risk function (expected loss, as a function of the underlying parameters) in the limit experiment is

$$r(S, \delta) = \mathrm{E}\, L(S, \delta) = \omega^{\text{t}}(I - G_S)\delta\delta^{\text{t}}(I - G_S)^{\text{t}}\omega + \omega^{\text{t}}G_S Q G_S^{\text{t}}\omega, \qquad (6.29)$$

where the term τ_0^2, common to all candidate models, is subtracted out. As we have seen, the FIC evolves from this by estimating $\delta\delta^{\text{t}}$ with $DD^{\text{t}} - Q$ (modulo a truncation lest the squared bias estimate becomes negative), etc.

Bayesian methods that partly parallel the development that led to the FIC evolve when there is prior information about δ, the q-dimensional parameter that measures departure from the narrow model. Suppose δ has a prior π, with finite second moments, so that $\int \delta\delta^{\text{t}} \, \mathrm{d}\pi(\delta) = B$ is finite. The Bayes risk is then

$$\text{b-risk}(S) = \int r(S, \delta) \, \mathrm{d}\pi(\delta) = \omega^{\text{t}}(I - G_S)B(I - G_S)^{\text{t}}\omega + \omega^{\text{t}}\pi_S^{\text{t}} Q_S \pi_S \omega.$$

Of particular interest is the canonical prior $\delta \sim N_q(0, v^2 Q)$. This corresponds to an isotropic prior for the transformed parameter $a = Q^{-1/2} \delta$ in the transformed experiment where $Z = Q^{-1/2} D$ is $N_q(a, I_q)$, with v dictating the level of concentration around the centre point $\delta = 0$. The $\delta = 0$ position is the prior mean point since the narrow model assumes this value. For this prior, the total Bayes risk becomes

$$\text{b-risk}(S) = v^2 \omega^t (I - G_S) Q (I - G_S)^t \omega + \omega^t \pi_S^t Q_S \pi_S \omega$$
$$= v^2 \omega^t Q \omega + (1 - v^2) \omega^t \pi_S^t Q_S \pi_S \omega.$$

Interestingly, this is easily minimised over S, with a result depending on whether v is above or below 1:

$$\widehat{S} = \begin{cases} \text{wide} & \text{if } v \geq 1, \\ \text{narr} & \text{if } v < 1. \end{cases}$$

This is the rather simple strategy that minimises the overall Bayesian risk.

The real Bayes solution to the focussed model selection problem is not quite that simplistic, however, and is found by minimising expected loss given data. From

$$\begin{pmatrix} \delta \\ D \end{pmatrix} \sim N_{2q} \left(\begin{pmatrix} 0 \\ 0 \end{pmatrix}, \begin{pmatrix} v^2 Q & v^2 Q \\ v^2 Q & (v^2 + 1) Q \end{pmatrix} \right)$$

follows

$$(\delta \mid D) \sim N_q(\rho D, \rho Q), \quad \text{where } \rho = v^2 / (1 + v^2). \tag{6.30}$$

Computing the expected loss given data, then, again subtracting out the immaterial constant term τ_0^2, gives

$$E\{L(S, \delta) \mid D\} = E[\{\omega^t (\delta - G_S D)\}^2 \mid D]$$
$$= \omega^t (\rho I - G_S) D D^t (\rho I - G_S)^t \omega + \rho \omega^t Q \omega.$$

Minimising expected loss given data therefore amounts to finding the S for which

$$|\omega^t (\rho I - G_S) D| = |\rho \omega^t D - \omega^t G_S D|$$

is smallest. That is, the best S is found by getting $\omega^t G_S D$ as close to the full unrestricted Bayes solution $\rho \omega^t D$ as possible. For small v, with ρ close to zero, this will be $S = \text{narr}$; while for large v, with ρ close to one, this will be $S = \text{wide}$. For intermediate ρ values other submodels will be preferred.

It is useful to supplement this method, which requires a fully specified prior for δ, with an empirical Bayes version that inserts a data-dictated value for v, and hence ρ. Since the statistic $D^t Q^{-1} D$ has mean $(v^2 + 1) q$, we suggest using $(v^*)^2 = \max(D^t Q^{-1} D / q - 1, 0)$ as empirical Bayes-type estimate of the prior spread parameter v^2. This leads to

$$\rho^* = \frac{(v^*)^2}{1 + (v^*)^2} = \begin{cases} 1 - q/(D^t Q^{-1} D) & \text{if } D^t Q^{-1} D \geq q, \\ 0 & \text{if } D^t Q^{-1} D \leq q. \end{cases}$$

We have reached a Bayesian focussed information criterion: compute

$$\text{BFIC}_{\text{lim}}(S) = |\rho^* \omega^t D - \omega^t G_S D| \qquad (6.31)$$

for each candidate model S, and select in the end the model with smallest such score. We observe that for $D^t Q^{-1} D \leq q$, the narrow model is best, while the wide model is preferred for large values of $D^t Q^{-1} D$. We term this Bayesian information criterion 'focussed' since it has been constructed with the specific focus parameter μ in mind; different foci lead to different versions of BFIC and potentially to different model rankings.

The derivation above used the framework of the limit experiment, where only δ is unknown and where $D \sim N_q(\delta, Q)$. For finite-sample data, we as usual insert parameter estimates where required, and define

$$\text{BFIC}(S) = |\widehat{\rho}\widehat{\omega}^t D_n - \widehat{\omega}^t \widehat{G}_S D_n|,$$

where $D_n = \widehat{\delta}$ and $\widehat{\rho} = \max\{1 - q/(D_n^t \widehat{Q}^{-1} D_n), 0\}$.

The BFIC of (6.31) emerged by minimising the expected mean squared error loss given data, as per the general Bayesian paradigm (and then inserting an estimate for the ρ parameter, in an empirical Bayes modus). This scheme may be followed also with other relevant loss functions, as we now briefly demonstrate by finding two more Bayesian-type focussed model selection criteria, the $\pi^*(S)$ method and the $\lambda^*(S)$ method.

The first of these starts from considering the loss function

$$L_{n,1}(S, \theta_0, \gamma_0 + \delta/\sqrt{n}) = \begin{cases} 0 & \text{if } r_n(S, \delta) \text{ is the lowest of all } r_n(S', \delta), \\ 1 & \text{if there is another } S' \text{ with lower } r_n(S', \delta), \end{cases}$$

writing $r_n(S, \delta)$ for the more complete $r_n(S, \theta_0, \gamma_0 + \delta/\sqrt{n})$ above. This random loss tends in distribution to

$$L_1(S, \delta) = I\{r(S, \delta) \text{ is not the lowest risk}\} = 1 - I\{r(S, \delta) \text{ is the smallest}\},$$

with $r(S, \delta)$ as in (6.29). The Bayes solution, working with this loss function instead of squared error loss, emerges by minimising expected loss given data, which is the same as finding the submodel S that maximises the probabilities

$$\pi^*(S) = P\{r(S, \delta) < \text{all the other } r(S', \delta) \mid D\}. \qquad (6.32)$$

With the $N_q(0, \nu^2 Q)$ prior for δ, for which the posterior distribution is $N_q(\rho D, \rho Q)$, we may actually compute each of the $\pi^*(S)$ probabilities via simulation of δ vectors from the $N_q(\rho D, \rho Q)$ distribution; see the application of Example 6.15. The candidate model with highest $\pi^*(S)$ is selected.

We should not confuse the probabilities (6.32) with prior or posterior model probabilities (such are involved in Bayesian model averaging, see Chapter 7). For the application of Example 6.15 we find that the $S = \{1, 2\}$ submodel, among eight competing candidates, has $\pi^*(S)$ equal to 0.358, for example. This is not the probability that $\{1, 2\}$ is 'the correct model', but rather the posterior probability that the still unknown parameter $\delta = (\delta_1, \delta_2, \delta_3)^t$ in that situation lies in the well-defined region $R_{1,2}$ of δ values for which $r(\{1, 2\}, \delta)$ is the smallest of the eight risk functions involved.

The second recipe uses the loss function

$$L_{n,2}(S, \theta_0, \gamma_0 + \delta/\sqrt{n}) = r_n(S, \delta) - \min_{S'} r_n(S', \delta).$$

The previous loss function penalises each non-optimal choice with the same sword, whereas the present variant perhaps more naturally concentrates on the actual distance in risk to the lowest possible risk. As above, the random loss $L_{n,2}(S)$ has a limit in distribution, namely $L_2(S, \delta) = r(S, \delta) - \min_{S'} r(S', \delta)$. The Bayes scheme is again to minimise the posterior expected loss, i.e.

$$\lambda^*(S) = \mathrm{E}[\{r(S, \delta) - \min_{S'} r(S', \delta)\} \mid D]. \tag{6.33}$$

For the prior $\delta \sim N_q(0, \nu^2 Q)$ we again use (6.30) when evaluating $\lambda^*(S)$, which we do by simulation.

Example 6.15 Counting bird species

This illustration of the BFIC is concerned with the data on bird species abundance on islands outside Ecuador, taken from Hand *et al.* (1994, case #52). For the present purposes let Y denote the number of species recorded on an island's paramos, along with

- the distance x_1 from Ecuador (in thousands of km, ranging from 0.036 to 1.380);
- the island's area z_1 (in thousands of square km, ranging from 0.03 to 2.17);
- the elevation z_2 (in thousands of m, ranging from 0.46 to 2.28); and
- distance z_3 from the nearest island (in km, ranging from 5 to 83).

Such data are recorded for each of 14 islands. The ecological questions relating to such data include identifications of those background parameters that may contribute more significantly to the number of species in different island habitats. Such knowledge may be used both for estimating bird populations on other islands and for saving threatened species through intervention.

The model takes Y to be Poisson with parameter $\xi = \exp(\beta_0 + \beta_1 x_1 + \gamma_1 z_1 + \gamma_2 z_2 + \gamma_3 z_3)$. We treat x_1 as a protected covariate, and ask which of the three extra covariates should be included in a good model. The focus parameter for this example is ξ, or equivalently the linear predictor $\log \xi$, at centre position in the covariate space, which here corresponds to $x_1 = 0.848$ and $(z_1, z_2, z_3) = (0.656, 1.117, 36.786)$. We find $D_n = (0.879, 0.725, 0.003)^t$ and $D_n^t \widehat{Q}_n^{-1} D_n = 20.471$, which also leads to $\widehat{\nu} = 2.413$, to $\widehat{\rho} = 0.853$, and to $\omega = (0.216, 0.131, 1.265)^t$.

Table 6.4 summarises our findings, using the model selection criteria BFIC, $\pi^*(S)$ and $\lambda^*(S)$. For comparison also AIC and the FIC are included. There is reasonable

Table 6.4. *Birds abundance data: selection of covariates z_1, z_2, z_3 via the BFIC and the $\pi^*(S)$ and $\lambda^*(S)$ criteria. Here '1,3' indicates the model with covariates z_1, z_3, and so on. For each model selection criterion the four best models are indicated.*

Submodel	AIC		FIC		BFIC		$\pi^*(S)$		$\lambda^*(S)$	
0	−112.648		0.0789		0.2459		0.0004		60.148	
1	−96.834	1	0.0042	4	0.0064	1	0.2172	2	1.086	4
2	−102.166		0.0116		0.0587		0.0418		7.479	
3	−113.619		0.0735		0.2359		0.0008		56.040	
1,2	−96.844	2	0.0041	1	0.0412	4	0.3583	1	0.232	1
1,3	−98.817	3	0.0042	3	0.0066	2	0.1737	3	1.076	3
2,3	−103.193		0.0089		0.0411	3	0.0704		5.146	
1,2,3	−98.768	4	0.0041	2	0.0422		0.1375	4	0.234	2

agreement about which models are good and which are not good, though there are some discrepancies about the top ranking. For this example we simulated 100,000 δ vectors from the posterior distribution $N_3(0.853\, D_n, 0.853\, \widehat{Q}_n)$, for estimating each of the $\pi^*(S)$ probabilities and $\lambda^*(S)$ expectations accurately. ■

6.11 Notes on the literature

The focussed information criterion is introduced in Claeskens and Hjort (2003), building on theoretical results in local misspecified models as in Hjort and Claeskens (2003). Hand and Vinciotti (2003) also advocate using a 'focus' when constructing a model, and write 'in general, it is necessary to take the prospective use of the model into account when building it'. Other authors have also expressed opinions echoing or supporting the 'focussed view'; see e.g. Hansen (2005) who work with econometric time series, Vaida and Blanchard (2005) on different selection aims in mixed models, and the editorial Longford (2005).

Claeskens *et al.* (2006) construct other versions of the FIC that do not aim at minimising the estimated mse, but rather allowing risk measures such as the one based on L_p error ($p \geq 1$), with $p = 1$ leading to a criterion that minimises the mean absolute error. When prediction of an event is important, they construct a FIC using the error rate as a risk measure to be minimised. In these situations it is typically not possible to construct unbiased estimators of limiting risk, as we have done in this chapter for the squared error loss function. For more on the averaged version of the FIC, see Claeskens and Hjort (2008). Applications to order selection in time series can be found in Hansen (2005) and Claeskens *et al.* (2007). The FIC for Cox models worked with in Section 6.8 has parallels in Hjort (2007b), where FIC methods are worked out for Aalen's nonparametric linear hazard regression model. Lien and Shrestha (2005) use the focussed information criterion for estimating optimal hedge ratios.

Exercises

6.1 *Submatrices as matrix products:* Define π_S as the projection matrix that maps $v = (v_1, \ldots, v_q)^t$ to the subvector $\pi_S v = v_S$ of components v_j with $j \in S$.

(a) Verify that $(0\ 1\ 0 \ldots 0)v = v_2$, while

$$\begin{pmatrix} 1 & 0 & 0 & 0 & \ldots & 0 \\ 0 & 0 & 1 & 0 & \ldots & 0 \end{pmatrix} v = \begin{pmatrix} v_1 \\ v_3 \end{pmatrix} = v_{\{1,3\}}.$$

Thus the matrix π_S is of size $|S| \times q$ with $|S|$ being the size of S.

(b) Show that

$$J_S = \begin{pmatrix} J_{00} & J_{01,S} \\ J_{10,S} & J_{11,S} \end{pmatrix} = \begin{pmatrix} J_{00} & J_{01}\pi_S^t \\ \pi_S J_{10} & \pi_S J_{11}\pi_S^t \end{pmatrix}.$$

(c) Show that the inverse $J_S^{-1} = (J_S)^{-1}$ of this matrix is likewise partitioned with blocks

$$J^{11,S} = (\pi_S Q^{-1}\pi_S^t)^{-1} = Q_S,$$
$$J^{01,S} = -J_{00}^{-1} J_{01}\pi_S^t Q_S,$$
$$J^{00,S} = J_{00}^{-1} + J_{00}^{-1} J_{01}\pi_S^t Q_S \pi_S J_{10} J_{00}^{-1}.$$

(d) Use this to show that $G_S = \pi_S^t Q_S \pi_S Q^{-1}$, and that $\mathrm{Tr}(G_S) = |S|$.

(e) Show finally that $\omega^t G_S Q G_S^t \omega = \omega^t Q_S^0 \omega = \omega_S^t Q_S \omega_S$.

6.2 *Estimating* $\delta\delta^t$: From $D \sim N_q(\delta, Q)$, show that DD^t has mean $\delta\delta^t + Q$. Similarly, show that $(\omega^t D)^2$ has mean $(\omega^t \delta)^2 + \omega^t Q\omega$. Study some competitors to the unbiased estimator $(\omega^t D)^2 - \omega^t Q\omega$, e.g. of the form $a(\omega^t D)^2 - b\omega^t Q\omega$, and investigate what values of (a, b) can be expected to do as well as or better than $(1, 1)$.

6.3 *Models with no common parameters:* Assume that no variables are common to all models, that is, there is no θ parameter in the model, only the γ part remains. We place a 'tilde' above matrices to distinguish them from similar matrices in the (θ, γ) situation. For the following points the local misspecification setting is assumed to hold.

(a) Show that the matrix $\widetilde{J}_{n,\mathrm{wide}}^{-1}$ is the same as Q_n (as defined for the (θ, γ) case) and $\widetilde{\omega} = -\frac{\partial\mu}{\partial\gamma}(\gamma_0)$.

(b) Show that $\mu(\widehat{\gamma}_S) - \mu(\gamma_{\mathrm{true}})$ may be represented as

$$(\tfrac{\partial\mu(\beta,\gamma)}{\partial\gamma_S})^t(\widehat{\gamma}_S - \gamma_0) + (\tfrac{\partial\mu(\beta,\gamma)}{\partial\gamma})^t(-\delta/\sqrt{n}) + o_P(1/\sqrt{n}).$$

(c) Show first that

$$\sqrt{n}(\widehat{\mu}_S - \mu_{\mathrm{true}}) \xrightarrow{d} \Lambda_S \sim N(\widetilde{\omega}^t(I_q - \widetilde{G}_S)\delta, \widetilde{\omega}^t(\widetilde{J}_S^{-1})^0\widetilde{\omega}),$$

and go on to prove that the limiting mean squared error corresponds to

$$\mathrm{mse}(S, \delta) = \widetilde{\omega}^t(I_q - \widetilde{G}_S)\delta\delta^t(I_q - \widetilde{G}_S^t)\widetilde{\omega} + \widetilde{\omega}^t(\widetilde{J}_S^{-1})^0\widetilde{\omega}.$$

(d) Proceed by inserting in the mean squared error the estimator $\widehat{\delta\delta^t} - \widetilde{J}^{-1}$ for $\delta\delta^t$ and replacing unknown parameters by their maximum likelihood estimators in the full model to verify that when we substitute $\widetilde{\omega}$, $\widetilde{G}_S$ and $\widetilde{J}$ for ω, G_S and J, the FIC formula (6.1) remains valid for the case of no common parameters.

6.4 *The 10,000-m:* We use the speedskating models discussed in Section 5.6 and refer to Exercise 5.2. Construct a FIC model selector for the focus parameter $\rho = \rho(x)$, the expected ratio Y/x for given x, where x is the 5000-m time and Y the 10,000-m time of a top skater, and find the best model, for different types of skaters. Implement a version of the AFIC to select the best model for an interval of x values.

6.5 *The 5000-m:* Repeat Exercise 6.4 with the data of the Ladies Adelskalenderen where Y is the time on the 5000-m and x the time on the 3000-m.

6.6 *FIC with polynomial regression:* Consider the model selection context of Example 6.5. Assume that the variables x_is have average equal to zero, and let $m_j = n^{-1} \sum_{i=1}^n x_i^j$ for $j = 2, 3, 4, 5, 6$. The focus parameter is $\mu(x) = \mathrm{E}(Y \mid x)$. Show first that ω has the two components $\omega_2 = m_2 - x^2 + (m_3/m_2)x$ and $\omega_3 = m_3 - x^3 + (m_4/m_2)x$, and that

$$
Q_n = \sigma^2 \begin{pmatrix} m_4 - m_2^2 - m_3^2/m_2 & m_5 - m_2 m_3 - m_3 m_4/m_2 \\ m_5 - m_2 m_3 - m_3 m_4/m_2 & m_6 - m_3^2 - m_4^2/m_2 \end{pmatrix}^{-1}.
$$

Next show that the four FIC values may be written

$$
\begin{aligned}
\widehat{\mathrm{FIC}}_0(x) &= (\omega_2 \widehat{\delta}_2 + \omega_3 \widehat{\delta}_3)^2, \\
\widehat{\mathrm{FIC}}_2(x) &= \{(\omega_3 - \omega_2 Q_{n,1} Q_n^{01}) \widehat{\delta}_3\}^2 + 2\omega_2^2 Q_{n,1}, \\
\widehat{\mathrm{FIC}}_3(x) &= \{(\omega_2 - \omega_3 Q_{n,2} Q_n^{10}) \widehat{\delta}_2\}^2 + 2\omega_3^2 Q_{n,2}, \\
\widehat{\mathrm{FIC}}_{23}(x) &= 2\omega^t Q_n \omega.
\end{aligned}
$$

6.7 *Risk functions for estimating a squared normal mean:* Consider the problem of Section 6.4, where the squared mean $\kappa = \xi^2$ is to be estimated based on having observed a single $X \sim \mathrm{N}(\xi, \sigma^2)$, where σ is known. Risk functions below are with respect to squared error loss.

(a) Show that the maximum likelihood estimator is $\widehat{\kappa}_1 = X^2$, and that its risk function is $\sigma^4(3 + 4\xi^2/\sigma^2)$.

(b) Show that the uniformly minimum variance unbiased estimator is $\widehat{\kappa}_2 = X^2 - \sigma^2$, and that its risk function is $\sigma^4(2 + 4\xi^2/\sigma^2)$.

(c) To find risk expressions for the modified estimator

$$
\widehat{\kappa}_3 = (X^2 - \sigma^2) I\{|X| \geq \sigma\}
$$

of (6.9), introduce first the functions $A_j(a, b) = \int_a^b y^j \phi(y)\, \mathrm{d}y$, and show that

$$
\begin{aligned}
A_1(a, b) &= \phi(a) - \phi(b), \\
A_2(a, b) &= A_0(a, b) + a\phi(a) - b\phi(b), \\
A_3(a, b) &= 2A_1(a, b) + a^2\phi(a) - b^2\phi(b), \\
A_4(a, b) &= 3A_3(a, b) + a^3\phi(a) - b^3\phi(b).
\end{aligned}
$$

Find expressions for the first and second moments of $\widehat{\kappa}_3$ in terms of the $A_j(a, b)$ functions, with $(a, b) = (-1 - \xi/\sigma, 1 - \xi/\sigma)$, and use this to calculate the risk function of $\widehat{\kappa}_3$. In particular, show that the risk improvement over the unbiased estimator is $2\phi(1) + 2\{1 - \Phi(1)\} = 0.8012$ at the origin.

6.8 *More general submodels:* Consider the situation of the football example in Section 6.6.4. One of the submodels sets two parameters equal ($b = c$), rather than including or excluding either parameter. Let A be an invertible $q \times q$ matrix, where subset models are considered in terms of the new parameterisation $\Upsilon = A(\gamma - \gamma_0)$, or $\gamma = \gamma_0 + A^{-1}\Upsilon$.

(a) Show that the new model has score vector

$$\begin{pmatrix} \widetilde{U}(y) \\ \widetilde{V}(y) \end{pmatrix} = \begin{pmatrix} U(y) \\ (A^{-1})^{t}V(y) \end{pmatrix},$$

leading to the information matrix

$$\widetilde{J}_{\text{wide}} = \begin{pmatrix} J_{00} & J_{01}A^{-1} \\ (A^{-1})^{t}J_{10} & (A^{-1})^{t}J_{11}A^{-1} \end{pmatrix}.$$

(b) Show that the Q matrix in the new model is related to that of the original model via $\widetilde{Q} = AQA^{t}$, that $\widetilde{\omega} = (A^{-1})^{t}\omega$, and $\widetilde{D}_n = \sqrt{n}\widehat{\Upsilon} = A\sqrt{n}(\widehat{\gamma} - \gamma_0)$ has limit distribution AD.

(c) In the submodel S, the focus parameter $v(\theta, \Upsilon) = \mu(\theta, \gamma_0 + A^{-1}\Upsilon)$ is estimated by $\widehat{v}_S = v(\widehat{\theta}_S, \widehat{\Upsilon}_S, 0_{S^c})$, with maximum likelihood estimators in the (θ, Υ_S) model. With $\widetilde{G}_S = \widetilde{Q}_S^0 \widetilde{Q}^{-1}$, prove that

$$\sqrt{n}(\widehat{v}_S - v_{\text{true}}) \xrightarrow{d} \widetilde{\Lambda}_S = \Lambda_0 + \widetilde{\omega}^{t}(A\delta - \widetilde{G}_S AD) = \Lambda_0 + \omega^{t}(\delta - A^{-1}\widetilde{G}_S AD).$$

(d) Show that the limit distribution in (c) has mean $\omega^{t}(I_q - A^{-1}\widetilde{G}_S A)\delta$ and variance

$$\tau_0^2 + \omega^{t}A^{-1}\widetilde{G}_S AQA^{t}\widetilde{G}_S^{t}(A^{-1})^{t}\omega = \tau_0^2 + \omega^{t}A^{-1}\widetilde{Q}_S^0(A^{-1})^{t}\omega.$$

Obtain the mean squared error, and as a next step an expression for the FIC that generalises the FIC of (6.1) where $A = I_q$.

7

Frequentist and Bayesian model averaging

In model selection the data are used to select one of the models under consideration. When a parameter is estimated inside this selected model, we term it *estimation-post-selection*. An alternative to selecting one model and basing all further work on this one model is that of *model averaging*. This amounts to estimating the quantity in question via a number of possible models, and forming a weighted average of the resulting estimators. Bayesian model averaging computes posterior probabilities for each of the models and uses those probabilities as weights. The methods we develop accept quite general random weighting of models, and these may for example be formed based on the values of an information criterion such as AIC or the FIC.

7.1 Estimators-post-selection

Let $\mathcal{A}$ be the collection of models S considered as possible candidates for a final model. The estimator of a parameter μ inside a model S is denoted $\widehat{\mu}_S$. A model selection procedure picks out one of these models as the final model. Let S_{aic} be the model selected by AIC, for example. The final estimator in the selected model is $\widehat{\mu}_{\text{aic}} = \widehat{\mu}_{S_{\text{aic}}}$. We may represent this estimator-post-selection as

$$\widehat{\mu}_{\text{aic}} = \sum_{S \in \mathcal{A}} I\{S = S_{\text{aic}}\}\widehat{\mu}_S, \tag{7.1}$$

that is, as a weighted sum over the individual candidate estimators $\widehat{\mu}_S$, in this case with weights of the special form 1 for the selected model and 0 for all other models. We shall more generally investigate estimators of the model average form

$$\widehat{\mu} = \sum_{S \in \mathcal{A}} c(S)\widehat{\mu}_S, \tag{7.2}$$

where the weights $c(S)$ sum to one and are allowed to be random, as with the estimator-post-selection class. A useful metaphor might be to look at the collection of estimates as having come from a list of statisticians; two statisticians (using the same data) will, via submodels S and S', deliver (correlated, but different) estimates $\widehat{\mu}_S$ and $\widehat{\mu}_{S'}$, etc. With model selection one attempts to find the best statistician for the given purpose, and then relies on his answer. With model averaging, on the other hand, one actively seeks a smooth compromise, perhaps among those statisticians that appear to be best equipped for the task.

Studying the behaviour of estimators of the type (7.2) is difficult, particularly when the weights are determined by the data, for example through a model selection procedure like AIC. Thus a proper study of the distribution of $\widehat{\mu}$ needs to take the randomness of the weights into account. This is different from applying classical distribution theory conditional on the selected model. We term the weights used in (7.1) 'AIC indicator weights'. Similarly, other model selection criteria define their own set of weights, leading to BIC indicator weights, FIC indicator weights, and so on. Since each of these sets of weights points to a possibly different model, the resulting distribution of the post-selection-estimator $\widehat{\mu}$ will be different for each case. In Section 7.3 a 'master theorem' is provided that describes the large-sample behaviour of estimators of the (7.2) form. In particular, this theorem provides the limit distribution of estimators-post-selection.

7.2 Smooth AIC, smooth BIC and smooth FIC weights

Model averaging provides a more general way of weighting models than just by means of indicator functions. Indeed, the weights $\{c(S): S \in \mathcal{A}\}$ of (7.2) can be any values summing to 1. They will usually be taken to be non-negative, but even estimators like $(1 + 2\widehat{w})\widehat{\mu}_1 - \widehat{w}(\widehat{\mu}_2 + \widehat{\mu}_3)$, with an appropriately constructed $\widehat{w} \geq 0$, are worthy of consideration. Here we give some examples of weighting schemes.

Example 7.1 Smooth AIC weights
Buckland *et al.* (1997) suggest using weights proportional to $\exp(\frac{1}{2}\mathrm{AIC}_S)$, where AIC_S is the AIC score for candidate model S. This amounts to

$$c_{\mathrm{aic}}(S) = \frac{\exp(\frac{1}{2}\mathrm{AIC}_S)}{\sum_{S' \in \mathcal{A}} \exp(\frac{1}{2}\mathrm{AIC}_{S'})} = \frac{\exp(\frac{1}{2}\Delta_{\mathrm{AIC},S})}{\sum_{S' \in \mathcal{A}} \exp(\frac{1}{2}\Delta_{\mathrm{AIC},S'})},$$

where $\Delta_{\mathrm{AIC},S} = \mathrm{AIC}_S - \max_{S'} \mathrm{AIC}_{S'}$. The point of subtracting the maximum AIC value is merely computational, to avoid numerical problems with very high or very low arguments inside the exp function. The sum in $c_{\mathrm{aic}}(S)$ extends over all models of interest. In some applications $\mathcal{A}$ would be the set of all subsets of a set of variables, while in other cases $\mathcal{A}$ consists of a sequence of nested models. Other examples are possible as well. We refer to $c_{\mathrm{aic}}(\cdot)$ as the smooth AIC weights, to distinguish them from the indicator AIC weights in post-AIC model selection. ∎

Example 7.2 Smooth BIC weights
In Bayesian model averaging, posterior probabilities for the models S are used as weights. An approximation to the posterior probability of model S being correct is given by the

BIC; see Chapter 3. The smooth BIC weights are defined as

$$c_{\text{bic}}(S) = \frac{\exp(\frac{1}{2}\text{BIC}_S)}{\sum_{S' \in \mathcal{A}} \exp(\frac{1}{2}\text{BIC}_{S'})}.$$

As for the smooth AIC weights, for numerical stability it is advisable to use $\Delta_{\text{BIC},S} = \text{BIC}_S - \max_{S'} \text{BIC}_{S'}$ instead of BIC_S. ■

Example 7.3 Smooth FIC weights
It is also attractive to smooth across estimators using the information of the FIC values. Define

$$c_{\text{fic}}(S) = \exp\left(-\tfrac{1}{2}\kappa \frac{\text{FIC}_S}{\widehat{\omega}^{\text{t}}\widehat{Q}\widehat{\omega}}\right) \bigg/ \sum_{S' \in \mathcal{A}} \exp\left(-\tfrac{1}{2}\kappa \frac{\text{FIC}_{S'}}{\widehat{\omega}^{\text{t}}\widehat{Q}\widehat{\omega}}\right) \quad \text{with } \kappa \geq 0.$$

Here κ is an algorithmic parameter, bridging from uniform weighting (κ close to zero) to the hard FIC (which is the case of large κ). The factor $\widehat{\omega}^{\text{t}}\widehat{Q}\widehat{\omega}$ in the scaling for κ is the constant risk of the minimax estimator $\widehat{\delta} = D$. The point of the scaling is that κ values used in different contexts now can be compared directly. One may show (see Exercise 7.1) that for the one-dimensional case $q = 1$, the value $\kappa = 1$ makes the weights of the smoothed FIC agree with those for the smoothed AIC. ■

Example 7.4 Interpolating between narrow and wide
Instead of including a large number of candidate models, we interpolate between the two extreme cases 'narrow' and 'wide' only,

$$\mu^* = (1 - \widehat{w})\widehat{\mu}_{\text{narr}} + \widehat{w}\widehat{\mu}_{\text{wide}}, \tag{7.3}$$

where again $\widehat{w}$ is allowed to depend on data. Using $\widehat{w} = I\{D_n^{\text{t}}\widehat{Q}^{-1}D_n \geq 2q\}$, for example, in the framework of Chapter 6, is large-sample equivalent to determining the choice between narrow and wide model by AIC. ■

Example 7.5 Averaging over linear regression models
Assume that response observations Y_i have concomitant regressors $x_i = (x_{i,1}, \ldots, x_{i,p})^{\text{t}}$ and possibly a further subset of additional regressors $z_i = (z_{i,1}, \ldots, z_{i,q})^{\text{t}}$. Which subset of these should best be included, and which ways are there of averaging over all models? The framework is that of $Y_i = x_i^{\text{t}}\beta + z_i^{\text{t}}\gamma + \varepsilon_i$ for $i = 1, \ldots, n$, where the ε_i are independent $N(0, \sigma^2)$ and an intercept may be included in the x_i variables. Suppose that the variables z_i have been made orthogonal to the x_i, in the sense that $n^{-1}\sum_{i=1}^{n} x_i z_i^{\text{t}} = 0$. Then

$$J_{n,\text{wide}} = \sigma^{-2}\text{diag}(2, \Sigma_{00}, \Sigma_{11}) \quad \text{with} \quad J_{n,\text{wide}}^{-1} = \sigma^2\text{diag}\left(\tfrac{1}{2}, \Sigma_{00}^{-1}, \Sigma_{11}^{-1}\right),$$

where $\Sigma_{00} = n^{-1} \sum_{i=1}^{n} x_i x_i^t$ and $\Sigma_{11} = n^{-1} \sum_{i=1}^{n} z_i z_i^t$. For the most important case of $\mu = E(Y \mid x, z)$ at some given location (x, z), model averaging estimators take the form

$$\widehat{\mu}(x, z) = \sum_{S \in \mathcal{A}} c(S \mid D_n)(x^t \widehat{\beta}_S + z_S^t \widehat{\gamma}_S) = x^t \widehat{\beta}^* + z^t \widehat{\gamma}^*.$$

The coefficients $\widehat{\beta}^* = \sum_{S \in \mathcal{A}} c(S \mid D_n) \widehat{\beta}_S$ and $\widehat{\gamma}^* = \sum_{S \in \mathcal{A}} c(S \mid D_n) \pi_S^t \widehat{\gamma}_S$ are in this case nonlinear regression coefficient estimates. For linear regression, as far as estimation of the mean response $E(Y \mid x, z)$ is concerned, model averaging is the same as averaging the regression coefficients obtained by fitting different models. Model averaging for more complex estimands, like a probability or a quantile, does not admit such a representation. ∎

7.3 Distribution of model average estimators

Throughout this chapter we work inside the local misspecification framework that has also been used in Chapters 5 and 6. Hence the real data-generating density in i.i.d. settings is of the form

$$f_{\text{true}}(y) = f(y, \theta_0, \gamma_0 + \delta/\sqrt{n}),$$

with a similar $f(y_i \mid x_i, \theta_0, \gamma_0 + \delta/\sqrt{n})$ for regression frameworks; see Sections 5.4, 5.7 and 6.3 for details and background discussion. Remark 5.3 is also of relevance for the following. The model densities we consider are of the form $f(y, \theta, \gamma_S, \gamma_{0,S^c})$, where the parameter vector (θ, γ) consists of two parts; θ corresponds to that parameter vector of length p which is present in all of the models, and γ represents the vector of length q of which only a subset might be included in the final used models. The smallest model only contains θ as unknown parameter vector and assumes known values γ_0 for γ. The true value of θ in the smallest model is equal to θ_0. In the biggest model all γ parameters are unknown. Models S in between the smallest and biggest model assume some of the γ components unknown, denoted γ_S; the other γ components, denoted γ_{0,S^c}, are assumed equal to the corresponding components of γ_0. Of interest is the estimation of $\mu_{\text{true}} = \mu(\theta_0, \gamma_0 + \delta/\sqrt{n})$. Since the true model is somewhere in between the narrow model, which assumes $\gamma = \gamma_0$, and the wide model, where the full vector γ is included, the distance δ between the true model and the narrow model is an important quantity. The estimator of δ, and its limit distribution, are given in Theorem 6.1.

In the theorem below, we allow the weight functions to depend on the data through the estimator D_n. The class of compromise or model average methods to work with has estimators of the form

$$\widehat{\mu} = \sum_{S \in \mathcal{A}} c(S \mid D_n) \widehat{\mu}_S,$$

with random weights typically summing to one. For the following result, let conditions be as for Theorem 6.1, which is valid both for i.i.d. and regression situations, modulo mild regularity conditions. It involves independent normal limit variables $\Lambda_0 \sim N(0, \tau_0^2)$ and $D \sim N_q(\delta, Q)$.

Theorem 7.1 *Assume that the weight functions $c(S \mid d)$ sum to 1 for all d and have at most a countable number of discontinuities. Then, under the local misspecification assumption,*

$$\sqrt{n}\left\{\sum_{S \in \mathcal{A}} c(S \mid D_n)\widehat{\mu}_S - \mu_{\text{true}}\right\} \xrightarrow{d} \Lambda = \sum_{S \in \mathcal{A}} c(S \mid D)\Lambda_S = \Lambda_0 + \omega^{\text{t}}\{\delta - \widehat{\delta}(D)\},$$

where $\widehat{\delta}(D) = \sum_{S \in \mathcal{A}} c(S \mid D)G_S D$.

Proof. There is joint convergence of all $\Lambda_{n,S} = \sqrt{n}(\widehat{\mu}_S - \mu_{\text{true}})$ and random weights $c(S \mid D_n)$ to their respective $\Lambda_S = \Lambda_0 + \omega^{\text{t}}(\delta - G_S D)$ and $c(S \mid D)$, by arguments used to prove Theorem 6.1. Hence the continuity theorem of weak convergence yields

$$\sum_S c(S \mid D_n)\Lambda_{n,S} \xrightarrow{d} \sum_S c(S \mid D)\Lambda_S.$$

This implies the statement of the theorem. $\qquad\square$

Unlike the individual Λ_S, the limiting random variable Λ is no longer normally distributed, because of the random weights $c(S \mid D)$. A nonlinear combination of normals is not normal; the limit Λ has a purely normal distribution only if the weights $c(S \mid D_n)$ are nonrandom constants $c(S)$. General expressions for the mean and variance are

$$\mathrm{E}\Lambda = \omega^{\text{t}}\left[\delta - \sum_{S \in \mathcal{A}} \mathrm{E}\{c(S \mid D)G_S D\}\right],$$

$$\mathrm{Var}\,\Lambda = \tau_0^2 + \omega^{\text{t}}\mathrm{Var}\left\{\sum_{S \in \mathcal{A}} c(S \mid D)G_S D\right\}\omega.$$

Taken together this shows that Λ has mean squared error

$$\mathrm{mse}(\delta) = \mathrm{E}\,\Lambda^2 = \tau_0^2 + \mathrm{E}\{\omega^{\text{t}}\widehat{\delta}(D) - \omega^{\text{t}}\delta\}^2,$$

with $\widehat{\delta}(D)$ defined in the theorem. Different model average strategies, where model selection methods are special cases, lead to different risk functions $\mathrm{mse}(\delta)$, and perhaps with advantages in different parts of the parameter space of δ values. We note in particular that using the narrow model corresponds to $c(\emptyset \mid D_n) = 1$ and to $\widehat{\delta}(D) = 0$, with risk $\tau_0^2 + (\omega^{\text{t}}\delta)^2$, while using the wide model is associated with $c(\text{wide} \mid D_n) = 1$ and $\widehat{\delta} = D$, with constant risk $\tau_0^2 + \omega^{\text{t}}Q\omega$. This is precisely what was used when establishing Theorem 5.3.

The limit distribution found in Theorem 7.1 is often dramatically non-normal. This is a consequence of the nature of most model average estimators, because they are nonlinear averages of asymptotically normal estimators. A formula for the density h of Λ may be worked out conditioning on D. This conditional distribution is normal, since Λ_0 and D are independent, and

$$\Lambda \mid (D = x) \sim \omega^{\mathrm{t}}\{\delta - \widehat{\delta}(x)\} + \mathrm{N}(0, \tau_0^2).$$

Hence the density function h for Λ may be expressed as

$$
\begin{aligned}
h(z) &= \int h(z \mid D = x)\phi(x - \delta, Q)\,\mathrm{d}x \\
&= \int \phi\left(\frac{z - \omega^{\mathrm{t}}\{\delta - \widehat{\delta}(x)\}}{\tau_0}\right)\frac{1}{\tau_0}\phi(x - \delta, Q)\,\mathrm{d}x,
\end{aligned}
\tag{7.4}
$$

in terms of the density $\phi(u, Q)$ of a $\mathrm{N}_q(0, Q)$ distribution. We stress the fact that (7.4) is valid for the full class of model average estimators treated in Theorem 7.1, including estimators-post-selection as well as those using smooth AIC or smooth FIC weights, for example. The Λ density may be computed via numerical integration, for given model average methods, at specified positions δ. Numerical integration is easy for $q = 1$ and not easy for $q \geq 2$. The densities displayed in Figure 7.2, for which $q = 2$, have instead been found by simulation using half a million copies of Λ and using kernel density estimation to produce the curves.

Consider for illustration first the $q = 1$-dimensional case, where the compromise estimators take the form

$$\widehat{\mu} = \{1 - c(D_n)\}\widehat{\mu}_{\mathrm{narr}} + c(D_n)\widehat{\mu}_{\mathrm{wide}},$$

for suitable functions of $D_n = \sqrt{n}(\widehat{\gamma} - \gamma_0)$. The limit distribution is then that of

$$\Lambda = \Lambda_0 + \omega\{\delta - \widehat{\delta}(D)\},$$

where $\widehat{\delta}(D) = \{1 - c(D)\} \cdot 0 + c(D)D$, following Theorem 7.1. Figure 7.1 displays the resulting densities for Λ, stemming from four different model average estimators $\widehat{\mu}$. These correspond to

$$c_1(D) = \frac{D^2/\kappa^2}{1 + D^2/\kappa^2},$$

$$c_2(D) = I\{|D/\kappa| \geq \sqrt{2}\},$$

$$c_3(D) = \frac{\exp(\frac{1}{2}D^2/\kappa^2 - 1)}{1 + \exp(\frac{1}{2}D^2/\kappa^2 - 1)},$$

$$c_4(D) = 1,$$

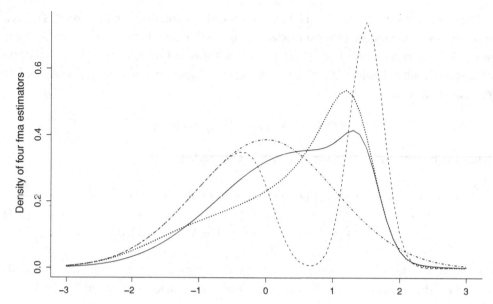

Fig. 7.1. Densities for Λ are shown for four model average estimators, in a situation where $\tau_0 = 0.25$, $\omega = 1$, $\kappa = 1$, at position $\delta = 1.5$ in the parameter space. The estimators correspond to weight functions c_1 (solid line, max density 0.415), c_2 (dotted line, max density 0.740), c_3 (dashed line, max density 0.535), and c_4 (dashed-dotted line, max density 0.387).

with $\kappa^2 = J^{11}$ as in Sections 5.2 and 5.3. The second and third of these are the AIC and smoothed AIC method, respectively, while the fourth is the wide-model minimax estimator $\widehat{\delta}(D) = D$.

Figure 7.2 shows Λ densities for three different estimation strategies in a $q = 2$-dimensional situation, at four different positions δ. The solid curve gives the density of an estimator selected by AIC, that is, of a post-selection-estimator. The dotted and dashed lines represent the density of a smoothed AIC and smoothed FIC estimator, which are both model averaged estimators. In this particular situation the smoothed FIC method is the winner in terms of performance, which is measured by tightness of the Λ distribution around zero.

Remark 7.1 More general model weights

The assumption given in the theorem that the weights should be directly dependent upon $D_n = \widehat{\delta}_{\text{wide}}$ can be weakened. The same limit distribution is obtained as long as $c_n(S) \rightarrow_d c(S \mid D)$, simultaneously with the $\sqrt{n}(\widehat{\mu}_S - \mu_{\text{true}})$ to Λ_S. Thus the weights do not need to be of the exact form $c(S \mid D_n)$. This is important when it comes to analysing the post-AIC and smoothed AIC estimators; see Example 7.1. The weights can in fact be dependent upon data in even more general ways, as long as we can accurately describe the simultaneous limit distribution of the $c_n(S)$ and the $\Lambda_{n,S}$. ∎

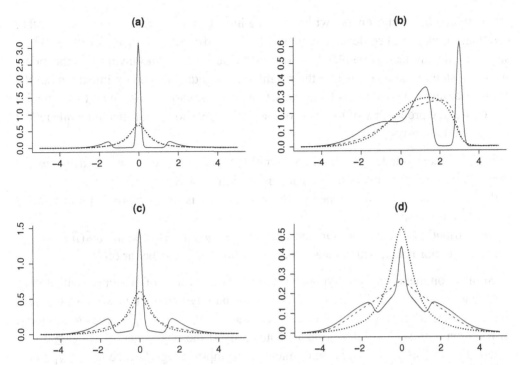

Fig. 7.2. Density function of the limiting distribution Λ of $\sqrt{n}(\widehat{\mu} - \mu_{\text{true}})$, for three model averaged estimators, at four positions in the parameter space. The situation studied has $q = 2$, $Q = \text{diag}(1, 1)$, $\omega = (1, 1)^t$ and $\tau_0 = 0.5$, and the four positions are (a) $(0, 0)$, (b) $(1.5, 1.5)$, (c) $(1, -1)$, (d) $(2, -2)$ for $\delta = (\delta_1, \delta_2)$. The densities have been found via density estimation based on half a million simulations per situation. The estimators are post-AIC (solid line), smoothed AIC (dotted line) and smoothed FIC (dashed line).

7.4 What goes wrong when we ignore model selection?

The applied statistics community has picked up on model selection methods partly via the convenience of modern software packages. There are certain negative implications of this, however. 'Standard practice' has apparently become to use a model selection technique to find a model, after which this part of the analysis is conveniently forgotten, and inference is carried out as if the selected model had been given a priori. This leads to too optimistic tests and confidence intervals, and generally to biased inference statements. In this section we quantify precisely how over-optimistic confidence intervals and significance values are, for these naive methods that ignore the model selection step.

A typical confidence interval, intended to have coverage probability 95%, would be constructed as

$$\mu \in \text{CI}_n = \widehat{\mu}_{\widehat{S}} \pm 1.96 \, \widehat{\tau}_{\widehat{S}} / \sqrt{n}, \tag{7.5}$$

where $\widehat{S}$ is the chosen model by the information criterion and $\widehat{\tau}_S / \sqrt{n}$ denotes any suitable estimator of the standard deviation of $\widehat{\mu}_S$. Using (7.5) indirectly amounts to conditioning on the final $\widehat{S}$ being chosen, but without taking into account the fact that any of the other candidate models in $\mathcal{A}$ could have been selected instead of $\widehat{S}$. Ignoring the model selection step in particular means completely setting aside the *uncertainties* involved in the model selection step of the analysis and the results are consequently too optimistic about the confidence level attained by such intervals. Our analysis below, which leads to the precise limit coverage probability of intervals of the (7.5) type, shows that such intervals reflect naivety in three ways:

- $\widehat{\tau}_{\widehat{S}}/\sqrt{n}$ underestimates the standard deviation of $\widehat{\mu}^* = \widehat{\mu}_{\widehat{S}}$, even when each separate $\widehat{\tau}_S/\sqrt{n}$ is a good estimator of the standard deviation of $\widehat{\mu}_S$;
- the $\widehat{\mu}^*$ estimator has a non-negligable bias, which is not captured by the interval construction; and
- the distribution of $\widehat{\mu}^*$ is often far from normal, even when each separate $\sqrt{n}(\widehat{\mu}_S - \mu_{\text{true}})$ is close to normality, which means that the '1.96' factor can be far off.

Similar comments apply to hypothesis tests and other forms of inference (with appropriate variations and modifications). Suppose the null hypothesis that we wish to test is $H_0: \mu = \mu_0$, with μ_0 a given value. The fact that $\sqrt{n}(\widehat{\mu}_S - \mu_{\text{true}})/\widehat{\tau}_S$ tends to a normal distribution with standard deviation 1 invites tests of the type 'reject if $|T_{n,S}| \geq 1.96$', where $T_{n,S} = \sqrt{n}(\widehat{\mu}_S - \mu_0)/\widehat{\tau}_S$, with intended asymptotic significance level (type I error) 0.05. The statistical practice of letting such a test follow a model selection step corresponds to

$$\text{reject if } |T_{n,\text{real}}| \geq 1.96, \quad \text{where } T_{n,\text{real}} = T_{n,\widehat{S}}. \tag{7.6}$$

The real type I error can be much bigger than the intended 0.05, as we demonstrate below.

7.4.1 The degree of over-optimism *

In Theorem 7.1 we obtained the precise asymptotic distribution of the post-selection-estimator $\widehat{\mu}_{\widehat{S}}$. With some additional details, provided below, this provides the exact limit probability for the event that μ_{true} belongs to the (7.5) interval. The task is to find the limit of

$$\text{P}(\mu_{\text{true}} \in \text{CI}_n) = \text{P}(\widehat{\mu}_{\widehat{S}} - 1.96\,\widehat{\tau}_{\widehat{S}}/\sqrt{n} \leq \mu_{\text{true}} \leq \widehat{\mu}_{\widehat{S}} + 1.96\,\widehat{\tau}_{\widehat{S}}/\sqrt{n})$$

and similar probabilities; the 1.96 number corresponds to the intention of having an approximate 95% confidence level. Introduce the random variable

$$V_n = \sqrt{n}(\widehat{\mu}_{\widehat{S}} - \mu_{\text{true}})/\widehat{\tau}_{\widehat{S}};$$

then the probability we study is $\text{P}(-1.96 \leq V_n \leq 1.96)$. Since V_n has a random denominator, the calculations need to be performed carefully. The estimator $\widehat{\tau}_S$ estimates the standard deviation of the limit of $\sqrt{n}(\widehat{\mu}_S - \mu_{\text{true}})$, which equals $\tau_S = (\tau_0^2 + \omega^{\text{t}} G_S Q G_S^{\text{t}} \omega)^{1/2}$. Since we cannot know beforehand which model S will be selected in the end, we need to consider all possibilities. Define the regions $R_S = \{x : c(S \mid x) = 1\}$. These define a partition of the q-dimensional space $\mathbb{R}^q$ of $D = x$ values, pointing to areas where each given S is chosen. Next let

$$\tau(x)^2 = \tau_0^2 + \sum_{S \in \mathcal{A}} I\{x \in R_S\} \omega^t Q_S^0 \omega.$$

The sum on the right-hand side contains only one nonzero value: there is one and only one set $\tilde{S}$ for which $x \in R_{\tilde{S}}$, and for this value of x, $\tau^2(x)$ is equal to $\tau_0^2 + \omega^t Q_{\tilde{S}}^0 \omega$. We now have

$$V_n = \frac{\sqrt{n}(\hat{\mu}_{\tilde{S}} - \mu_{\text{true}})}{\hat{\tau}_{\tilde{S}}} \xrightarrow{d} \frac{\Lambda}{\tau(D)} = \frac{\Lambda_0 + \omega^t\{\delta - \hat{\delta}(D)\}}{\tau(D)},$$

where $\hat{\delta}(D) = \sum_S c(S \mid D) G_S D$. This follows from Theorem 7.1, along with a supplementary argument that the numerator and denominator converge jointly in distribution to $(\Lambda, \tau(D))$. Consequently, for the real coverage probability $p_n(\delta)$ of a naive confidence interval that ignores model selection, we have derived that

$$p_n = P(|V_n| \leq 1.96) \to p(\delta) = P\left(\left|\frac{\Lambda_0 + \omega^t\{\delta - \hat{\delta}(D)\}}{\tau(D)}\right| \leq 1.96\right).$$

Since conditional on a value $D = x$,

$$\Lambda \mid \{D = x\} \sim N(\omega^t\{\delta - \hat{\delta}(x)\}, \tau_0^2) = \tau_0 N + \omega^t\{\delta - \hat{\delta}(x)\},$$

with N a standard normal, we arrive at an easier-to-compute formula for $p(\delta)$. This probability may be expressed as

$$p(\delta) = \sum_{S \in \mathcal{A}} \int_{R_S} P\left\{-1.96 \leq \frac{\tau_0 N + \omega^t\{\delta - \hat{\delta}(x)\}}{\tau_S} \leq 1.96\right\} \phi(x - \delta, Q) \, dx,$$

in which $\phi(v, Q)$ is the density function of a $N_q(0, Q)$ random variable. This derivation holds for *any* post-selection-estimator.

Example 7.6 Confidence coverage for the post-AIC-method

For $q = 1$ we choose between a narrow and a wide model. In this case the coverage probability of a naive confidence interval is easily calculated via numerical integration. The matrix Q is a number denoted now by κ^2, which is the limit variance of $\sqrt{n}(\hat{\gamma} - \gamma_0)$. For the narrow model $G_{\text{narr}} = 0$, $\tau(x) = \tau_{\text{narr}} = \tau_0$, and $R_{\text{narr}} = \{x: |x/\kappa| \leq \sqrt{2}\}$. For the full model $G_{\text{wide}} = 1$, $\tau(x) = \tau_{\text{wide}} = (\tau_0^2 + \omega^2\kappa^2)^{1/2}$ and $R_{\text{wide}} = \{x: |x/\kappa| > \sqrt{2}\}$. The general findings above lead to

$$p(\delta) = \int P\left(\left|\frac{\Lambda_0 + \omega\{\delta - \hat{\delta}(x)\}}{\tau(x)}\right| \leq z_0\right) \phi\left(\frac{x - \delta}{\kappa}\right) \frac{1}{\kappa} \, dx$$

$$= \int_{|x/\kappa| \leq \sqrt{2}} P\left(\left|\frac{\Lambda_0 + \omega(\delta - 0)}{\tau_0}\right| \leq z_0\right) \phi\left(\frac{x - \delta}{\kappa}\right) \frac{1}{\kappa} \, dx$$

$$+ \int_{|x/\kappa| \geq \sqrt{2}} P\left(\left|\frac{\Lambda_0 + \omega(\delta - x)}{\tau_{\text{wide}}}\right| \leq z_0\right) \phi\left(\frac{x - \delta}{\kappa}\right) \frac{1}{\kappa} \, dx,$$

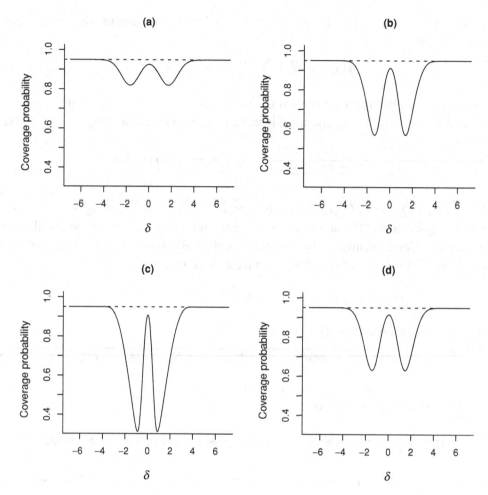

Fig. 7.3. True coverage probability for the naive AIC procedure, when the choice is between two models, the narrow and the wide. The four situations correspond to (ω, κ) equal to respectively $(1, 1)$, $(2, 1)$, $(4, 1)$, $(1, \sqrt{3})$; for each of them, $\tau_0 = 1$, and the intended coverage is 0.95.

with z_0 adjusted to match the intended coverage level, e.g. 1.96 for intended level 95%. The real coverage $p(\delta)$ is often significantly smaller than the intended level. Figure 7.3 displays the true coverage probability as a function of δ, for the naive AIC confidence interval, when the model choice is between the narrow and the wide models. ∎

Example 7.7 Confidence coverage when AIC chooses between four models

We have also carried out such computations for the case of $q = 2$ where we choose between four models. Here we used simulation to compute the coverage probabilities. Figure 7.4 depicts the coverage probability for AIC choice amongst four models in a situation where $\omega = (1, 1)^t$ and $Q = \mathrm{diag}(1, 1)$. We note that the correct (or intended) coverage probability is obtained when δ gets far enough away from $(0, 0)$. ∎

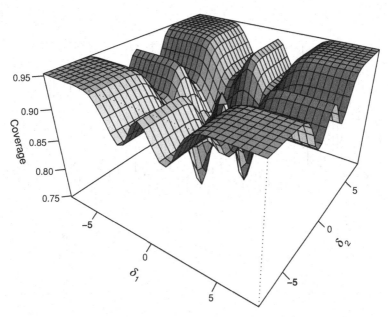

Fig. 7.4. True coverage probability for the naive AIC-based procedure with intended coverage 0.95, when AIC selects among four models, for $q = 2$. The situation corresponds to $\omega = (1, 1)^t$ and $Q = \text{diag}(1, 1)$.

Example 7.8 Confidence intervals for a Weibull quantile

Assume i.i.d. data $Y_1, \ldots, Y_n$ come from the Weibull distribution $1 - \exp\{-(\theta y)^\gamma\}$, as in Examples 2.5 and 5.6. One wishes to construct a confidence interval for the 0.10-quantile $\mu = A^{1/\gamma}/\theta$; here $A = -\log(1 - 0.10)$. We examine how the intervals-post-AIC selection behave; these are as in (7.5) with two candidate models, the narrow exponential ($\gamma = 1$) and the wide Weibull ($\gamma \neq 1$). When $\gamma = 1 + \delta/\sqrt{n}$, we obtained in Section 2.8 that the asymptotic probability that AIC selects the Weibull is $P(\chi_1^2(\delta^2/\kappa^2) \geq 2)$, where $\kappa = \sqrt{6}/\pi$ (see Example 5.6). We choose the values such that $\delta = 1.9831^{1/2}\kappa$ corresponds to AIC limit probabilities $\frac{1}{2}$ and $\frac{1}{2}$ for the two models.

Figure 7.5 shows 100 simulated post-AIC confidence intervals of intended coverage level 90%, thus using 1.645 instead of 1.96 in (7.5), for sample size $n = 400$. It illustrates that about half of the intervals, those corresponding to AIC selecting the narrow model, are too short, in addition to erring on the negative side of the true value (marked by the horizontal line). The other half of the intervals are appropriate. The confidence level of the narrow half is as low as 0.07, while the confidence level of the wide half is 0.87, acceptably close to the asymptotic limit 0.90; these figures are found from simulating 10,000 replicas of the described Weibull model. The overall combined coverage level is 0.47, which is of course far too low. The importance of this prototype example is that behaviour similar to that exposed by Figure 7.5 may be expected in most applied statistics situations where an initial model selection step is used before computing one's standard confidence interval. ∎

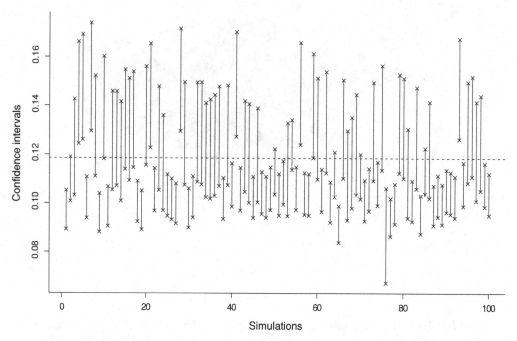

Fig. 7.5. 100 simulated confidence intervals for the 0.10-quantile of a Weibull distribution, with the true quantile indicated by the horizontal line. The intervals are of post-AIC-selection type (7.5) with AIC choosing between the exponential and the Weibull model. The sample size is $n = 400$, and the Weibull parameters are set up to give AIC selection probabilities $\frac{1}{2}$ and $\frac{1}{2}$ for the two models. Half of the intervals are too short and in addition err on the negative side.

7.4.2 The inflated type I error *

We consider testing point hypotheses, of the type H_0: $\mu = \mu_0$. We saw above that the real coverage level p_n of an intended 95% confidence interval can be drastically lower if the interval is naively constructed after model selection. Similarly, for a hypothesis test with a significance level α_n, the probability of a false rejection can be expected to be higher than the intended level if the test statistic is constructed naively after model selection.

As introduced in connection with (7.6), let $T_{n,S} = \sqrt{n}(\widehat{\mu}_S - \mu_0)/\widehat{\tau}_S$. Statistical practice has partly been to reject if $|T_{n,\text{real}}| = |T_{n,\widehat{S}}| \geq 1.96$, with $\widehat{S}$ the submodel arrived at via model selection, e.g. AIC. Let us examine

$$\alpha_n = \mathrm{P}(|T_{n,\text{real}}| \geq 1.96 \mid H_0 \text{ is true}),$$

under circumstances where the η in $\theta_n = \theta_0 + \eta/\sqrt{n}$ is arranged such that $\mu_{\text{true}} = \mu(\theta_0 + \eta/\sqrt{n}, \gamma_0 + \delta/\sqrt{n})$ is equal to μ_0. This assumes that the interest parameter $\mu = \mu(\theta, \gamma)$ is not a function of γ alone. We have

$$T_{n,\text{real}} \xrightarrow{d} T = \sum_S c(S \mid D)T_S,$$

where $T_S = \{\Lambda_0 + \omega^t(\delta - G_S D)\}/\tau_S$ and with $c(S \mid D)$ the indicator for S winning the model selection in the limit experiment. As earlier, $\tau_S = (\tau_0^2 + \omega^t G_S Q G_S^t \omega)^{1/2}$. Note that each T_S is a normal, with standard deviation 1, but with nonzero means $\omega^t(I - G_S)\delta$. Thus the limit distribution variable T is a nonlinear mixture of normals with nonzero means and variance one. For the type I error or false rejection rate we have

$$\alpha_n \to \alpha(\delta) = P(|T| \geq 1.96),$$

as a consequence of Theorem 7.1. The probability on the right-hand side is fully determined by the usual ingredients $\Lambda_0 \sim N(0, \tau_0^2)$ and $D \sim N(\delta, Q)$. The limit type I error may be computed in quite general cases by conditioning on D and performing numerical integration, parallelling calculations in Section 7.4.1. For the particular case of testing $\mu = \mu_0$ via (7.6) in a $q = 1$-dimensional case in combination with AIC; one finds

$$\alpha(\delta) = \int_{|x/\kappa| \leq \sqrt{2}} P\left(\frac{|\Lambda_0 + \omega\delta|}{\tau_0} \geq 1.96\right) \phi(x, \kappa^2) \, dx$$
$$+ \int_{|x/\kappa| > \sqrt{2}} P\left(\frac{|\Lambda_0 + \omega(\delta - x)|}{\tau_{\text{wide}}} \geq 1.96\right) \phi(x, \kappa^2) \, dx,$$

which can be much larger than the naively intended 0.05.

As an example, consider the following continuation of Example 7.8, with data from the Weibull distribution, and focus parameter the 0.10-quantile $\mu = A^{1/\gamma}/\theta$, with $A = -\log(1 - 0.10)$. We wish to test the null hypothesis that μ is equal to a specified value μ_0, say $\mu_0 = 1$. For this illustration we set the parameters $\gamma = 1 + \delta_0/\sqrt{n}$ and $\theta = \theta_0 + \eta_0/\sqrt{n}$ such that the AIC model selection limit probabilities are $\frac{1}{2}$ and $\frac{1}{2}$, which requires $\delta_0 = 1.9831^{1/2}\kappa$, and such that the true 0.10-quantile is in fact identical to μ_0. The test statistic used is

$$T_{n,\text{real}} = \begin{cases} T_{n,\text{narr}} = \sqrt{n}(\widehat{\mu}_{\text{narr}} - \mu_0)/\widehat{\tau}_0 & \text{if AIC prefers exponential,} \\ T_{n,\text{wide}} = \sqrt{n}(\widehat{\mu}_{\text{wide}} - \mu_0)/\widehat{\tau}_{\text{wide}} & \text{if AIC prefers Weibull,} \end{cases}$$

and the null hypothesis is rejected when $|T_{n,\text{real}}| \geq 1.96$, as per (7.6). Here $\widehat{\mu}_{\text{narr}} = A/\widehat{\theta}_{\text{narr}}$ and $\widehat{\mu}_{\text{wide}} = A^{1/\widehat{\gamma}}/\widehat{\theta}$, while $\widehat{\tau}_0$ and $\widehat{\tau}_{\text{wide}}$ are estimators of $\tau_0 = A/\theta$ and $\tau_{\text{wide}} = (A/\theta)\{1 + (\log A - b)^2\kappa^2\}^{1/2}$.

We simulated 10,000 data sets of size $n = 400$ from this Weibull model, to see how the test statistics behaved. Figure 7.6 displays histograms of $T_{n,\text{narr}}$ and $T_{n,\text{wide}}$ separately. Both are approximately normal, and both have standard deviations close enough to 1, but $T_{n,\text{narr}}$ is seriously biased to the left. The real test statistic $T_{n,\text{real}}$ has a very non-normal distribution, portrayed via the 10,000 simulations in Figure 7.7. The conditional type I errors are an acceptable 0.065 for the wide-based test but an enormous 0.835 for the narrow-based test, giving an overall type I error of 0.46. So $|T_{n,\text{real}}|$ exceeds the naive 1.96 threshold in almost half the cases, even though the null hypothesis $\mu = \mu_0$ is perfectly true in all of the 10,000 simulated situations. The reason for the seriously inflated type I error is that $T_{n,\text{narr}}$, which is the chosen test statistic in half of the cases, is so severely

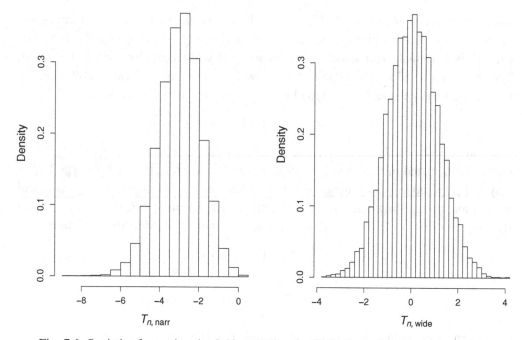

Fig. 7.6. Statistics for testing the 0.10-quantile of a Weibull distribution, with sample size $n = 400$. Histograms are shown for $T_{n,\text{narr}}$ and $T_{n,\text{wide}}$ separately, based on 10,000 simulations. Both are approximately normal and have standard deviations close to 1, but $T_{n,\text{narr}}$ is seriously biased to the left.

biased. The naive testing-after-selection strategy does a bad job, since one with high probability is led to a too simple test statistic.

7.5 Better confidence intervals

In this section we examine ways of constructing confidence intervals that do not run the over-optimism danger spelled out above. This discussion will cover not only post-selection-estimators that use AIC or the FIC, but also smoothed versions thereof. Results and methods below have as point of departure Theorem 7.1, which gives the precise description of the limit Λ of $\Lambda_n = \sqrt{n}(\widehat{\mu} - \mu_{\text{true}})$, for the wide class of model average estimators of the type $\widehat{\mu} = \sum_S c(S \,|\, \widehat{\delta}_{\text{wide}})\widehat{\mu}_S$. The challenge is to construct valid confidence intervals for μ_{true}.

7.5.1 Correcting the standard error

Buckland *et al.* (1997) made an explicit suggestion for correcting for the model uncertainty in the calculation of the variance of post-selection-estimators. This leads to modified confidence intervals. Their method has later been recommended by Burnham and Anderson (2002, Section 4.3), particularly when used for the smoothed AIC weights

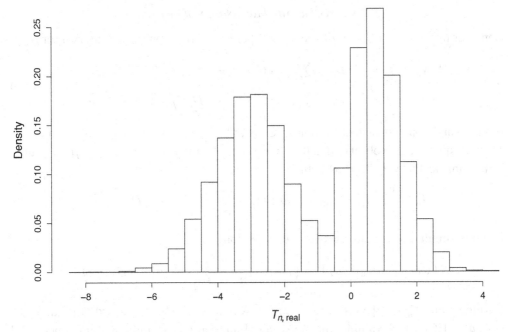

Fig. 7.7. For the situation described in the text, the histogram is shown over the 10,000 simulated values of the real test statistic $T_{n,\text{real}}$; this is equal to $T_{n,\text{narr}}$ or $T_{n,\text{wide}}$ according to which model is preferred by AIC. In 50.7% of cases AIC chose the Weibull, in 49.3% the exponential. Conditional type I errors were 0.834 and 0.065; total type I error is 0.46.

for $c(S \mid D_n)$. The method consists of using $\widehat{\mu} \pm u \, \widehat{\text{se}}_n$ as confidence intervals, with u the appropriate normal quantile and formula (9) in Buckland *et al.* (1997) for the estimated standard error $\widehat{\text{se}}_n$. Rephrased using our notation,

$$\widehat{\text{se}}_n = \sum_S c(S \mid D_n)(\widehat{\tau}_S^2/n + \widehat{b}_S^2)^{1/2},$$

where $\widehat{\tau}_S$ is a consistent estimator of the standard deviation $\tau_S = (\tau_0^2 + \omega^{\text{t}} G_S Q G_S^{\text{t}} \omega)^{1/2}$ and the bias estimator $\widehat{b}_S = \widehat{\mu}_S - \widehat{\mu}$. The resulting coverage probability p_n is not studied accurately in the references mentioned, but it is claimed that it will be close to the intended $P(-u \leq N(0, 1) \leq u)$. Hjort and Claeskens (2003) study p_n in more detail and find that $p_n = P(-u \leq B_n \leq u)$, where $B_n = (\widehat{\mu} - \mu_{\text{true}})/\widehat{\text{se}}_n$ has a well-defined limit distribution:

$$B_n \xrightarrow{d} B = \frac{\Lambda}{\widehat{\text{se}}} = \frac{\Lambda_0 + \omega^{\text{t}}\{\delta - \sum_{S \in \mathcal{A}} c(S \mid D) G_S D\}}{\sum_S c(S \mid D)\{\tau_S^2 + [\omega^{\text{t}}\{\sum_{S' \in \mathcal{A}} c(S' \mid D) G_{S'} - G_S\} D]^2\}^{1/2}}.$$

This variable has a normal distribution, for given D, but is clearly not standard normal when averaged over the distribution of D, and neither is it centred at zero, so the coverage probability p_n can be quite different from the intended value.

7.5.2 Correcting the bias using wide variance

Consider instead the following lower and upper bound of a confidence interval for μ_{true},

$$\text{low}_n = \widehat{\mu} - \widehat{\omega}^{\text{t}}\Big\{D_n - \sum_{S \in \mathcal{A}} c(S \mid D_n)G_S D_n\Big\}\Big/\sqrt{n} - u\widehat{\kappa}/\sqrt{n},$$
$$\text{up}_n = \widehat{\mu} - \widehat{\omega}^{\text{t}}\Big\{D_n - \sum_{S \in \mathcal{A}} c(S \mid D_n)G_S D_n\Big\}\Big/\sqrt{n} + u\widehat{\kappa}/\sqrt{n},$$

(7.7)

where $\widehat{\omega}$ and $\widehat{\kappa}$ are consistent estimators of ω and $\kappa = \tau_{\text{wide}} = (\tau_0^2 + \omega^{\text{t}}Q\omega)^{1/2}$, and u is a normal quantile. We observe that the coverage probability $p_n = \text{P}(\text{low}_n \le \mu_{\text{true}} \le \text{up}_n)$ is the same as $\text{P}(-u \le T_n \le u)$, where

$$T_n = \Big[\sqrt{n}(\widehat{\mu} - \mu_{\text{true}}) - \widehat{\omega}^{\text{t}}\Big\{D_n - \sum_{S \in \mathcal{A}} c(S \mid D_n)G_S D_n\Big\}\Big]\Big/\widehat{\kappa}.$$

There is simultaneous convergence in distribution

$$\Big(\sqrt{n}(\widehat{\mu} - \mu_{\text{true}}), D_n\Big) \xrightarrow{d} \Big(\Lambda_0 + \omega^{\text{t}}\Big\{\delta - \sum_{S \in \mathcal{A}} c(S \mid D)G_S D\Big\}, D\Big).$$

It follows that $T_n \to_d \{\Lambda_0 + \omega^{\text{t}}(\delta - D)\}/\kappa$, which is simply a standard normal. Thus, with $u = 1.645$, for example, the constructed interval has asymptotic confidence level precisely the intended 90% level. This method is first-order equivalent to using the wide model for confidence interval construction, with a modification for location.

7.5.3 Simulation from the Λ distribution

For any given model averaging scheme we have a precise description of the limit variable

$$\Lambda = \Lambda_0 + \omega^{\text{t}}\Big\{\delta - \sum_S c(S \mid D)G_S D\Big\},$$

involving as before $\Lambda_0 \sim \text{N}(0, \tau_0^2)$ independent of $D \sim \text{N}_q(\delta, Q)$. The model averaging distribution Λ may in particular be simulated at each position δ, using the consistently estimated quantities τ_0, ω, K, G_S. Such simulation is easily carried out via a large number B of $\Lambda_{0,j} \sim \text{N}(0, \tau_0^2)$ and $D_j \sim \text{N}_q(\delta, Q)$. For each of the vectors D_j we recalculate the model averaging weights, leading to the matrix $\sum_S c(S \mid D_j)G_S$. The observed distribution of $\Lambda_{0,j} + \omega^{\text{t}}\{\delta - \sum_S c(S \mid D_j)G_S^{\text{t}} D_j\}$ can be used to find $a = a(\delta)$ and $b = b(\delta)$ with

$$\text{P}\{a(\delta) \le \Lambda(\delta) \le b(\delta)\} = 0.95.$$

(7.8)

The sometimes highly non-normal aspects of the distribution of $\Lambda = \Lambda(\delta)$ were illustrated in Figure 7.2, for given positions δ in the parameter space; the densities in that figure were in fact arrived at via a million simulations of the type just described. The distribution of Λ may also fluctuate when viewed as a function of δ, for a fixed model averaging scheme. This is illustrated in Figure 7.8, pertaining to a situation where $q = 1$,

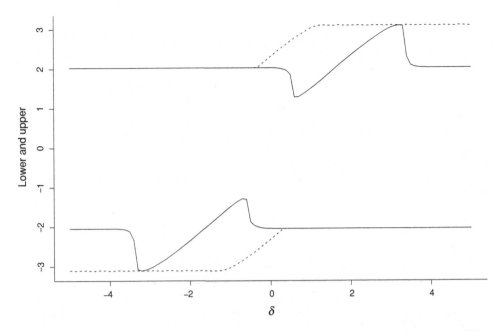

Fig. 7.8. Displaying the pointwise $a(\delta)$ and $b(\delta)$ (solid lines) that for each δ gives 0.95 level coverage for $\Lambda(\delta)$, along with the conservative $a_0(\delta)$ and $b_0(\delta)$ (dashed lines). The situation is that of $q = 1$, with $\tau_0 = 0.25$, $\omega = 1$, $\kappa = 1$, and $\widehat{\delta}(D) = c(D)D$ as in the AIC threshold method.

$\tau_0 = 0.25$, $\omega = 1$, $\kappa = 1$, for the AIC selector. In other words,

$$\Lambda = \Lambda_0 + \omega\{\delta - c(D)D\},$$

where $D \sim \mathrm{N}(\delta, \kappa^2)$ and $c(D) = I\{D^2/\kappa^2 \geq 2\}$. Simulations gave for each δ the appropriate $a(\delta)$ to $b(\delta)$ interval as in (7.8); these are the two full lines in Figure 7.8.

If we know the correct numerical values for the $[a(\delta), b(\delta)]$ interval, we can infer from

$$\mathrm{P}\big(a(\delta) \leq \sqrt{n}(\widehat{\mu} - \mu_{\mathrm{true}}) \leq b(\delta)\big) \to \mathrm{P}\big(a(\delta) \leq \Lambda(\delta) \leq b(\delta)\big) = 0.95$$

that $[\widehat{\mu} - b(\delta)/\sqrt{n}, \widehat{\mu} - a(\delta)/\sqrt{n}]$ is an interval covering μ_{true} with asymptotic confidence level 0.95. One somewhat naive method is now to plug in $D_n = \widehat{\delta}_{\mathrm{wide}}$ here, carry out the required Λ simulations at this position in the parameter space, and use

$$\mathrm{CI}_n = [\widehat{\mu} - \widehat{b}/\sqrt{n}, \widehat{\mu} - \widehat{a}/\sqrt{n}].$$

This corresponds to using $\widehat{a} = a(D_n)$ and $\widehat{b} = b(D_n)$ above. It is reasonably simple to carry out in practice and may lead to satisfactory coverage levels, but not always, however. The limit of the coverage probability $p_n(\delta)$ here can be described accurately by again turning to the limit experiment. The limit becomes

$$p(\delta) = \mathrm{P}\{a(D) \leq \Lambda(\delta) \leq b(D)\},$$

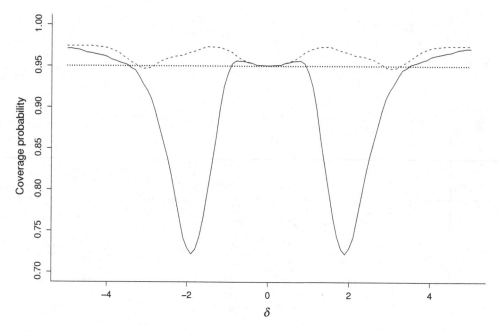

Fig. 7.9. For the situation of Figure 7.8, we display the confidence level attained by the naive method using $[a(D), b(D)]$ (solid line), along with the two-stage conservative method (dashed line).

which can be computed (typically, via simulations) in given situations. Figure 7.9 illustrates that the coverage for this method can become far too low, for the same situation as in Figure 7.8.

7.5.4 A two-stage confidence procedure

A better but more laborious idea is as follows. Instead of fixing the value of δ we first make a confidence ellipsoid for δ. Consider the event $A_n = \{\rho_n(D_n, \delta) \le z\}$, where $z = (\chi^2_{q,0.95})^{1/2}$, and the distance function used is

$$\rho_n(D_n, \delta) = \{(D_n - \delta)^t \widehat{Q}^{-1} (D_n - \delta)\}^{1/2}.$$

The limit probability of A_n is 0.95, giving a confidence ellipsoid for δ based on D_n. In conjunction with (7.8), define

$$\widehat{a}_0 = \min\{a(\delta): \rho_n(D_n, \delta) \le z\},$$
$$\widehat{b}_0 = \max\{b(\delta): \rho_n(D_n, \delta) \le z\}.$$

The construction is illustrated in Figure 7.8, where one can read the required $\widehat{a}_0(D_n)$ and $\widehat{b}_0(D_n)$ from any value of D_n. Our proposed two-stage confidence interval construction is

$$\text{CI}_n^* = [\widehat{\mu} - \widehat{b}_0/\sqrt{n}, \widehat{\mu} - \widehat{a}_0/\sqrt{n}]. \tag{7.9}$$

We claim that its limit coverage level is always above 0.90.

To see this, we again use simultaneous convergence of all relevant variables to those in the limit experiment, to find that the coverage probability $r_n(\delta)$ that $\mu_{\text{true}} \in \text{CI}_n^*$ converges to is

$$r(\delta) = \text{P}\{a_0(D) \le \Lambda(\delta) \le b_0(D)\},$$

where

$$a_0(D) = \min\{a(\delta): \rho(D, \delta) \le z\} \quad \text{and} \quad b_0(D) = \max\{b(\delta): \rho(D, \delta) \le z\},$$

and $\rho(D, \delta)^2$ is the χ_q^2 distance $(D - \delta)^t Q^{-1}(D - d)$. Letting A be the event that $\rho(D, \delta) \le z$, we find

$$0.95 = \text{P}\{a(\delta) \le \Lambda(\delta) \le b(\delta), A\} + \text{P}\{a(\delta) \le \Lambda(\delta) \le b(\delta), A^c\}$$
$$\le \text{P}\{a_0(D) \le \Lambda(\delta) \le b_0(D)\} + \text{P}(A^c).$$

This shows that $q(\delta) \ge 0.90$ for all δ. For the AIC illustration with $q = 1$ the limit coverage probability is depicted in Figure 7.9. As we see, the procedure can be quite conservative; it is here constructed to be always above 0.90, but is in this case almost entirely above 0.95.

This two-stage construction can be fine-tuned further, by adjusting the confidence ellipsoid radius parameter z above, so as to reach the shortest possible interval of the type CI_n^* with limit probability equal to some prespecified level. Such calibration would depend on the type of model average procedure being used.

7.6 Shrinkage, ridge estimation and thresholding

Model selection in regression amounts to setting some of the regression coefficients equal to zero (namely for those covariates that are not in the selected model). On the other hand, model averaging for linear regression corresponds to keeping all coefficients but averaging the coefficient values obtained across different models, as we saw in Example 7.5. An alternative to weighting or averaging the regression coefficients is to use so-called shrinkage methods. These may shrink some, or most, or all of the coefficients towards zero. We focus on such methods in this section and go on to Bayesian model averaging in the next; these methodologies are in fact related, as we shall see.

7.6.1 Shrinkage and ridge regression

Ridge regression (Hoerl and Kennard, 1970) has precisely the effect of pulling the regression coefficients towards zero. Ridge estimation is a common technique in linear regression models, particularly in situations with many and perhaps positively correlated covariates. In such cases of multicolinearity it secures numerical stability and has often more robustness than what the ordinary least squares methods can deliver. In the linear regression model $Y = X\beta + \varepsilon$, with β of length q, the ridge regression estimators are obtained by

$$\widehat{\beta}_{\text{ridge}} = (X^{\text{t}}X + \lambda I_q)^{-1}X^{\text{t}}Y.$$

The parameter λ controls the amount of ridging, with $\lambda = 0$ corresponding to the usual least squares estimator $\widehat{\beta}$. This method is particularly useful when the matrix X contains highly correlated variables such that $X^{\text{t}}X$ is nearly singular (no longer invertible). Adding a constant λ to the diagonal elements has an effect of stabilising the computations. Since the least squares estimators (with $\lambda = 0$) are unbiased for β, the ridge estimator $\widehat{\beta}_{\text{ridge}}$ with $\lambda \neq 0$ is a biased estimator. It can be shown that adding bias in this way has an advantageous effect on the variance: while the bias increases, the corresponding variance decreases. In fact, there always exist values of λ such that the mean squared error of $\widehat{\beta}_{\text{ridge}}$ is smaller than that of the least squared estimator $\widehat{\beta}$. The value λ is often chosen in a subjective way, as attempts at estimating such a favourable λ from data make the risk bigger again.

An equivalent way of presenting the ridge regression estimator is through the following constraint estimation problem. The estimator $\widehat{\beta}_{\text{ridge}}$ minimises the sum of squares $\|Y - X\beta\|^2$ subject to the constraint that $\sum_{j=2}^{q} \beta_j^2 < c$, with c some constant. Note that the intercept is not restricted by the constraint in this method. Using a Lagrange multiplier, another equivalent representation is that $\widehat{\beta}_{\text{ridge}}$ minimises the penalised least squares criterion

$$\|Y - X\beta\|^2 + \lambda \sum_{j=2}^{q} \beta_j^2.$$

Such a method of estimation is nowadays also used for flexible model estimation with penalised regression splines, see Ruppert *et al.* (2003). In that method only part of the regression coefficients, those corresponding to spline basis functions, are penalised. This way of setting the penalty, with only penalising some but not all coefficients, allows much flexibility.

Other choices for the penalty lead to other types of estimators. The lasso (Tibshirani, 1996) uses instead an ℓ_1 penalty and obtains $\widehat{\beta}_{\text{lasso}}$ as the minimiser of

$$\|Y - X\beta\|^2 + \lambda \sum_{j=2}^{q} |\beta_j|.$$

The abbreviation lasso stands for 'least absolute shrinkage and selection operator'. Although quite similar to the ridge regression estimator at first sight, the properties of the lasso estimator are very different. Instead of only shrinking coefficients towards zero, this penalty actually allows some of the coefficients to become identically zero. Hence, the lasso performs shrinkage and variable selection simultaneously. The combination of an ℓ_1 (as in the lasso) and an ℓ_2 penalty (as in ridge regression) is called the elastic net

(Zou and Hastie, 2005). Efron *et al.* (2004) develop least angle regression (LARS), a computationally efficient algorithm for model selection that encompasses the lasso as a special case. Another related method is the non-negative garotte (Breiman, 1995). This method applies a different shrinkage factor to each regression coefficient.

7.6.2 Thresholding in wavelet smoothing

Selection or shrinkage of nonzero coefficients in wavelet models is obtained via thresholding methods. As we will explain below, hard thresholding performs the selection of nonzero coefficients, while soft thresholding corresponds to shrinkage.

Wavelet methods are widely used in signal and image processing. Consider the following regression model:

$$Y_i = f(x_i) + \varepsilon_i,$$

where the function $f(\cdot)$ is not specified, and needs to be estimated from the data. This is a nonparametric estimation problem, since we do not have a parametric formula for f. The situation is now more complex since there are no 'parameters' θ_j to estimate as there were in the linear regression model $Y_i = \theta_1 + \theta_2 X_i + \varepsilon_i$, for example. In vector notation, we write the above nonparametric model as $Y = f + \varepsilon$, all vectors are of length n, the sample size. A forward wavelet transformation multiplies (in a smart and fast way) these vectors with a matrix $\widetilde{W}$ to arrive at

$$w = v + \eta, \text{ or, equivalently, } \widetilde{W}Y = \widetilde{W}f + \widetilde{W}\varepsilon.$$

If the error terms ε are independent, the same holds for the transformed errors η when the transformation matrix $\widetilde{W}$ is orthogonal. The values w are called the *wavelet coefficients*. Typically, only a few wavelet coefficients are large, those correspond to important structure in the signal f. As an example, consider the data in Figure 7.10. The true signal $f(\cdot)$ is a stepwise block function. The left panel shows the observations y_i versus x_i; this is the signal with noise. In the right panel we plot the wavelet coefficients obtained using the R function wd in the library wavethresh, using the default arguments. Only a few of these wavelet coefficients are large, the majority take values close to zero. Since there are as many wavelet coefficients as there are data, $n = 512$ in the example, this shows that we can probably set many of them equal to zero, without being afraid that too much structure gets lost when reconstructing the signal. This important property is called *sparsity*. Many fewer wavelet coefficients than the number of data points can then be used to reconstruct a smooth signal.

One way of reducing the number of wavelet coefficients is coefficient selection. This is called *thresholding*, and more in particular, *hard thresholding*. Only those wavelet coefficients are kept that are larger than a user-set threshold, all others are set equal to zero. Hard thresholding is a form of variable selection. Shrinkage is another way, and one of the most important examples is *soft thresholding*. In this case, all wavelet coefficients

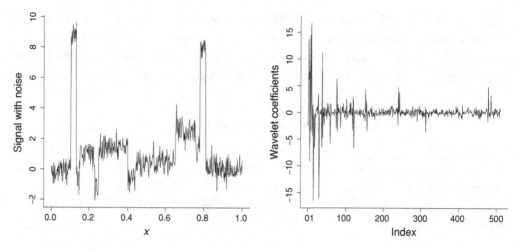

Fig. 7.10. Plot of the signal observations y_i versus x_i and the corresponding wavelet coefficients. The sample size is $n = 512$.

below the value of the threshold are set equal to zero, and all wavelet coefficients that exceed the threshold are shrunk towards zero with the value of the threshold. For an extensive treatment of thresholding for noise reduction, see Jansen (2001).

An important and thoroughly studied question is how to set the value of the threshold. By applying the backward wavelet transformation, we obtain the reconstructed signal $\widehat{f} = W\widehat{v}$, where $W = \widetilde{W}^{-1}$ and $\widehat{v}$ is an estimator of v (see below). A small threshold will lead to a reproduced signal that is close to the original, noisy, signal. A large value of the threshold will produce a signal that is much smoother, as compared to the original signal with noise.

The threshold assessment methods can be roughly grouped into three classes:

(1) A first class of methods tries to find the threshold value such that the mean squared error of the wavelet estimator is minimised. Stein's (1981) unbiased risk estimator (SURE) is constructed with the goal of doing exactly that. SURE(λ) is an unbiased estimator of the mean squared error and the threshold value λ_{sure} is that value which minimises SURE(λ). If this is combined with soft thresholding, we arrive at the SURE shrinkage estimator of Donoho and Johnstone (1995). An equivalent representation for soft thresholding is by finding the wavelet coefficients w to minimise

$$\|Y - Ww\|^2 + 2\lambda \sum |w_j|,$$

which takes the same form as for the lasso estimator.

For SURE shrinkage, the solution is the vector of shrunk wavelet coefficients

$$\widehat{v} = \text{sign}(w)\max(|w| - \lambda_{\text{sure}}, 0).$$

Other mean squared error-based threshold finding methods are cross-validation (Nason, 1996) and generalised cross-validation (Jansen *et al.*, 1997; Jansen and Bultheel, 2001),

$$\text{GCV}(\lambda) = \frac{n^{-1}\|y - \widehat{y}\|^2}{\{n_0(\lambda)/n\}^2},$$

where $n_0(\lambda)$ counts the number of wavelet coefficients that are smaller than the threshold λ and hence are set to zero by thresholding. GCV can be computed in a fast way, without actually having to go through the process of leaving out an observation and refitting. GCV is shown to be an asymptotically optimal threshold estimator in a mean squared error sense. Denote by λ^* the value of λ that minimises the expected mse, and by λ_{gcv} that value obtained by minimising GCV, then the ratio $\text{Emse}(\lambda_{\text{gcv}})/\text{Emse}(\lambda^*) \rightarrow 1$ (Jansen, 2001, Chapter 4). A comparison with SURE yields that $\text{GCV}(\lambda) \approx \text{SURE}(\lambda) + \sigma^2$, where σ^2 is the error variance.

(2) A second class of thresholding method considers the problem as an instance of a multiple testing problem. The focus is on overfitting, or false positives. Examples of such methods are the universal threshold (Donoho and Johnstone, 1994) $\lambda_{\text{univ}} = \sigma\sqrt{2\log n}$ and the false discovery rate FDR (Benjamini and Hochberg, 1995; Abramovich and Benjamini, 1996).

(3) Finally, a third class consists of Bayesian methods. Wavelet coefficients are assigned a prior which is typically a mixture of a point-mass at zero and a heavy-tailed distribution. This again expresses sparsity, many coefficients are zero, and only few are large. Several people have worked on this topic, amongst them are Chipman *et al.* (1997), Abramovich *et al.* (1998), Vidakovic (1998) and Ruggeri and Vidakovic (1999). It can be shown that in a model with a point-mass at zero, taking the posterior mean as an estimator leads to shrinkage but no thresholding. This corresponds to ridge regression. On the other hand, working with the posterior median corresponds to a thresholding procedure. That means, when a coefficient is below an implicit threshold, that coefficient is replaced by zero. When the coefficient is above that threshold, it is shrunk. The transition between thresholded and shrunk coefficients is in general smoother than in soft thresholding. Empirical Bayesian methods have been studied by Johnstone and Silverman (2004, 2005).

Figure 7.11 gives first the noisy data together with the true underlying signal, and contains next several estimates using different thresholding schemes. For this example, the universal threshold takes the value 2.136. Because this value is larger than all other threshold values, the smoothest fit is found here. SURE corresponds to a threshold with value 0.792. Then we have two pictures with cross-validation. The first one uses soft thresholding which results in $\lambda = 0.402$, hard thresholding uses $\lambda = 0.862$. The last one applies soft thresholding with GCV and finds the threshold value 0.614. As cross-validation and SURE concentrate on the mse of the output, rather than on smoothness, post-processing

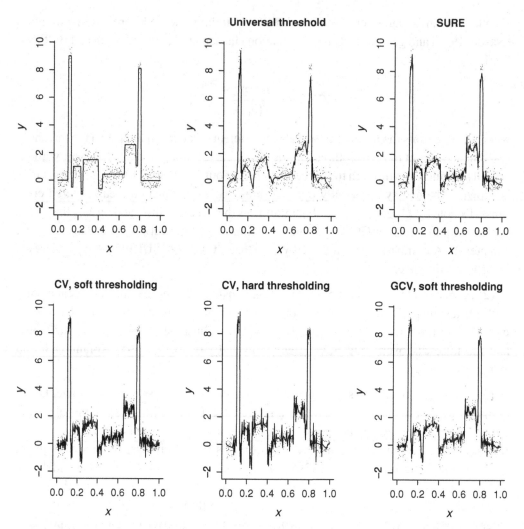

Fig. 7.11. Test data, with the true block function and various wavelet estimates using different thresholding schemes.

should remove, or at least reduce, the false positives. This post-processing typically looks at coefficients at successive scales. In this way, a low threshold preserves the true jumps in the underlying data and yet arrives at a smooth reconstruction in between.

7.7 Bayesian model averaging

Suppose there is a set of models which are all 'reasonable' for estimating a quantity μ from the set of data y. Consider a parameter of interest μ that is defined and has a common interpretation for all of the considered models $M_1, \ldots, M_k$. Instead of using one single model for reaching inference for μ, Bayesian model averaging constructs $\pi(\mu \mid y)$, the posterior density of μ given the data, not conditional on any model. This is arrived at via the Bayes formula. Bayesian model averaging essentially starts from specifying

- prior probabilities $P(M_j)$ for all models $M_1, \ldots, M_k$ under consideration,
- prior densities $\pi(\theta_j \mid M_j)$ for the parameters θ_j of the M_j model.

This is as in Section 3.2. As before, denote the likelihood in model M_j by $\mathcal{L}_{n,j}$. Then, given the prior information on the parameter given the model, the integrated likelihood of model M_j is given by

$$\lambda_{n,j}(y) = \int \mathcal{L}_{n,j}(y, \theta_j)\pi(\theta_j \mid M_j)\,\mathrm{d}\theta_j.$$

The $\lambda_{n,j}(y)$ is also the marginal density at the observed data. Using the Bayes theorem, see also (3.2), the posterior density of the model is obtained as

$$P(M_j \mid y) = \frac{P(M_j)\lambda_{n,j}(y)}{\sum_{j'=1}^{k} P(M_{j'})\lambda_{n,j'}(y)}.$$

Next we compute for each model the posterior density of μ assuming that model M_j is true. This we denote by $\pi(\mu \mid M_j, y)$. The above ingredients combine to express the posterior density of the quantity of interest as

$$\pi(\mu \mid y) = \sum_{j=1}^{k} P(M_j \mid y)\pi(\mu \mid M_j, y).$$

Instead of using a single conditional posterior density $\pi(\mu \mid M_j, y)$ assuming model M_j to be true, the posterior density $\pi(\mu \mid y)$ is a weighted average of the conditional posterior densities, where the weights are the posterior probabilities of each model. By not conditioning on any given model, Bayesian model averaging does not make the mistake of ignoring model uncertainty. The posterior mean is likewise a weighted average of the posterior means in the separate models,

$$\mathrm{E}(\mu \mid y) = \sum_{j=1}^{k} P(M_j \mid y)\mathrm{E}(\mu \mid M_j, y),$$

from properties of mixture distributions. Similarly the posterior variance may be expressed via the formula

$$\mathrm{Var}(\mu \mid y) = \sum_{j=1}^{k} P(M_j \mid y)[\mathrm{Var}(\mu \mid M_j, y) + \{\mathrm{E}(\mu \mid M_j, y) - \mathrm{E}(\mu \mid y)\}^2],$$

see Exercise 7.5.

In a subjective Bayesian analysis, the prior probabilities $P(M_j)$ and prior distributions for the parameters are chosen to reflect the investigator's degree of belief in the various models and the parameters therein. A Bayesian analysis independent of the investigator's own beliefs requires noninformative priors. This is not an easy issue and is particularly challenging in model averaging situations, since noninformative priors tend to be improper, and then the normalising constants need calibration across models. Kass and Wasserman (1996) discuss these and related issues. A possible prior for the models is the uniform one that gives each model equal probability. For some choices of models it is debatable whether assigning them equal probabilities is really noninformative. When the models are nested, one could argue that it is more natural to put smaller prior probabilities on the models of larger dimension. Jeffreys (1961) proposed using the improper prior $p_j = 1/(j+1)$ for $j = 0, 1, \ldots$, for such nested models. A proper noninformative prior for the positive integers was proposed by Rissanen (1983). Sometimes one may consider more than one model having a given dimension; the singleton model-averaging schemes that average models each containing only a single of the variables, are such examples. A scheme which gives equal probability to each individual model of a certain dimension has been proposed by Berger and Pericchi (1996). Numerical integration or use of Markov chain Monte Carlo methods is typically needed to compute the posterior probabilities. For more information on the basics of Bayesian model averaging, see Draper (1995) and Hoeting *et al.* (1999).

Example 7.9 Fifty years survival with Bayesian model averaging

Example 3.11 used data from Dobson (2002, Chapter 7) on 50 years survival after graduation for groups of students, and dealt with AIC and DIC rankings of three models about four survival probabilities; model M_1 takes $\theta_1 = \theta_2 = \theta_3 = \theta_4$; model M_2 takes $\theta_1 = \theta_2$ and $\theta_3 = \theta_4$; while model M_3 operates with four different probabilities. Presently we shall use a Bayesian model-averaging approach to analyse the data, which in particular means not being forced to select one of the three models and then discard the other two. As for Example 3.11, the priors used are of the Beta $(\frac{1}{2}c, \frac{1}{2}c)$ type, for each unknown probability, where we use the Jeffreys prior with $c = 1$ in our numerical illustrations. In the spirit of questions discussed by Dobson, we shall study two parameters for our illustration,

$$\mu_1 = \mathrm{odds}(\theta_{\mathrm{ave}}) \quad \text{and} \quad \mu_2 = \mathrm{odds}(\theta_{\mathrm{men}})/\mathrm{odds}(\theta_{\mathrm{women}}). \tag{7.10}$$

Here $\mathrm{odds}(p) = p/(1-p)$, θ_{ave} is $\sum_{j=1}^{4}(m_j/m)\theta_j$, with $m = 25 + 19 + 18 + 7 = 69$, while θ_{men} and θ_{women} are respectively $(m_1\theta_1 + m_2\theta_2)/(m_1 + m_2)$ and $(m_3\theta_3 + m_4\theta_4)/(m_3 + m_4)$.

The likelihood is as in (3.17). This leads to marginal likelihoods

$$\lambda_1 = A\frac{\Gamma\left(\frac{1}{2}c + \sum_{j=1}^{4} y_j\right)\Gamma\left(\frac{1}{2}c + \sum_{j=1}^{4}(m_j - y_j)\right)}{\Gamma\left(c + \sum_{j=1}^{4} m_j\right)},$$

$$\lambda_2 = A\frac{\Gamma\left(\frac{1}{2}c + \sum_{j=1,2} y_j\right)\Gamma\left(\frac{1}{2}c + \sum_{j=1,2}(m_j - y_j)\right)}{\Gamma\left(c + \sum_{j=1,2} m_j\right)}$$
$$\times \frac{\Gamma\left(\frac{1}{2}c + \sum_{j=3,4} y_j\right)\Gamma\left(\frac{1}{2}c + \sum_{j=3,4}(m_j - y_j)\right)}{\Gamma\left(c + \sum_{j=3,4} m_j\right)},$$

$$\lambda_3 = A\prod_{j=1}^{4}\frac{\Gamma\left(\frac{1}{2}c + y_j\right)\Gamma\left(\frac{1}{2}c + m_j - y_j\right)}{\Gamma\left(c + m_j\right)}.$$

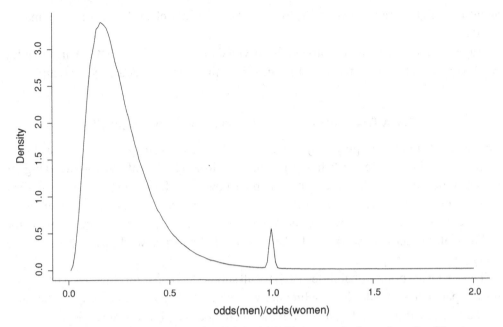

Fig. 7.12. The density of the ratio odds(men)/odds(women) given data, for 50 years survival after graduation in 1940, for the Australian data.

The factor A cancels out when computing the posterior probabilities for the three models, $P(M_j \mid \text{data}) \propto P(M_j)\lambda_j$ for $j = 1, 2, 3$. With equal prior probabilities $1/3, 1/3, 1/3$ for the three models, we find posterior model probabilities 0.0138, 0.3565, 0.6297.

The posterior distributions of $(\theta_1, \theta_2, \theta_3, \theta_4)$ are already worked out and described in connection with Example 3.11. We now obtain the full posterior distribution of the parameter $\mu = \mu(\theta_1, \theta_2, \theta_3, \theta_4)$. The operationally simplest numerical strategy is to simulate say a million four-vectors from the three-mixture distribution, and then display a density estimate for each of the μ parameters of interest, along with summary numbers like the posterior 0.05, 0.50, 0.95 quantiles. For μ_1 and μ_2 of (7.10) we find respectively 1.179, 1.704, 2.511 and 0.080, 0.239, 0.618 for these quantiles. Figure 7.12 displays the density of the odds(men)/odds(women) ratio; we used kernel density estimation of the $\log \mu_2$ values, followed by back-transformation, in order to avoid the boundary problem at zero. Note the bump at the value 1, corresponding to the small but positive posterior probability 0.0138 that all four probabilities are equal. ∎

We note that Bayesian model averaging, although in many respects a successful enterprise, has some inner coherence problems that are mostly downplayed. These relate to the fact that when different priors are used in different models to quantify the implied prior for a given interpretable parameter of interest μ, then conflicting priors emerge about a single parameter. The single Bayesian statistician who applies Bayesian model averaging with say k different models has been willing to quantify his own prior opinion about the single parameter μ in k different ways. It is not clear that this always makes good sense.

Bayesian model averaging software is provided in R in the library BMA. This includes BMA functions for linear regression, generalised linear models and survival data.

7.8 A frequentist view of Bayesian model averaging *

The candidate models employed by Bayesian model averaging may in principle be almost arbitrary and do not necessarily have to correspond to submodels of a well-defined bigger model. Often they do, however, and our plan now is to look more carefully at such procedures and their performances in the context where the candidate models represent submodels $S \subset \{1, \ldots, q\}$, with q the dimension of the nonprotected parameter vector γ. We shall furthermore work inside the local modelling framework where $\|\gamma - \gamma_0\| = O(1/\sqrt{n})$, as in Chapter 6.

Thus there are prior probabilities $P(S)$ for all submodels and prior densities $\pi(\theta, \delta_S \mid S)$ for the parameters inside the S submodel writing as earlier $\gamma = \gamma_0 + \delta/\sqrt{n}$. The integrated likelihood of model S, involving the likelihood $\mathcal{L}_{n,S}$ for this model is

$$\lambda_{n,S}(y) = \int \mathcal{L}_{n,S}(y, \theta, \gamma_0 + \delta_S/\sqrt{n}) \pi(\theta, \delta_S \mid S) \, d\theta \, d\delta_S. \qquad (7.11)$$

In this framework, the posterior density of the parameters may be expressed as

$$\pi(\theta, \delta \mid y) = \sum_S P(S \mid y) \pi(\theta, \delta_S \mid S, y). \qquad (7.12)$$

Here $\pi(\theta, \delta_S \mid S, y)$ is the posterior calculated under the model indexed by S (with $\delta_j = 0$ for $j \notin S$) while

$$P(S \mid y) = P(S)\lambda_{n,S}(y) \Big/ \sum_{S'} P(S')\lambda_{n,S'}(y)$$

is the probability of model S given data.

We now derive an approximation to and limit distribution results for $\lambda_{n,S}(y)$ of (7.11), under the local alternatives framework. We saw in Chapter 3 that the familiar BIC stems in fact from an approximation to this quantity. Let as before $\widehat{\theta}_S$ and $\widehat{\delta}_S = \sqrt{n}(\widehat{\gamma}_S - \gamma_{0,S})$ be the maximum likelihood estimators inside the S model. Then one possible approximation is by using the Laplace approximation as in Section 3.2.2,

$$\lambda_{n,S}(y) \doteq \mathcal{L}_{n,S}(y, \widehat{\theta}_S, \widehat{\gamma}_S) n^{-(p+|S|)/2} (2\pi)^{(p+|S|)/2} |J_{n,S}|^{-1/2} \pi(\widehat{\theta}_S, \widehat{\delta}_S \mid S), \quad (7.13)$$

with as before $J_{n,S}$ the observed information matrix of size $(p + |S|) \times (p + |S|)$. The quantity

$$2 \log \lambda_{n,S}(y) \approx 2 \max \log \mathcal{L}_{n,S} - (p + |S|) \log n$$

is precisely the BIC approximation (cf. Chapter 3 and Hoeting *et al.* (1999), equation (13), modulo an incorrect constant).

The asymptotic approximation (7.13), which underlies the BIC, is valid in the framework of fixed models $f(y, \theta, \gamma)$ and a fixed $g_{\text{true}}(y)$, where in particular $\delta = \sqrt{n}(\gamma - \gamma_0)$ grows with n. In this framework the candidate model S_0 with smallest Kullback–Leibler distance to the true density will have $p_n(S_0) \to 1$ as n grows, or in other words the best model will win in the end. This follows since the dominant term of $\max \ell_{n,S}$ will be $n \max \int g_{\text{true}}(y) \log f(y, \theta, \gamma) \, \mathrm{d}y$. In the framework of local alternative models the magnifying glass is set up on the $\sqrt{n}(\gamma - \gamma_0)$ scale, and different results apply. Maximised log-likelihoods are then not $O_P(n)$ apart, as under the fixed models scenario, but have differences related to noncentral chi-squared distributions. Secondly, the $n^{-|S|/2}$ ingredient above, crucial to the BIC, disappears.

Theorem 7.2 *Let the prior for the S subset model be of the form $\pi_0(\theta)\pi_S(\delta_S)$, with π_0 continuous in a neighbourhood around θ_0. Then, under standard regularity conditions, when n tends to infinity,*

$$\lambda_{n,S}(y) \doteq \mathcal{L}_{n,S}(\widehat{\theta}_S, \widehat{\gamma}_S) n^{-p/2} (2\pi)^{(p+|S|)/2} \pi_0(\widehat{\theta}_S) |J_{n,S}|^{-1/2} \kappa_n(S),$$

where $\kappa_n(S) = \int \phi(\delta_S - \widehat{\delta}_S, J_{n,S}^{11}) \pi_S(\delta_S) \, \mathrm{d}\delta_S$. This approximation holds in the sense that $\log \lambda_{n,S}(y)$ is equal to the logarithm of the right-hand side plus a remainder term of size $O_P(n^{-1/2})$.

Proof. We work with the case of the full model, where $S = \{1, \ldots, q\}$, and write $\widehat{\theta}$ and $\widehat{\delta}$ for $\widehat{\theta}_{\text{wide}}$ and $\widehat{\delta}_{\text{wide}}$, and so on. The general case can be handled quite similarly. Define the likelihood ratio

$$F_n(s, t) = \frac{\mathcal{L}_n(\widehat{\theta} + s/\sqrt{n}, \gamma_0 + (\widehat{\delta} + t)/\sqrt{n})}{\mathcal{L}_n(\widehat{\theta}, \gamma_0 + \widehat{\delta}/\sqrt{n})} = \frac{\mathcal{L}_n(\widehat{\theta} + s/\sqrt{n}, \widehat{\gamma} + t/\sqrt{n})}{\mathcal{L}_n(\widehat{\theta}, \widehat{\gamma})}.$$

Then, with Taylor expansion analysis, one sees that

$$\log F_n(s, t) = -\frac{1}{2} \begin{pmatrix} s \\ t \end{pmatrix}^{\mathrm{t}} J_n \begin{pmatrix} s \\ t \end{pmatrix} + O_P\left(n^{-1/2} \left\| \begin{pmatrix} s \\ t \end{pmatrix} \right\|^3 \right).$$

For a calculation needed in a moment we shall need the following integration identity which follows from properties of the multivariate normal density. For a symmetric positive definite $(p + q) \times (p + q)$ matrix A,

$$\int \exp\left\{-\frac{1}{2} \begin{pmatrix} s \\ t \end{pmatrix}^{\mathrm{t}} A \begin{pmatrix} s \\ t \end{pmatrix}\right\} \mathrm{d}s = \frac{(2\pi)^{p/2}}{|A|^{1/2}|A^{11}|^{1/2}} \exp\{-\frac{1}{2} t^{\mathrm{t}}(A^{11})^{-1} t\},$$

where A^{11} is the $q \times q$ lower right-hand submatrix of A^{-1}. Substituting $\theta = \widehat{\theta} + s/\sqrt{n}$ and $\delta = \widehat{\delta} + t$ in the $\lambda_{n,S}(y)$ integral (7.11) and using the integration identity now leads to

$$\lambda_{n,S}(y) = \mathcal{L}_n(\widehat{\theta}, \widehat{\gamma}) n^{-p/2} \int F_n(s,t) \pi_0(\widehat{\theta} + s/\sqrt{n}) \pi(\widehat{\delta} + t) \, ds \, dt$$

$$\doteq \mathcal{L}_n(\widehat{\theta}, \widehat{\gamma}) n^{-p/2} \pi_0(\widehat{\theta}) (2\pi)^{p/2} |J_n|^{-1/2} |J_n^{11}|^{-1/2}$$

$$\times \int \pi(\widehat{\delta} + t) \exp\{-\tfrac{1}{2} t^{\mathrm{t}} (J_n^{11})^{-1} t\} \, dt.$$

This proves the claims made. For further details and technicalities related to this proof, see Hjort (1986a) and Hjort and Claeskens (2003, Appendix). □

When n tends to infinity we have that $J_{n,S} \to_p J_S$, and the limit of $J_{n,S}^{11}$ is Q_S, cf. Section 6.1. Combining this with some previous results, we find that

$$\lambda_{n,S}(y) \doteq k \exp(\tfrac{1}{2} \widehat{\delta}_S^{\mathrm{t}} Q_S^{-1} \widehat{\delta}_S)(2\pi)^{|S|/2} |J_S|^{-1/2} \int \phi(\delta_S - \widehat{\delta}_S, Q_S) \pi_S(\delta_S) \, d\delta_S,$$

where the constant k equals $n^{-p/2}(2\pi)^{p/2} \pi_0(\widehat{\theta})$.

This leads to a precise description of posterior probabilities for the different models in the canonical limit experiment where all quantities have been estimated with full precision except δ, for which we use the limit $D \sim \mathrm{N}_q(\delta, Q)$ of $D_n = \sqrt{n}(\widehat{\gamma}_{\mathrm{wide}} - \gamma_0)$. This also points to a relation between AIC and the limit density λ_S, in view of the result

$$\mathrm{AIC}_{n,S} - \mathrm{AIC}_{n,\emptyset} \xrightarrow{d} \mathrm{aic}(S, D) = D^{\mathrm{t}} Q^{-1} Q_S^0 Q^{-1} D - 2|S|$$

of Theorem 5.4. It furthermore holds that

$$\widehat{\delta}_S^{\mathrm{t}} Q_S^{-1} \widehat{\delta}_S \xrightarrow{d} D^{\mathrm{t}} Q^{-1} Q_S^0 Q^{-1} D,$$

from Hjort and Claeskens (2003, Section 3.2). Hence, the limit version of the factor $\exp(\tfrac{1}{2} \widehat{\delta}_S^{\mathrm{t}} Q_S^{-1} \widehat{\delta}_S)$, appearing in the expression of $\lambda_{n,S}(y)$, can be rewritten as $\exp\{\tfrac{1}{2} \mathrm{aic}(S, D)\} \exp(|S|)$. Thus, with $P(S \mid D) \propto P(S) \lambda_S$, we have that

$$\lambda_S = \exp(\tfrac{1}{2} D^{\mathrm{t}} (Q_S^{-1})^0 D)(2\pi)^{|S|/2} |J_S|^{-1/2} \int \phi(\delta_S - D_S, Q_S) \pi_S(\delta_S) \, d\delta_S$$

$$= \exp(\tfrac{1}{2} \mathrm{AIC}_S) \exp(|S|)(2\pi)^{|S|/2} |J_S|^{-1/2} \qquad (7.14)$$

$$\times \int \phi(\delta_S - D_S, Q_S) \pi_S(\delta_S) \, d\delta_S,$$

with $D_S = Q_S(Q^{-1}D)_S$.

The methods above also yield first-order approximations to posteriors in submodels. The posterior of δ_S in model S will in fact converge to

$$\pi_S(\delta_S \mid \widehat{\delta}_S) = \mathrm{const} \cdot \pi_S(\delta_S) \phi(\delta_S - \widehat{\delta}_S, Q_S), \qquad (7.15)$$

with $\phi(u, Q_S)$ the density of a $\mathrm{N}_{|S|}(0, Q_S)$; see Hjort and Claeskens (2003, Section 9). This further entails that the Bayesian submodel estimator $\widetilde{\mu}_S = \mathrm{E}_S(\mu \mid \mathrm{data})$ is asymptotically equivalent to the simpler estimator $\bar{\mu}_S = \mathrm{E}\{\mu(\widehat{\theta}_S, \gamma_{0,S} + \delta_S/\sqrt{n}) \mid \widehat{\delta}_S\}$, where the distribution of δ_S is that of (7.15).

Finally, these methods lead to clear limit distribution results for $\sqrt{n}(\widetilde{\mu}_S - \mu_{\mathrm{true}})$, that can be compared directly with those found in Section 6.3 for the maximum likelihood estimators. It turns out that these Bayesian submodel limit distributions are exactly as for a class of frequentist generalised ridging methods; cf. Hjort and Claeskens (2003, Sections 8 and 9). This may be used to construct simpler numerical approximations to the often intricate Bayesian model average estimators and to compare performances with likelihood estimators.

7.9 Bayesian model selection with canonical normal priors *

The most important special case is when the prior for δ_S is $\mathrm{N}_{|S|}(0, \tau_S^2 Q_S)$. This corresponds to equally spread-out and independent priors around zero for the transformed parameters $a_S = (Q^{-1/2}\delta)_S$. Then

$$\lambda_S = \exp\left(\tfrac{1}{2}\frac{\tau_S^2}{1+\tau_S^2} D_S^{\mathrm{t}} Q_S^{-1} D_S\right)(1 + \tau_S^2)^{-|S|/2}|J_{00}|^{-1/2}.$$

The last determinant is independent of $|S|$, and is arrived at via $|J_S|^{-1/2}|Q_S|^{-1/2}$, since $|J_S| = |J_{00}||Q_S|^{-1}$. For the narrow model with $S = \emptyset$, result (7.14) is also valid, and $\lambda_\emptyset = |J_{00}|^{-1/2}$.

This leads to a Bayesian information criterion, which we term the BLIC, with L for 'local', as in local neighbourhood asymptotics. This construction of the criterion is similar to that of the BIC, but relates to a different statistical magnifying glass, more particularly that of using $\gamma = \gamma_0 + \delta/\sqrt{n}$. From (7.14) the criterion reads

$$\mathrm{BLIC} = \frac{\tau_S^2}{1+\tau_S^2} D_S^{\mathrm{t}} Q_S^{-1} D_S - |S|\log(1 + \tau_S^2) + 2\log p(S),$$

since the posterior model probability is close to being proportional to $P(S)\lambda_S$. For the narrow model $\mathrm{BLIC} = 2\log P(\emptyset)$. The value τ_S is meant to be a spread measure for δ_S in submodel S. We select the candidate model with largest BLIC since this is the most probable one, given data, in the Bayesian formulation.

The formula above is valid for the limit experiment. For real data we use $\widehat{\delta}_S$ for D_S, which leads to

$$\widehat{\mathrm{BLIC}} = \frac{\tau_S^2}{1+\tau_S^2} n(\widehat{\gamma}_S - \gamma_{0,S})^{\mathrm{t}} \widehat{Q}_S^{-1}(\widehat{\gamma}_S - \gamma_{0,S}) - |S|\log(1 + \tau_S^2) + 2\log P(S).$$

For the estimation of the spread, we first have that $D_S^{\mathrm{t}} Q_S^{-1} D_S$ given δ is a noncentral chi-squared with parameter $\delta_S^{\mathrm{t}} Q_S^{-1} \delta_S$. The mean of $|S| + \delta_S^{\mathrm{t}} Q_S^{-1} \delta_S$ equals $|S|(1 + \tau_S^2)$. We thus suggest, in an empirical Bayes fashion, to estimate $1 + \tau_S^2$, by $D_S^{\mathrm{t}} Q_S^{-1} D_S/|S|$.

This gives

$$\text{BLIC}^* = |S|\{\widehat{\tau}_S^2 - \log(1 + \widehat{\tau}_S^2)\} + 2\log P(S),$$

with $\widehat{\tau}_S^2 = \max\{D_S^t Q_S^{-1} D_S / |S| - 1, 0\}$. There are various alternatives to the procedure above which may also be considered.

7.10 Notes on the literature

Problems associated with inference after model selection have been pointed out by Hurvich and Tsai (1990), Pötscher (1991), Chatfield (1995) and Draper (1995) amongst others. Sen and Saleh (1987) studied effects of pre-testing in linear models. Buckland *et al.* (1997) made the important statement that the uncertainty due to model selection should be incorporated into statistical inference. Burnham and Anderson (2002) work further on this idea and suggest in their chapter 4 a formal method for inference from more than one model, which they term 'multimodel inference'. Pötscher (1991) studied effects of model selection on inference for the case of a nested model search. Kabaila (1995) builds on this work to study the effect of model selection on confidence and prediction regions. In Kabaila (1998), for the setting of normal linear regression, critical values are chosen such that the constructed interval is as short as possible but still the desired coverage probability is guaranteed. For normal linear regression models, Kabaila and Leeb (2006) study upper bounds for the large-sample limit minimal coverage probability of naively constructed confidence intervals. The main parts of this chapter are based on the paper about frequentist model average estimators by Hjort and Claeskens (2003).

Foster and George (1994) studied the performance of estimators post-selection in a linear regression setting and proposed the risk inflation criterion (RIC). The risk inflation is defined as the ratio of the risk of the estimator post-selection to the risk of the best possible estimator which uses the true variables in the model. An out-of-bootstrap method for model averaging with model weights obtained by the bootstrap is constructed by Rao and Tibsirani (1997). Yang (2001) uses sample-splitting for determining the weights.

Ridge regression has a large literature, with Hoerl and Kennard (1970) an early important contribution. These methods have essentially been confined to linear regression models, though, sometimes with a high number of covariates; see Frank and Friedman (1993) and Hastie *et al.* (2001) for discussion. Ridge regression in neural networks is also called the weight principle, or weight decay (Saunders *et al.*, 1998; Smola and Schölkopf, 1998). Generalised ridging by shrinking subsets of maximum likelihood parameters appears in Hjort and Claeskens (2003). The bias-reduced estimators of Firth (1993) for likelihood models, and in particular for generalised linear models, also possess the property of shrinkage. George (1986a,b) studies multiple shrinkage estimation in normal models.

Bayesian model averaging has seen literally hundreds of journal papers over the past decade or so; see e.g. Draper (1995), Hoeting *et al.* (1999), Clyde (1999), Clyde and George (2004). The literature has mostly been concerned with issues of interpretation and computation. Results about the large-sample behaviour of Bayesian model-averaging schemes were found in Hjort and Claeskens (2003). Methods and results of the present chapter may be generalised to proportional hazard regression models, where algorithmic and interpretation aspects have been discussed by Volinsky *et al.* (1997).

Exercises

7.1 *Smoothing the FIC:* Consider the smoothed FIC estimator with weights as in Example 7.3. Show that in the $q = 1$-dimensional case, using smoothing parameter $\kappa = 1$ is equivalent to using the smoothed AIC weights.

7.2 *The Ladies Adelskalenderen:* Refer to Exercise 2.9. We study the times on the 3000-m and 5000-m. Fit a linear regression model with response variable Y to the 5000-m time and the 3000-m time as covariate x. Next, fit the model with an added quadratic variable and allowing for heteroscedasticity as in the data analysis of Section 5.6. Compute a model-averaged estimator of $E(Y \mid x)$ over these two models, give its asymptotic distribution function and graph this function, inserting estimates for unknowns. Use an appropriate set of weights.

7.3 *The low birthweight data:* For a set of relevant candidate models, (i) obtain the post-model-selection weights for AIC, BIC and FIC, (ii) obtain the smooth-AIC, smooth-BIC and smooth-FIC weights and use these to construct model-averaged estimators.

7.4 *Exponential or Weibull, again:* Suppose that two models are considered for life-length data $Y_1, \ldots, Y_n$, the Weibull one, with cumulative distribution $1 - \exp\{-(\theta y)^\gamma\}$, and the exponential one, where $\gamma = 1$.

(a) Write out precise formulae for a Bayesian model average method for estimating the median $\mu = (\log 2)^{1/\gamma}/\theta$, with the following ingredients: the prior probabilities are $\frac{1}{2}$ and $\frac{1}{2}$ for the two models; the prior for θ is a unit exponential, in both cases; and the prior for γ, in the case of the Weibull model, is such that $\delta = \sqrt{n}(\gamma - 1)$ is a N$(0, \tau^2)$, say with $\tau = 1$.

(b) For the two priors involved in (a), find the two implied priors for μ, and compare them, e.g. by histograms from simulations.

(c) Implement the method and compare its performance to that of other model average estimators of μ.

7.5 *Mean and variance given data:* This exercise leads to formulae for the posterior mean and variance of a focus parameter.

(a) Suppose X is drawn from a mixture distribution of the type $\sum_{j=1}^{k} p_j g_j$, where $p_1, \ldots, p_k$ are probabilities and $g_1, \ldots, g_k$ are densitites. Then X may be represented as X_J, with J taking values $1, \ldots, k$ with probabilities $p_1, \ldots, p_k$. Use this to show that

$$\mathrm{E}X = \widetilde{\xi} = \sum_{j=1}^{k} p_j \xi_j \quad \text{and} \quad \mathrm{Var}\, X = \sum_{j=1}^{k} \{p_j \sigma_j^2 + p_j (\xi_j - \widetilde{\xi})^2\},$$

where ξ_j and σ_j are the mean and standard deviation for a variable drawn from g_j.

(b) Then use this to derive the formulae for posterior mean and posterior variance of the interest parameter μ in Section 7.7.

7.6 *Mixing over expansion orders for density estimation:* In Example 2.9 we worked with density estimators of the form

$$\widehat{f}_m(x) = f_0(x) \exp\left\{\sum_{j=1}^{m} \widehat{a}_j(m)\psi_j(x)\right\} \Big/ c_m(\widehat{a}(m)),$$

where $\widehat{a}(m)$ with components $\widehat{a}_1(m), \ldots, \widehat{a}_m(m)$ is the maximum likelihood estimator inside the mth-order model. We used AIC to select the order, say within an upper bound $m \le m_{\mathrm{up}}$.

(a) Consider the density estimator that mixes over orders,

$$\widehat{f}(x) = \sum_{m=1}^{m_{\mathrm{up}}} c_n(m \mid D_n)\widehat{f}_m(x),$$

where the weights depend on $D_n = \sqrt{n}\widehat{a}(m_{\mathrm{up}})$. Work out the characteristics of the limit distribution for $\sqrt{n}(\widehat{f} - f)$, assuming that the real f is within the class studied, of order m_{up}.

(b) Implement this method, with f_0 the standard normal, and try it out on simulated data, for a suitable list of true densities. Use AIC and smoothed AIC weights.

(c) Then compare performance with the mixture estimator that uses FIC and smoother FIC weights for $c_n(m \mid D_n)$, where the focus parameter is the position $x_0 = x_0(a)$ at which the density is maximal.

8

Lack-of-fit and goodness-of-fit tests

We build on order selection ideas to construct statistics for testing the null hypothesis that a function takes a certain parametric form. Traditionally two labels have been applied for such tests. Most often, the name lack-of-fit test is reserved for testing the fit of a function in a regression context, for example statements about the functional form of the mean response. On the other hand, tests for the distribution function of random variables are called goodness-of-fit tests. Examples include testing for normality and testing Weibull-ness of some failure time data. The main difference with the previous chapters is that there is a well-defined model formulated under a null hypothesis, which is to be tested against broad alternative classes of models. The test statistics actively employ model selection methods to assess adequacy of the null hypothesis model. The chapter also includes a brief discussion of goodness-of-fit monitoring processes and tests for generalised linear models.

8.1 The principle of order selection

General approaches for nonparametric testing of hypotheses consist of computing both a parametric and a nonparametric estimate of the hypothesised curve and constructing a test statistic based on some measure of discrepancy. One example is the sum of squared differences $\sum_{i=1}^{n}\{\widehat{\mu}(x_i) - \mu_{\widehat{\theta}}(x_i)\}^2$, where $\widehat{\mu}$ can be any nonparametric estimator, for example a local polynomial, spline or wavelet estimator of the function μ. Other examples of tests are based on the likelihood ratio principle. Our emphasis is on tests which use a model selection principle under the alternative hypothesis. Two such classes of tests are the order selection test in the lack-of-fit setting and the Neyman smooth-type tests in the goodness-of-fit framework.

Example 8.1 Testing for linearity
Consider the regression model

$$Y_i = \mu(x_i) + \varepsilon_i \quad \text{where } \varepsilon_i \sim N(0, \sigma^2) \text{ for } i = 1, \ldots, n.$$

A linear least squares regression is easily fit to the $(x_1, y_1), \ldots, (x_n, y_n)$ data and has a straightforward interpretation. In order to be comfortable that this model provides a reasonable fit to the data, we wish to test the null hypothesis

H_0: there exist values (θ_1, θ_2) such that $\mu(x) = \theta_1 + \theta_2 x$ for all x.

Since the exact value of the coefficients (θ_1, θ_2) is not known beforehand, we can equivalently write this hypothesis as

$$H_0: \mu(\cdot) \in \{\mu_\theta(\cdot): \theta = (\theta_1, \theta_2) \in \Theta\},$$

where $\mu_\theta(x) = \theta_1 + \theta_2 x$ and the parameter space Θ in this case is $\mathbb{R}^2$.

In parametric testing problems an alternative could consist, for example, of the set of all quadratic functions of the form $\theta_1 + \theta_2 x + \theta_3 x^2$. If this is the alternative hypothesis we test the linear null model against the quadratic alternative model. In nonparametric testing we are not satisfied with such an approach. Why would the model be quadratic? Why not cubic, or quartic, or something completely different? It might be better for our purposes to not completely specify the functional form of the function under the alternative hypothesis, but instead perform a test of H_0 against

$$H_a: \mu(\cdot) \notin \{\mu_\theta: \theta = (\theta_1, \theta_2) \in \Theta\}.$$

This is the type of testing situation that we consider in this chapter. ∎

We test whether a certain function g belongs to a parametric family,

$$H_0: g(\cdot) \in \{g_\theta(\cdot): \theta \in \Theta\}. \tag{8.1}$$

The parameter space Θ is a subset of $\mathbb{R}^p$ with p a finite natural number. In a regression setting the function g can, for example, be the mean of the response variable Y, that is $g(x) = E(Y \mid x)$, or it can be the logit of the probability of success in case Y is a Bernoulli random variable $g(x) = \text{logit}\{P(Y = 1 \mid x)\}$. Yet another example is when g represents the standard deviation function, $g(x) = \sigma(x)$, where we might test for a certain heteroscedasticity pattern. This hypothesis is contrasted with the alternative hypothesis

$$H_a: g(\cdot) \notin \{g_\theta(\cdot): \theta \in \Theta\}. \tag{8.2}$$

Since there is no concrete functional specification of the function g under the alternative hypothesis, we construct a sequence of possible alternative models via a series expansion, starting from the null model. Write $g(x) = g_\theta(x) + \sum_{j=1}^{\infty} \gamma_j \psi_j(x)$. This series does not need to be orthogonal, although that might simplify practical computation and make some theoretical results easier to obtain. Some common examples for basis functions ψ_j are polynomial functions, in particular orthogonal Legendre polynomials, cosine functions or a trigonometric system with both sine and cosine functions, in which case the coefficients γ_j are called the Fourier coefficients. Wavelet basis functions are another possibility. The functions ψ_j are known functions that span a 'large' space of functions, since we wish to keep the space of alternative models reasonably large. We do not provide details here on properties of function approximation spaces, but rather explain the connection to model selection. It is understood that the function g_θ does not consist of a linear combination of functions ψ_j. If this is the case, we discard those ψ_j from the set of basis functions used in the series expansion.

Now consider several approximations to the function g, constructed by considering only finite contributions to the sum. A sequence of nested models is the following:

$$g(x; m) = g_\theta(x) + \sum_{j=1}^{m} \gamma_j \psi_j(x) \quad \text{for } m = 1, 2, \ldots$$

The function $g(x; 0)$ corresponds to the model g_θ under the null hypothesis. Thus a nested set of models $g(\cdot; 0), g(\cdot; 1), \ldots$ are available, usually with an upper bound m_n for the order m. A model selection method can pick one of these as being the best approximation amongst the constructed series. An order selection test exploits this idea. It uses a model selection criterion to choose one of the available models. If this model is $g(\cdot; 0)$, the null hypothesis is not rejected. If the selected model is not the null model, there is evidence to reject the null hypothesis. Hence, order selection tests are closely linked to model selection methods. Let $\widehat{m}$ be the order chosen by a model selection mechanism. A test of H_0 in (8.1) versus the nonparametric alternative H_a in (8.2) can now simply be

reject H_0 if and only if $\widehat{m} \geq 1$.

This is called an order selection test (Eubank and Hart, 1992; Hart, 1997).

8.2 Asymptotic distribution of the order selection test

We consider first order selection tests based on AIC. The log-likelihood function of the data under the model with order m is $\ell_n(\theta; \gamma_1, \ldots, \gamma_m)$. AIC, see Chapter 2, takes the form

$$\text{AIC}(m) = 2\,\ell_n(\widehat{\theta}_{(m)}; \widehat{\gamma}_1, \ldots, \widehat{\gamma}_m) - 2(p + m),$$

with p the dimension of θ, the vector of all parameters in the null model. The subscript (m) indicates that the estimator of θ is obtained in the model with order m for the γ_j. Again, $m = 0$ corresponds to the null model. The test rejects the null hypothesis if and only if $\widehat{m}_{\text{AIC}} \geq 1$. The limiting distribution results are easier to understand when working with AIC differences

$$\begin{aligned} \text{aic}_n(m) &= \text{AIC}(m) - \text{AIC}(0) \\ &= 2\{\ell_n(\widehat{\theta}_{(m)}; \widehat{\gamma}_1, \ldots, \widehat{\gamma}_m) - \ell_n(\widehat{\theta}_{(0)})\} - 2m = L_{m,n} - 2m. \end{aligned}$$

We recognise the $L_{m,n}$ term as the log-likelihood ratio statistic for the parametric testing problem

$$H_0: g(x) = g_\theta(x) \text{ versus } H_a: g(x) = g_\theta(x) + \sum_{j=1}^{m} \gamma_j \psi_j(x) \text{ for all } x. \qquad (8.3)$$

The penalty term in AIC is twice the difference in degrees of freedom for these two models. The model order chosen by AIC is the argument that maximises AIC, or equivalently, $\mathrm{aic}_n(m)$:

$$\widehat{m}_{\mathrm{aic}} = \arg \max_m \mathrm{aic}_n(m).$$

Rejecting the null hypothesis when $\widehat{m}_{\mathrm{aic}} \geq 1$ is equivalent to rejecting when

$$T_{n,\mathrm{OS}} = \sup_{m \geq 1} L_{m,n}/m > 2.$$

The order selection test statistic $T_{n,\mathrm{OS}} = \max_{m \leq r} L_{m,n}/m$ with upper bound r has limiting null distribution equal to that of

$$T_{\mathrm{OS},r} = \max_{m \leq r}(Z_1^2 + \cdots + Z_m^2)/m,$$

with $Z_1, \ldots, Z_r$ independent N(0, 1).

Lemma 8.1 *Consider the nested model sequence indexed by $m = 0, 1, \ldots, r$ for some upper threshold r, and assume the null hypothesis is true. Then, under asymptotic stability conditions on the sequence of covariate vectors, there is joint convergence in distribution of $(L_{1,n}, \ldots, L_{r,n})$ to that of $(L_1, \ldots, L_r)$, which are partial sums of $Z_1^2, \ldots, Z_r^2$ with $Z_1, \ldots, Z_r$ independent N(0, 1).*

Proof. It is clear that $L_{m,n} \to_d \chi_m^2$ under H_0, for each separate m, but the statement we are to prove is stronger. There is joint convergence of the collection of $L_{m,n}$ to that of

$$L_m = D^{\mathrm{t}} Q^{-1} Q_m^0 Q^{-1} D \quad \text{for } m = 1, \ldots, r,$$

where we use notation parallelling that of Section 6.1. Thus Q is the usual $r \times r$ lower-right submatrix of the limiting $(p + r) \times (p + r)$ inverse information matrix, and $D \sim \mathrm{N}_r(0, Q)$; $Q_m = (\widetilde{\pi}_m Q \widetilde{\pi}_m^{\mathrm{t}})^{-1}$ is the $m \times m$ lower-right submatrix of the $(p + m) \times (p + m)$ inverse information matrix in the $(p + m)$-dimensional submodel, writing $\widetilde{\pi}_m$ for the projection matrix $\pi_{\{1,\ldots,m\}}$; and $Q_m^0 = \widetilde{\pi}_m^{\mathrm{t}} Q_m \widetilde{\pi}_m$ is the $r \times r$ extension of Q_m that adds zeros. The statement of the lemma follows from this in case of a diagonal Q.

The nontrivial aspect of the lemma is that it holds also for general nondiagonal Q. Let $N = Q^{-1/2} D \sim \mathrm{N}_r(0, I_r)$ and write $L_m = N^{\mathrm{t}} H_m N$ with projection matrices $H_m = Q^{-1/2} Q_m^0 Q^{-1/2}$. The point is now that the differences $H_1, H_2 - H_1, H_3 - H_2, \ldots$ are orthogonal; $(H_3 - H_2)(H_4 - H_3) = 0$, etc. This means that differences $L_m - L_{m-1} = N^{\mathrm{t}}(H_m - H_{m-1})N$ are independent χ_1^2 variables, proving the lemma. $\square$

The limiting distribution of $T_{\mathrm{OS},r}$ can easily be simulated for any fixed r. Various papers deal with the limiting case where $r = m_n \to \infty$ with growing sample size. Aerts *et al.* (1999) list the needed conditions to prove in their theorem 1 for this case of $r \to \infty$

that

$$T_{n,\mathrm{OS}} \xrightarrow{d} T_{\mathrm{OS}} = \sup_{m \geq 1} \frac{1}{m} \sum_{j=1}^{m} Z_j^2.$$

Woodroofe (1982) used a combinatorial lemma of Spitzer (1956) to show that this variable has distribution function

$$P(T_{\mathrm{OS}} \leq x) = \exp\left\{-\sum_{j=1}^{\infty} \frac{P(\chi_j^2 > jx)}{j}\right\}. \tag{8.4}$$

The limiting type I error is $P(T_{\mathrm{OS}} > 2)$. Hart (1997) has shown that this cumulative distribution function can be well approximated by

$$F(x, M) = \exp\left\{-\sum_{j=1}^{M} \frac{P(\chi_j^2 > jx)}{j}\right\}.$$

The error of approximation $|F_T(x) - F(x, M)|$ is bounded by $(M + 1)^{-1}\epsilon_x^{M+1}/(1 - \epsilon_x)$ where $\epsilon_x = \exp(-\{(x - 1) - \log x\}/2)$. It is now easy to calculate (see Exercise 8.1) that the critical level of this test is equal to about 0.288.

By most standards of hypothesis testing, this cannot be accepted. Table 7.1 of Hart (1997) shows for Gaussian data with known variance the percentiles of the order selection statistic for various sample sizes. A simple remedy to the large type I error is to change the penalty constant in AIC. For any level $\alpha \in (0, 1)$ we can select the penalty constant c_n such that under the null hypothesis $P(T_{\mathrm{OS}} > c_n) = \alpha$. Via the approximation above, it follows that for $\alpha = 0.05$ the penalty constant is $c_\infty = 4.179$ for the limit distribution T_{OS}. Hence, the test which rejects when the maximiser $\widehat{m}_{AIC;c_\infty}$ of $L_{m,n} - 4.179\,m$, over $m = 0, 1, \ldots$, is strictly positive, has asymptotic level 0.05 (Aerts *et al.*, 1999).

Example 8.2 Low-iron rat teratology data

We consider the low-iron rat teratology data (Shepard *et al.*, 1980). A total of 58 female rats were given different amounts of iron supplements, ranging from none to normal levels. The rats were made pregnant and after three weeks the haemoglobin level of the mother (a measure for the iron intake), as well as the total number of foetuses (here ranging between 1 and 17), and the number of foetuses alive, was recorded. Since individual outcomes (dead or alive) for each foetus might be correlated for foetuses of the same mother animal, we use the beta-binomial model, which allows for intra-cluster correlation. This full-likelihood model has two parameters: π, the proportion of dead foetuses and ρ, the correlation between outcomes of the same cluster. The model allows for clusters of different sizes, as is the case here. We condition on the cluster sizes n_i. The beta-binomial likelihood can be written

$$f(y_i, \pi_i, \rho_i) = \binom{n_i}{y_i} \frac{\Gamma(k_i)}{\Gamma(k_i\pi_i)\Gamma(k_i(1 - \pi_i))} \frac{\Gamma(k_i\pi_i + y_i)\Gamma(k_i(1 - \pi_i) + n_i - y_i)}{\Gamma(k_i + n_i)},$$

where k_i is the strength parameter of the underlying Beta distribution $(k_i \pi_i, k_i(1 - \pi_i))$ for the probabilities. It is related to the correlation parameter via $\rho_i = 1/(k_i + 1)$; see Exercise 8.2. One may check that when Y_i has the beta-binomial distribution with parameters (n_i, π_i, ρ_i), as here, then $E\, Y_i = n_i \pi_i$, as for binomial data, but that the variance is increased,

$$\text{Var}\, Y_i = n_i \pi_i (1 - \pi_i)\left(1 + \frac{n_i - 1}{k_i + 1}\right) = n_i \pi_i (1 - \pi_i)\{1 + (n_i - 1)\rho_i\},$$

see again Exercise 8.2. It also follows that if $\rho_i = 0$, then $Y_i \sim B(n_i, \pi_i)$.

Let $\pi(x)$ denote the expected proportion of dead foetuses for female rats with haemoglobin level x. Suppose we wish to test whether

$$H_0 \colon \pi(x) = \frac{\exp(\theta_1 + \theta_2 x)}{1 + \exp(\theta_1 + \theta_2 x)} \quad \text{for each } x.$$

The alternative model is not parametrically specified. We apply the order selection test using (a) cosine basis functions $\psi_j(x) = \cos(\pi j x)$ (where x is rescaled to the $(0,1)$ range) and (b) polynomial functions $\psi_j(x) = x^{j+1}$. For the cosine basis functions the value of $T_{n,\text{OS}}$ equals 4.00, with a corresponding p-value of 0.06. For the polynomial basis, the test statistic has value 6.74, with p-value 0.01. For computational reasons an upper bound of $r = 14$ is used for both tests. Both show some evidence of lack of fit of the linear model on logit scale. ∎

8.3 The probability of overfitting *

AIC is often blamed for choosing models with too many parameters. Including these superfluous parameters is called overfitting. For nested models, this phenomenon has been studied in detail, and leads to a study of generalised arc-sine laws, as we explain below. It turns out that the effect is far less dramatic than often thought. Other criteria such as Mallows's C_p and the BIC will be discussed as well.

Example 8.3 Lack-of-fit testing in generalised linear models

We let AIC choose the order of the polynomial regression model in a generalised linear model. In a regression setting, the model is of the form

$$f(y_i; \theta_i, \eta) = \exp[\{y_i \theta_i - b(\theta_i)\}/a(\phi) + c(y_i, \phi)],$$

where $b(\cdot)$ and $c(\cdot)$ are known functions, corresponding to the type of exponential family considered. The lack-of-fit perspective is that we are not sure whether a simple parametric model fits the data well. Therefore we extend this simple model with some additional terms. As approximators to θ_i we take

$$\theta_i = \theta(x_i; \beta_1, \ldots, \beta_{p+m}) = \sum_{j=1}^{p} \beta_j x_{i,j} + \sum_{j=1}^{m} \beta_{p+j} \psi_j(x_i), \quad m = 1, 2, \ldots \quad (8.5)$$

It is understood that the functions $\psi_j(\cdot)$ are not linear combinations of any of the variables x_j for $j = 1, \ldots, p$. A polynomial basis $\psi_j(x) = x^j$ (in one dimension), or an orthogonalised version of this, is one common choice. In Fourier series expansions, we might take $\psi_j(x) = \sqrt{2}\cos(j\pi x)$, for example. If the simple model $\theta_i = x_i^t\beta$ fits the data well, the number of additional components should be $m = 0$. If not, a better value for the order will be some $m \geq 1$. The question here is to select the number of additional terms in the sequence of nested models. The models are indeed nested since the model with $m = 0$ is a subset of that with $m = 1$, which in its turn is a subset of the model with $m = 2$, etc. The question we wish to answer is: what is the probability that a model selection method such as AIC will select a value $\widehat{m}_{\mathrm{aic}} \geq 1$, while in fact $m = 0$? ∎

Example 8.4 Density estimation via AIC

Consider again the situation of Examples 2.9 and 3.5. It is actually similar to that of the previous example. We start with a known density function f_0, and build a more complex density function via a log-linear expansion. The density function, after adding m additional terms, has the form $f_m = f_0\exp(\sum_{j=1}^m a_j\psi_j)/c_m(a)$. See Examples 2.9 and 3.5 for more details. The sequence of density functions $f_0, f_1, \ldots$ induces a nested sequence of models. What is the probability that a model selection criterion will select more log-linear terms than really necessary? ∎

A detailed calculation of the limiting probabilities of AIC selecting particular model orders in nested models is facilitated by working with AIC differences. Since $\mathrm{AIC}(m) = 2\ell_n(\widehat{\theta}_{(m)}, \widehat{\gamma}_1, \ldots, \widehat{\gamma}_m) - 2(p + m)$, the difference is

$$\mathrm{aic}_n(m) = \mathrm{AIC}(m) - \mathrm{AIC}(0) = 2L_{m,n} - 2m,$$

with $L_{m,n}$ the likelihood ratio statistic for testing (8.3). Under the null hypothesis, the limiting AIC differences may be written as

$$\mathrm{aic}(m) = \sum_{j=1}^m Z_j^2 - 2m = \sum_{j=1}^m (Z_j^2 - 2),$$

where $Z_1, Z_2, \ldots$ are independent standard normals.

Overfitting means that more parameters than strictly necessary are included in the selected model. Via the representation of limiting AIC differences via χ^2 random variables we can calculate the probability that this happens. The combinatorial lemma of Spitzer (1956) and the generalised arc-sine laws obtained by Woodroofe (1982) give us the probability distribution needed to determine that AIC in the limit picks a certain model order. The result is based on properties of random walks. Let $Y_1, Y_2, \ldots$ be independent and identically distributed random variables. Then define the random walk $\{W_j : j \geq 0\}$ with $W_0 = 0$ and $W_j = Y_1 + \cdots + Y_j$. Working with the limiting χ^2 random variables, this can be applied as follows.

We now write $Y_j = Z_j^2 - 2$, and define the random walk accordingly. The index m maximising AIC is also the maximising index of the random walk $\{W_j\}$:

$$\widehat{m}_{\mathrm{aic}} = \text{the } m \text{ for which } W_m \text{ equals } \max_{1 \le j \le r} W_j,$$

with r the upper bound of number of coefficients considered. Under the present null hypothesis, $\widehat{m}_{\mathrm{aic}}$ is the number of superfluous parameters. Note that this maximum is well defined since the random walk has a negative drift, $EY_j = -1$. To calculate $P(\widehat{m}_{\mathrm{aic}} = m)$ we use the generalised arc-sine probability distributions as in Woodroofe (1982), to obtain that for $m = 0, \ldots, r$,

$$P(\widehat{m}_{\mathrm{aic}} = m) = P(W_1 > 0, \ldots, W_m > 0)P(W_1 \le 0, \ldots, W_{r-m} \le 0),$$

where for $m = 0$ the first probability on the right-hand side is defined to be 1, and for $m = r$ the second factor is defined to be 1. Conditions under which this result holds can be found in Woodroofe (1982), and are essentially the same as those used by Aerts *et al.* (1999), the latter conditions which were also used to derive the limiting distribution of $T_{n,\mathrm{OS}}$. Figure 8.1 depicts the generalised arc-sine probability distribution $P(\widehat{m}_{\mathrm{aic}} = m)$ for three situations of the upper bound, respectively $r = 5, 10$ and $r = \infty$.

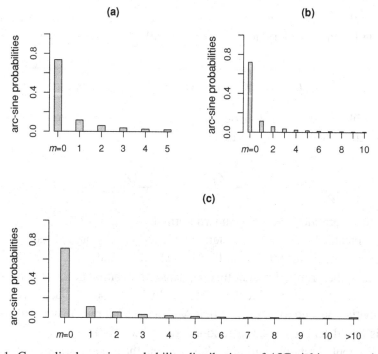

Fig. 8.1. Generalised arc-sine probability distributions of AIC picking a certain model order when the zero model order is the true one. Maximum order considered is for (a) $r = 5$, (b) $r = 10$ and (c) $r = \infty$.

Woodroofe (1982) calculates the probability of superfluous parameters for these choices of upper bounds and finds that for the limiting situation when $r \to \infty$ the expected number of superfluous parameters is 0.946, while the probability of correctly identifying the null model is 0.712. This precisely quantifies the amount of overfitting to be expected, and shows that the problem is less dramatic than often thought. We expect on average less than one superfluous parameter to be selected.

Let us ask the same question for Mallows's (1973) C_p criterion. For linear regression models, C_p approximates the sum of squared prediction errors, divided by an unbiased estimator of σ^2, usually defined in the biggest model. Writing SSE_p for the residual sum of squares in the model using p regression variables, see Section 4.4,

$$C_p = \mathrm{SSE}_p / \widehat{\sigma}^2 - (n - 2p),$$

where $\widehat{\sigma}^2 = \mathrm{SSE}_r / (n - r)$, with r indicating the largest number of regression variables considered. The index p which minimises C_p is preferred by the criterion, and we let $\widehat{p}(C_p)$ denote this chosen index. Woodroofe (1982) finds that the expected number of superfluous parameters is equal to

$$\mathrm{E}\{\widehat{p}(C_p)\} = \sum_{m=1}^{r} \mathrm{P}(F_{m,n-r} > 2),$$

with $F_{m,n-r}$ denoting a random variable distributed according to an F distribution with degrees of freedom $(m, n - r)$. This probability depends both on the sample size and on the maximum number of parameters r.

Figure 8.2 depicts the expected number of superfluous parameters as a function of the maximum number of coefficients in the model for two sample sizes, $n = 50$ and

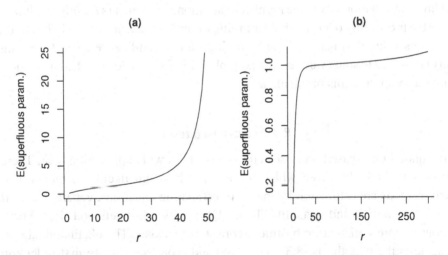

Fig. 8.2. Expected number of superfluous parameters selected by Mallows's C_p criterion for sample size (a) 50, (b) 500, as a function of r the maximum number of coefficients in the model.

$n = 500$. The overfitting problem gets worse if the number of parameters increases, for fixed sample size. Keeping r fixed, the limiting values as $n \to \infty$ are, for respectively $r = 5, 10, 20$ and 30, given by $0.571, 0.791, 0.915$ and 0.937. Convergence to the limit is quite slow. In the limiting situation of a sample of infinite size, there is less than one superfluous parameter to be expected. For finite samples the situation might be much worse, as the figure shows.

A completely different story is to be told about the BIC. Theorems 4.2 (weak consistency) and 4.3 (strong consistency) show, under the stated conditions, that the BIC selects the most simple correct model with probability one (Theorem 4.2), or almost surely (Theorem 4.3), when sample size increases. This means, if there is a correct model in our search list, that the BIC will eventually correctly identify this model, and hence that in the large-sample limit there is zero probability of selecting too many parameters.

Example 8.5 (8.3 continued) Lack-of-fit testing in generalised linear models
Let the null hypothesis

$$H_0: \theta(x_i; \beta_1, \ldots, \beta_{p+m}) = \sum_{j=1}^{p} \beta_j x_{i,j}$$

be true. The calculation of the probability of superfluous parameters can be used to compute the significance level of a test of H_0 versus the alternative hypothesis that the structure takes a more complicated form for some $m \geq 1$, as in (8.5). The test can be performed by rejecting H_0 precisely when $\widehat{m}_{\text{aic}} \geq 1$. The significance level of this test is computed as

$$\alpha = P(\widehat{m}_{\text{aic}} \geq 1 \mid m = 0).$$

In the limiting situation where the number of additional parameters r tends to infinity, α is 1 minus the probability of correctly identifying the null model, and is $1 - 0.712 = 0.288$. Usually this value is considered much too high for a significance level. Changing the penalty constant 2 to some value c_n, for example 4.179 for a 5% level test, as in Section 8.2, provides a solution to this problem. ∎

8.4 Score-based tests

The likelihood ratio based order selection tests find a wide application area. There are estimation methods, however, which do not specify a full likelihood model. Classes of such methods are estimation by means of estimating equations, generalised estimating equations (GEE) and quasi-likelihood (QL) or pseudo-likelihood (PL). For those situations a score statistic or robustified score statistic exists. The likelihood ratio statistic $L_{m,n}$ tests the hypothesis (8.3). The Wald and score statistic are first-order equivalent approximations to the likelihood ratio statistic. Either one can be used instead of the likelihood ratio statistic to determine the appropriate model order and construct an order selection lack-of-fit test. Below we give the tests in terms of score statistics, which might be preferred above Wald statistics since the latter are known to not be invariant to reparameterisations of nonlinear restrictions (Phillips and Park, 1988).

Let $U_m(\widehat{\theta}, 0_m)$ be the score vector, evaluated at the null model estimates. In full-likelihood models U_m consists of the first partial derivatives of the log-likelihood with respect to the parameters. More generally, suppose we start with estimating equations of the form

$$U_n(\theta, \gamma_1, \ldots, \gamma_m) = \sum_{i=1}^{n} \Psi_m(Y_i; \theta, \gamma_1, \ldots, \gamma_m) = 0.$$

The vector $(\widehat{\theta}, \widehat{\gamma}_1, \ldots, \widehat{\gamma}_m)$ solves these equations (one for each parameter component). In our study of misspecified parametric models two matrices made their appearance, a matrix of minus the second partial derivatives of the log-likelihood function with respect to the parameters, and the matrix of products of first partial derivatives. The expectation of this matrix with respect to the true distribution of the data is equal to the Fisher information matrix in case the model is correctly specified. Working with estimating equations, we have a similar set-up. Define

$$J_n(\theta) = -n^{-1} \sum_{i=1}^{n} \frac{\partial}{\partial \theta} \Psi_m(Y_i; \theta), \quad K_n(\theta) = n^{-1} \sum_{i=1}^{n} \Psi_m(Y_i; \theta) \Psi_m(Y_i; \theta)^{\mathrm{t}}.$$

The score statistic is

$$S_{m,n}(\theta, 0_m) = n^{-1} U_n(\theta, 0_m)^{\mathrm{t}} J_n^{-1}(\theta, 0_m) U_n(\theta, 0_m),$$

while the robustified score statistic becomes

$$R_{m,n}(\theta, 0_m) = n^{-1} U_n(\theta, 0_m)^{\mathrm{t}} J_n^{-1}(\theta, 0_m) \left\{ J_n^{-1}(\theta, 0_m) K_n(\theta, 0_m) J_n^{-1}(\theta, 0_m) \right\}^{-1}$$
$$\times J_n^{-1}(\theta, 0_m) U_n(\theta, 0_m).$$

This leads to two new AIC-like model selection criteria, replacing $L_{m,n}$ by either $S_{m,n}(\widehat{\theta}, 0_m)$ or $R_{m,n}(\widehat{\theta}, 0_m)$. It also yields two model order selectors:

$$m_{\mathrm{aic},S} = \arg\max_{m \geq 0} \{ S_{m,n}(\widehat{\theta}, 0_m) - c_n m \},$$

where traditionally $c_n = 2$, and its robustified version

$$m_{\mathrm{aic},R} = \arg\max_{m \geq 0} \{ R_{m,n}(\widehat{\theta}, 0_m) - c_n m \}.$$

A nonparametric lack-of-fit test using the score statistic rejects the null hypothesis if and only if $\widehat{m}_{\mathrm{aic},S} \geq 1$; similarly, using the robustified score statistic, the test is instead to reject the null hypothesis if and only if $\widehat{m}_{\mathrm{aic},R} \geq 1$. Order selection statistics are similarly obtained as

$$T_{n,\mathrm{OS},S} = \sup_{m \geq 1} S_{m,n}(\theta, 0_m)/m \quad \text{and} \quad T_{n,\mathrm{OS},R} = \sup_{m \geq 1} R_{m,n}(\theta, 0_m)/m.$$

The critical value at the 5% level of significance is 4.179. For asymptotic distribution theory, we refer to Aerts *et al.* (1999).

8.5 Two or more covariates

While the construction of a nested model sequence was rather trivial for one covariate, a little more thought is needed when there are two or more covariates. Let us consider the case of two covariates and construct a model sequence, that is, a path in the alternative models space, leading to an omnibus test. Depending on how this path is chosen, some special tests can be obtained, for example tests for the presence of interaction, or tests about the form of the link function.

Let g be an unknown function of the covariates x_1 and x_2. We test the null hypothesis

$$H_0: g \in \{g(\cdot, \cdot; \theta): \theta \in \Theta\}.$$

A series expansion which uses basis functions ψ_j gives alternative models of the form

$$g(x_1, x_2) = g(x_1, x_2; \theta) + \sum \sum_{(j,k) \in \Lambda} \alpha_{j,k} \psi_j(x_1) \psi_k(x_2).$$

The definition of the index set Λ will, in general, depend on the specific model under the null hypothesis. For example, if we wish to test the null hypothesis that $g(x_1, x_2, \theta)$ has the form $\theta_1 + \theta_2 \psi_1(x_1) + \theta_2 \psi_2(x_2)$, then it is obvious that these two terms do not need to be included in the sequence representing the alternative model. To make notation simpler, we assume now that the function $g(x_1, x_2, \theta)$ is constant. This is a no-effect null hypothesis, where Λ is a subset of $\{(j, k): 0 \leq j, k \leq n, \; j + k \geq 1\}$. In analogy to the case of only one covariate, we define log-likelihood, score and robust score statistics $L_{\Lambda,n}$, $S_{\Lambda,n}$ and $R_{\Lambda,n}$, along with the corresponding information criteria

$$\text{AIC}_L(\Lambda; c_n) = L_{\Lambda,n} - c_n \, \text{length}(\Lambda),$$
$$\text{AIC}_S(\Lambda; c_n) = S_{\Lambda,n} - c_n \, \text{length}(\Lambda),$$
$$\text{AIC}_R(\Lambda; c_n) = R_{\Lambda,n} - c_n \, \text{length}(\Lambda),$$

respectively, where length(Λ) denotes the number of elements in Λ.

To carry out a test we maximise AIC($\Lambda; c_n$) over a collection of subsets Λ_1, $\Lambda_2, \ldots, \Lambda_{m_n}$, assumed to be nested. This means that $\Lambda_1 \subset \Lambda_2 \subset \cdots \subset \Lambda_{m_n}$, and we call such a collection of sets a *model sequence*. The challenge now is deciding on how to choose a model sequence, since obviously there are many possibilities. One important consideration is whether a given sequence will lead to a consistent test. To ensure consistency against virtually any alternative to the null hypothesis H_0, it is required that length(Λ_{m_n}) $\to \infty$ in such a way that, for each $(j, k) \neq (0, 0)$ $(j, k \geq 0)$, (j, k) is in Λ_{m_n} for all sample sizes n sufficiently large. The choice of a model sequence is further simplified if we consider only tests that place equal emphasis on the two covariates. In

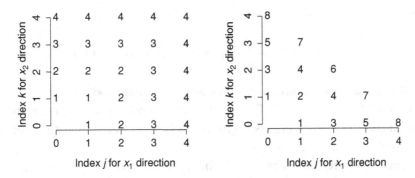

Fig. 8.3. Model sequences in two dimensions. The plotted number refers to the number of the step in which, for (j, k), the basis function $\psi_j(x_1)\psi_k(x_2)$ enters the model. The first sequence adds an increasing number of terms in every step. The second sequence only adds one or two new basis functions at a time.

other words, terms of the form $\psi_j(x_1)\psi_k(x_2)$ and $\psi_k(x_1)\psi_j(x_2)$ should enter the model simultaneously.

In Figure 8.3 we consider two different choices for the model sequence. The first few models in the sequences are graphically represented by plotting the number of the step in which the basis elements enter the model for each index (j, k). For the first model sequence, in step 1, we add $\psi_1(x_1)$, $\psi_1(x_2)$ and the interaction $\psi_1(x_1)\psi_1(x_2)$. In step 2 the following basis functions are added: $\psi_2(x_1)$, $\psi_2(x_2)$, $\psi_2(x_1)\psi_1(x_2)$, $\psi_1(x_1)\psi_2(x_2)$ and $\psi_2(x_1)\psi_2(x_2)$. In terms of a polynomial basis this reads: x_1, x_2, $x_1 x_2$ for step 1 and x_1^2, x_2^2, $x_1^2 x_2$, $x_1 x_2^2$ and $x_1^2 x_2^2$ for step 2. This model sequence adds $2j + 1$ terms to the previous model at step j. Here, the penalisation, which is linearly related to the number of parameters in the model, grows very fast. This implies that tests based on this sequence will in general have bad power properties, mainly caused by the heavy penalisation in the selection of the order. This problem is less severe in the second model sequence, where at most two new terms are added at each step. Other model sequences leading to omnibus tests are certainly possible. See, for example, Aerts *et al.* (2000).

Example 8.6 The POPS data

We consider the POPS data, from the 'project for preterm and small-for-gestational age' (Verloove and Verwey, 1983; le Cessie and van Houwelingen, 1991, 1993). The study consists of 1310 infants born in the Netherlands in 1983 (we deleted 28 cases with missing values). The following variables are measured: x_1, gestational age (≤ 32 weeks); x_2, birthweight ($\leq 1500\,\text{g}$); $Y = 0$ if the infant survived two years without major handicap, and $Y = 1$ if otherwise. We wish to test the null hypothesis

$$H_0: \text{logit}\{E(Y \mid x_1, x_2)\} = \theta_0 + \theta_1 x_1 + \theta_2 x_1^2 + \theta_3 x_2 + \theta_4 x_2^2,$$

against the alternative hypothesis that there is another functional form. Legendre polynomials $\psi_k(x_1)$ and $\psi_\ell(x_2)$ are used to represent all models. With $\psi_0(x) \equiv 1$, the null hypothesis can be written as

$$H_0: \text{logit}(EY) = \sum_{k=0}^{2} \alpha_{k,0} \psi_k(x_1) + \sum_{\ell=1}^{2} \alpha_{0,\ell} \psi_\ell(x_2).$$

We considered alternative additive models extending this null model by extra terms $\psi_k(x_1)$ and $\psi_\ell(x_2)$ with $k, \ell = 3, \ldots, 15$. For the alternative models allowing interaction terms we included the above main effects up to the sixth order, together with all interaction terms $\psi_k(x_1)\psi_\ell(x_2)$ where $2 \leq k + \ell \leq 6$.

The results are as follows. For $c_n = 2$, the traditional AIC penalty factor, the value of the score statistic at the AIC selected model equals 10.64, with an asymptotic p-value of 0.073. The corresponding value of the order selection statistic $T_{n,\text{OS}}$ which uses $c_n = 4.179$, is 3.55, with asymptotic p-value equal to 0.036. We can conclude that there is some evidence against the null hypothesis. The order selected by AIC using $c_n = 2$ is 3, while the BIC chose order 1. For AIC the chosen model includes covariates $x_1, x_1^2, x_2, x_2^2, x_1x_2, x_1x_2^2$. The BIC chooses the model that only includes x_1 and x_2. If we perform the same type of test, now taking $c_n = \log n$, as in the BIC, the value of the score statistic at the model selected by the BIC is equal to 0.42, with corresponding p-value 0.811, failing to detect a significant result. ∎

8.6 Neyman's smooth tests and generalisations

Order selection tests as presented in the previous section rely on AIC. In this section we consider a similar class of tests, based on the BIC principle, and mainly specified for goodness-of-fit testing. First, we return to the original source of the tests as developed by Neyman (1937). This introduction to Neyman's smooth-type tests follows partly sections 5.6 and 7.6 of Hart (1997).

Consider the goodness-of-fit testing situation where we test whether independent and identically distributed data $X_1, \ldots, X_n$ have distribution function $F = F_0$, where F_0 is continuous. Testing the null hypothesis H_0 that $F(x) = F_0(x)$ for all x is equivalent to testing that $F_0(X_i)$ has a uniform distribution on $(0, 1)$.

8.6.1 The original Neyman smooth test

Neyman proposed the following order m smooth alternative to H_0. Under the alternative hypothesis, the density function g of the random variable $F_0(X_i)$ is

$$g(x) = \exp\left\{\theta_0 + \sum_{j=1}^{m} \theta_j \psi_j(x)\right\} \quad \text{for } x \in (0, 1),$$

with ψ_j representing the jth Legendre polynomial, transformed to be orthonormal on $(0,1)$. The null hypothesis is therefore equivalent to

$$H_0: \theta_1 = \ldots = \theta_m = 0.$$

The original Neyman test statistic is a score statistic

$$N_m = \sum_{j=1}^{m} V_j^2 \quad \text{where } V_j = n^{-1/2} \sum_{i=1}^{n} \psi_j(F_0(X_i)).$$

In other words, the random variable V_j represents $\sqrt{n}$ times the average of the jth coefficient in the expansion. Under the null hypothesis $N_m^2 \to_d \chi_m^2$, see Exercise 8.3. The term 'smooth test' comes from Neyman's original article; he thought of the order m alternative as a smooth alternative to the uniform density. This test does not belong to the category of 'omnibus' tests, since the alternative hypothesis is parametrically specified via the choice of the Legendre polynomials and the order m, as indeed Neyman fixed the choice of m prior to performing the test.

Remark 8.1 Lack-of-fit in regression
In the regression setting, where data are modelled $Y_i = \mu(x_i) + \varepsilon_i$, a lack-of-fit statistic is constructed as follows. Consider the 'no-effect' null hypothesis $H_0: \mu(x) = \theta_0$. Smooth alternatives of order m are of the same form as considered earlier in Section 8.1,

$$\mu(x) = \theta_0 + \sum_{k=1}^{m} \theta_k \psi_k(x).$$

The basis functions may be chosen to be orthonormal over the design points. This means that $n^{-1} \sum_{i=1}^{n} \psi_j(x_i)\psi_k(x_i)$ is equal to one for $j = k$ and is zero otherwise. Define $\psi_0(x) = 1$. Least squares estimators of the regression coefficients are easily obtained, as

$$\widehat{\theta}_k = n^{-1} \sum_{i=1}^{n} Y_i \psi_k(x_i) \quad \text{for } k = 0, \ldots, m.$$

The test statistic is defined as $N_{m,n} = n \sum_{k=1}^{m} \widehat{\theta}_k^2 / \widehat{\sigma}^2$, with $\widehat{\sigma}^2$ a consistent estimator of the error variance. Variance estimators can be obtained under either the null model, or some alternative model (for example the biggest one), or can be constructed independently of the models used. The tests will be different for different choices. ∎

8.6.2 Data-driven Neyman smooth tests

A data-driven version of the Neyman smooth test can be constructed by letting any model selection mechanism choose the order $\widehat{m}$ from a set of possible orders. Ledwina (1994) used the BIC to determine a good value of m: let $\text{BIC}_0 = 0$ and define $\text{BIC}_m = n \sum_{k=1}^{m} \widehat{\theta}_k^2 / \widehat{\sigma}^2 - m \log n$. The data-driven Neyman smooth test statistic is $T_{n,\text{bic}} = n \sum_{k=1}^{\widehat{m}_{\text{bic}}} \widehat{\theta}_k^2 / \widehat{\sigma}^2$ when $\widehat{m}_{\text{bic}} \geq 1$ and is zero otherwise.

This test statistic based on the BIC has a peculiar limiting distribution. Since the BIC is consistent as a model selection criterion (see Chapter 4), under the null hypothesis $\widehat{m}_{\text{bic}}$ is consistently estimating 0. This implies that $T_{n,\text{bic}}$ equals zero with probability going to 1, as the sample size $n \to \infty$. As a consequence, the level of this test tends to 0 with growing n. A simple solution to this problem is provided by Ledwina (1994) in the setting of goodness-of-fit testing. The proposal is to omit the null model as one of the possibilities in the BIC. That is, we insist that $m \geq 1$. The test rejects the null hypothesis for large values of $T_{n,\text{bic}}$, where now $\widehat{m}_{\text{bic}}$ maximises BIC_m for $m \geq 1$. Since the BIC is not allowed to choose $\widehat{m} = 0$, $\widehat{m}_{\text{bic}}$ converges in probability to 1, leading to a χ_1^2 distribution.

Theorem 8.1 *Denote by $\widehat{m}_{\text{bic}}$ the maximiser of BIC_m for $1 \leq m \leq M$. Then, the Neyman smooth-type statistic $T_{n,\text{bic}}$ with model order $\widehat{m}_{\text{bic}}$ has under H_0 asymptotically a χ_1^2 distribution.*

The simplicity of this asymptotic distribution should be contrasted with the behaviour of the test under a local sequence of alternatives of the form

$$\mu(x) = \theta_0 + \sum_{k=1}^{m} \frac{b_k}{\sqrt{n}} \psi_k(x). \tag{8.6}$$

Assuming such an alternative model and a growth condition on m it can be shown that P(BIC chooses first component only) $\to 1$. This implies that $T_{n,\text{bic}} \to_d \chi_1^2(b_1^2)$. This spells trouble for the BIC test, since its limiting behaviour is only dictated by the first coefficient b_1, not by $b_2, \ldots, b_k$. In particular, if $b_1 = 0$ but the others are non-zero, then $T_{n,\text{bic}} \to_d \chi_1^2$. Hence its limiting local power is identical to the level of the test, reflecting asymptotic inability to discover important alternatives.

8.7 A comparison between AIC and the BIC for model testing *

The arguments in this section hold for both lack-of-fit and goodness-of-fit tests. Under the null hypothesis $H_0: m = 0$ and some growth condition on m_n

$$T_{n,\text{aic}} \overset{d}{\to} \sum_{m=0}^{m_n} I\{\widehat{m}_{\text{aic}} = m\} \sum_{j=1}^{m} Z_j^2,$$

where $Z_1, Z_2, \ldots$ are i.i.d. standard normal random variables. The mixing probabilities $P(\widehat{m}_{\text{aic}} = m)$ for the limiting distribution under the null hypothesis follow the generalised arc-sine distribution (see Section 8.3). The main difference with the BIC-type test arises because under H_0, one has $P(\widehat{m}_{\text{bic}} = 1) \to 1$ (when $m = 0$ is excluded).

Let us now step away from the nested model sequences, and consider an all-subsets model search. We keep the upper bound $m_n = m$ fixed and consider *all* subsets S of $\{1, \ldots, m\}$. There are 2^m possibilities,

$$\emptyset, \{\psi_1\}, \ldots, \{\psi_m\}, \{\psi_1, \psi_2\}, \ldots, \ldots, \{\psi_1, \ldots, \psi_m\},$$

and we let AIC find the best subset.

Under the null hypothesis H_0 and a growth condition on m_n, one can show that

$$T_{n,\text{aic}}^* \xrightarrow{d} \sum_S I\{S_{\text{aic}} = S\} \sum_{j \in S} Z_j^2.$$

An omnibus test rejects H_0 if S_{aic} is non-empty. AIC chooses the empty set if and only if all $Z_j^2 - C \le 0$ for $j = 1, \ldots, m$. This in particular implies

$$P(T_{n,\text{aic}}^* = 0) \to \prod_{j=1}^m \Gamma_1(C, b_j^2)$$

where $\Gamma_1(\cdot, b_j^2)$ is the cumulative distribution function of $\chi_1^2(b_j^2)$ under the local alternative model (8.6). The limiting local power equals $1 - \prod_{j=1}^m \Gamma_1(C, b_j^2)$. The level of this test can be adjusted by choosing the constant c such that $\Gamma_1(c, 0)^m = 1 - \alpha$.

Now, let us turn again to the BIC. Under a sequence of local alternatives, and a growth condition on m, P(BIC chooses set S with $|S| \ge 2$) $\to 0$, which is equivalent to P(BIC chooses a singleton) $\to 1$. Hence

$$T_{n,\text{bic}}^* \xrightarrow{d} \max_{j \le m_0} (b_j + Z_j)^2.$$

This shows that for large n, the all-subsets version of the BIC behaves precisely as the all-subsets version of AIC, both are based on individual Z_j variables. Hence, with an all-subsets model search, the disadvantages a BIC-based test experiences when using a nested model sequence disappear.

There are many other search strategies. One may, for example, consider all subsets of $\{1, \ldots, m_0\}$, followed by a nested sequence $\{1, \ldots, m\}$ for $m > m_0$. Another possibility is to allow the criteria: choose the two biggest components, or perhaps three biggest components, among $\{1, \ldots, m_0\}$. More information, along with proofs of various technical statements relating to performance of this type of test, are given in Claeskens and Hjort (2004).

8.8 Goodness-of-fit monitoring processes for regression models *

Methods presented and developed in this section are different in spirit from those associated with order and model selection ideas worked with above. Here the emphasis is on constructing 'monitoring processes' for regression models, for example of GLM type, to assess adequacy of various model assumptions, both visually (plots of functions compared to how they should have looked like if the model is correct) and formally (test statistics and p-values associated with such plots). Such model check procedures are important also from the perspective of using model selection and averaging methods, say for variable selection reasons with methods of earlier chapters, in that many of these build on the start assumption that at least the biggest of the candidate models is adequate. It is also our view that good strategies for evaluating model adequacy belong to the statistical toolbox of model selection and model averaging. If two different types of generalised linear models are being considered for a complicated set of data, say corresponding to two different link functions, then one strategy is to expose each to a goodness-of-fit screening check and then proceed with methods of selection and averaging using the best of the two alternatives. Sometimes a goodness-of-fit test shows that even the best of a list of candidate models does not fit the data well, which is motivation for looking for yet better models.

For concreteness of presentation we choose to develop certain general goodness-of-fit methods inside the realm of Poisson regression models; it will not be a difficult task to extend the methods to the level of generalised linear models. Suppose therefore that independent count data $Y_1, \ldots, Y_n$ are associated with p dimensional covariate vectors $x_1, \ldots, x_n$, and consider the model that takes $Y_i \sim \text{Pois}(\xi_i)$, with $\xi_i = \exp(x_i^t \beta)$, for $i = 1, \ldots, n$. We think of $s_i = x_i^t \beta$ as linear predictors in an interval $[s_{\text{low}}, s_{\text{up}}]$ spanning all relevant $s = x^t \beta$. Let also $\widehat{s}_i = x_i^t \widehat{\beta}$, in terms of the maximum likelihood estimator $\widehat{\beta}$, and write $\widehat{\xi}_i = \exp(\widehat{s}_i)$ for the estimated ξ_i. Now define

$$A_n(s) = n^{-1/2} \sum_{i=1}^{n} I\{x_i^t \widehat{\beta} \leq s\}(Y_i - \widehat{\xi}_i)x_i \quad \text{for } s \in [s_{\text{low}}, s_{\text{up}}],$$

to be thought of as a p-dimensional monitoring process. Note that A_n starts and ends at zero, since $\sum_{i=1}^{n}(Y_i - \widehat{\xi}_i)x_i = 0$ defines the maximum likelihood estimator.

Various visual and formalised tests may now be developed, based on empirical process theory for how the p-dimensional A_n process behaves under Poisson conditions. We do not go into these details here, but describe some of the easier-to-use tests that flow from this theory. First we consider a chi-squared test. Form m cells or windows $C_k = (c_{k-1}, c_k]$ in the linear risk interval of $\widehat{s}_i$ values, via cut-off values $s_{\text{low}} = c_0 < \cdots < c_m = s_{\text{up}}$, and define increments $\Delta A_{n,k} = A_n(c_k) - A_n(c_{k-1})$ and $\Delta J_{n,k} = n^{-1} \sum_{\text{wind } k} \widehat{\xi}_i x_i x_i^t$. Our test statistic is

$$T_n = \sum_{k=1}^{m} (\Delta A_{n,k})^t \Delta J_{n,k}^{-1} \Delta A_{n,k}.$$

The windows C_k need to be chosen big enough in order for each of $\Delta J_{n,k}$ to be invertible, which means that at least p linearly independent x_is need to be caught for each cell. Note also that $\Delta A_{n,k} = n^{-1/2}(O_k - E_k)$, where $O_k = \sum_{\text{wind } k} Y_i x_i$ and $E_k = \sum_{\text{wind } k} \widehat{\xi}_i x_i$ are the p-dimensional 'observed' and 'expected' quantities for window k. One may show that

$$T_n = n^{-1} \sum_{k=1}^{m} (O_k - E_k)^t \Delta J_{n,k}^{-1} (O_k - E_k) \xrightarrow{d} \chi_{(m-1)p}^2.$$

Example 8.7 The number of bird species
We apply the goodness-of-fit test to a data set on the number of different bird species found living on islands outside Ecuador; see Example 6.15 and Hand *et al.* (1994, case #52) for some more details. For each of 14 such islands, the number Y of bird species

living on the island's paramos is recorded (ranging from 4 to 37), along with various covariates. For the purposes of the present illustration, we include x_1, the island's area (in thousands of square km, ranging from 0.03 to 2.17) and x_2, the distance from Ecuador (in thousands of km, ranging from 0.036 to 1.380).

We have ordered the 14 $\widehat{s}_i = x_i^t \widehat{\beta}$ values, and placed cut points at number 1, 4, 7, 10, 14 of these (2.328, 2.612, 2.725, 3.290, 3.622), to form four windows with the required three $\widehat{s}_i$s inside each. The T_n test is then found to be the sum of values 7.126, 2.069, 7.469, 3.575 over the four windows, with a total of 20.239. This is to be compared to a χ^2 with $3 \cdot 3 = 9$ degrees of freedom, and gives a p-value of 0.016, indicating that the Poisson model is not fully adequate for explaining the influence of covariates x_1 and x_2 on the bird species counts Y. ∎

The $\chi^2_{(m-1)p}$ test above works in overall modus, with equal concern for what happens across the p monitoring components of A_n. One may also build tests that more specifically single out one of the components, or more generally a linear combination of these. Let f be some fixed p-vector, and consider

$$Z_n(s) = f^t A_n(s) = n^{-1/2} \sum_{i=1}^{n} I\{x_i^t \widehat{\beta} \le s\}(Y_i - \widehat{x}_i)f^t x_i \quad \text{for } s \in [s_{\text{low}}, s_{\text{up}}].$$

This is an ordinary one-dimensional process, and with $f = (1, 0, \ldots, 0)^t$, for example, Z_n is simply the first of A_n's components. Consider the process increments over m windows,

$$\Delta Z_{n,k} = Z_n(c_k) - Z_n(c_{k-1}) = f^t \Delta A_{n,k} = n^{-1/2}\{O_k(f) - E_k(f)\},$$

that compare the observed $O_k(f) = \sum_{\text{wind } k} Y_i f^t x_i$ with the expected $E_k(f) = \sum_{\text{wind } k} \widehat{\xi}_i f^t x_i$, We now construct a χ^2_{m-1} test $T_n(f)$, that is the estimated quadratic form in the collection of increments $\Delta Z_{n,k}$:

$$T_n(f) = \sum_{k=1}^{m} \frac{\Delta Z_{n,k}^2}{\widehat{\lambda}_k} + \left(\sum_{k=1}^{m} \frac{\Delta J_{n,k} f \Delta Z_{n,k}}{\widehat{\lambda}_k} \right)^t \widehat{R}^{-1} \left(\sum_{k=1}^{m} \frac{\Delta J_{n,k} f \Delta Z_{n,k}}{\widehat{\lambda}_k} \right),$$

where $\widehat{\lambda}_k$ and $\widehat{R}$ are estimates of $\lambda_k = f^t \Delta J_k f$ and $R = J - \sum_{k=1}^{m} \Delta J_k f f^t \Delta J_k / (f^t \Delta J_k f)$.

It is not difficult to generalise the methods exhibited here to form monitoring processes for the class of generalised linear models. With $f = (1, 0, \ldots, 0)^t$ above, we have a χ^2_{m-1} that resembles and generalises the popular Hosmer–Lemeshow goodness-of-fit test for logistic regression models. Studies of local power make it possible to outperform this test.

8.9 Notes on the literature

Woodroofe (1982) is an important early paper on probabilistic aspects of order selection methods, partly exploiting theory for random walks. The property of overfitting for

AIC is illustrated and discussed by McQuarrie and Tsai (1998). Hart (1997) gives an extensive treatment of lack-of-fit tests starting from nonparametric estimation methods. Examples and strategies for choosing the sequence of alternative models in case of multiple regression can be found in Aerts *et al.* (2000). Several papers have been devoted to aspects of order selection lack-of-fit tests in different connections, see e.g. Eubank and Hart (1992) and again Hart (1997). More information on order selection testing whether data come from a certain parametric density function (rather than in a regression context) is dealt with in Claeskens and Hjort (2004), where detailed proofs can be found and further developments are discussed for the statements in Sections 8.6–8.7. For more information on the Neyman smooth test, we refer to Ledwina (1994), Inglot and Ledwina (1996) and Inglot *et al.* (1997). Aerts *et al.* (2004) construct tests which are based on a BIC approximation to the posterior probability of the null model. The tests can be carried out in either a frequentist or a Bayesian way. Hart (2009) builds further on this theme but uses a Laplace approximation instead. The monitoring processes of Section 8.8 are related to those studied in Hjort and Koning (2002) and in Hosmer and Hjort (2002). These lead in particular to generalisations of the Hosmer–Lemeshow goodness-of-fit statistic to generalised linear models.

Exercises

8.1 *Order selection tests:*
  (a) Verify that for AIC with penalty constant 2, the limiting type I error of the order selection test $T_{n,\text{OS}}$ is about equal to 0.288.
  (b) Obtain the penalty constants for tests which have respectively levels 0.01, 0.05 and 0.10. *Hint.* Consider the following S-Plus/R code, with cn representing the cut-off point c_n, and where you may set e.g. m = 100:
  pvalue = 1 − exp(−sum((1 − pchisq((1 : m)*cn, 1 : m))/(1 : m))).
  Quantify the approximation error.
  (c) Show that the limiting null distribution of the order selection test (8.4) has support $[1, \infty)$, that is, that there is zero probability on $[0, 1]$. Compute its density function and display it.

8.2 *The beta-binomial model:* Assume that $Y_1, \ldots, Y_m$ are independent Bernoulli $(1, p)$ variables for given p, but that p itself has a Beta distribution with parameters $(kp_0, k(1 - p_0))$. Find the marginal distribution of $(Y_1, \ldots, Y_m)$, and of their $Y = \sum_{j=1}^m Y_j$. Show that Y has

$$\text{mean } mp_0 \quad \text{and} \quad \text{variance } mp_0(1 - p_0)\Big(1 + \frac{m - 1}{k + 1}\Big).$$

Finally demonstrate that the correlation between Y_j and Y_k is $\rho = 1/(k + 1)$. The binomial case corresponds to $k = \infty$.

8.3 *The Neyman smooth test:* Prove that the original Neyman smooth statistic defined in Section 8.6.1 has an asymptotic chi-squared distribution with m degrees of freedom under the null hypothesis.

8.4 *The Ladies Adelskalenderen:* For the ladies' speedskating data, test the null hypothesis that the 3000-m time is linearly related to the 1500-m time. Use both the AIC-based order selection test and the BIC-based data-driven Neyman smooth test.

8.5 *US temperature data:* Consider the US temperature data that are available in R via data(ustemp); attach(ustemp). The response variable Y is the minimum temperature, regression variables are latitude and longitude.

   (a) Fit both an additive model (without interaction) and a model with interaction between $x_1 =$ latitude and $x_2 =$ longitude. Compare the AIC values of both models. Which model is preferred?

   (b) Use the order selection idea to construct a test for additivity. The null hypothesis is that the model is additive, $E Y = \beta_0 + \beta_1 x_1 + \beta_2 x_2$. The alternative hypothesis assumes a more complicated model. Construct a nested model sequence by adding one variable at a time: $x_1 x_2$, x_1^2, x_2^2, $x_1^2 x_2$, $x_2^2 x_1$, x_1^3, $x_1^2 x_2^2$, x_1^3, x_2^3, etc. Use the order selection test to select the appropriate order and give the p-value.

   (c) Consider some alternatives to the testing scheme above by not starting from an additive model that is linear in (x_1, x_2), but rather with an additive model of the form $\beta_0 + \sum_{j=1}^{k}(\beta_{1,j} x_1^j + \beta_{2,j} x_2^j)$, for $k = 3, 4$ or 5. Investigate the effect of the choice of k on the order selection test.

9

Model selection and averaging schemes in action

In this chapter model selection and averaging methods are applied in some usual regression set-ups, like those of generalised linear models and the Cox proportional hazards regression model, along with some less straightforward models for multivariate data. Answers are suggested to several of the specific model selection questions posed about the data sets of Chapter 1. In the process we explain in detail what the necessary key quantities are, for different strategies, and how these are estimated from data. A concrete application of methods for statistical model selection and averaging is often a nontrivial task. It involves a careful listing of all candidate models as well as specification of focus parameters, and there might be different possibilities for estimating some of the key quantities involved in a given selection criterion. Some of these issues are illustrated in this chapter, which is concerned with data analysis and discussion only; for the methodology we refer to earlier chapters.

9.1 AIC and BIC selection for Egyptian skull development data

We perform model selection for the data set consisting of measurements on skulls of male Egyptians, living in different time eras; see Section 1.2 for more details. Our interest lies in studying a possible trend in the measurements over time and in the correlation structure between measurements.

Assuming the normal approximation at work, we construct for each time period, and for each of the four measurements, pointwise 95% confidence intervals for the expected average measurement of that variable and in that time period. The results are summarised in Figure 9.1. For maximal skull breadth, there is a clear upward trend over the years, while for basialveolar length the trend is downward.

The model selection of the Egyptian skull data starts by constructing a list of possible models. We use the normality assumption $Y_{t,i} \sim N_4(\xi_{t,i}, \Sigma_{t,i})$ and will consider several possibilities for modelling the mean vector and covariance structure. Within a time period (for fixed t) we assume the $n_t = 30$ four-dimensional skull measurements to be independent and identically distributed.

248

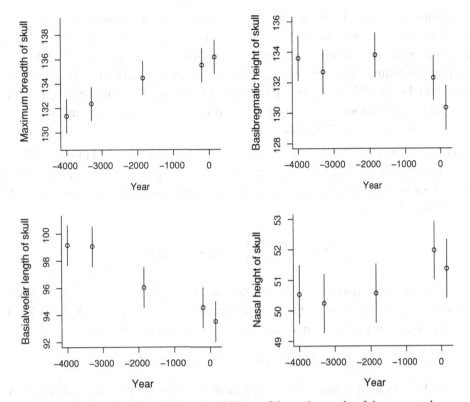

Fig. 9.1. Egyptian skull data. Pointwise 95% confidence intervals of the expected measurement in each of the five time periods.

Model 1. This model makes the least assumptions. For each time period t there is a possibly different mean vector ξ_t, and a possibly different covariance matrix Σ_t. Considering each of the four skull measurements separately, this model does not make any assumptions about how the mean profile of any of these measurements change over time. The corresponding likelihood is a product of four-dimensional normal likelihoods, leading to a likelihood of the form

$$\mathcal{L}_{M_1} = \prod_{t=1}^{5}\left\{\prod_{i=1}^{30}\phi(Y_{t,i} - \xi_t, \Sigma_t)\right\},$$

where as earlier $\phi(y, \Sigma)$ is the density of a N$(0, \Sigma)$. The log-likelihood at the maximum likelihood estimators

$$\widehat{\xi}_t = \bar{y}_{t,\bullet} \quad \text{and} \quad \widehat{\Sigma}_t = n_t^{-1}\sum_{i=1}^{n_t}(y_{t,i} - \bar{y}_{t,\bullet})(y_{t,i} - \bar{y}_{t,\bullet})^{\mathrm{t}}$$

can be written explicitly as

$$\ell_{M_1} = \tfrac{1}{2}\sum_{t=1}^{5}\{-n_t \log|\widehat{\Sigma}_t| - 4n_t - 4n_t \log(2\pi)\}.$$

To compute the values of AIC and the BIC we need to count the number of parameters. There are five four-dimensional mean vectors, resulting in 20 parameters. Each 4×4 covariance matrix is symmetric, leading to another five times ten parameters. In total, this model has 70 estimated parameters. For AIC this leads to $\text{AIC}(M_1) = 2\ell_{M_1} - 2 \cdot 70 = -3512.960$. For the BIC, the penalty is equal to $\log n$ times the number of parameters, with n the total number of observations, in this case 150. We compute $\text{BIC}(M_1) = 2\ell_{M_1} - \log(150) \cdot 70 = -3723.704$.

Model 2. We now consider several simplifications of the unstructured model 1. Model 2 makes the simplification that all five covariance matrices Σ_t are equal, without specifying any structure for this matrix, and without any assumptions about the mean vectors. This gives the likelihood function

$$\mathcal{L}_{M_2} = \prod_{t=1}^{5}\left\{\prod_{i=1}^{30} \phi(Y_{t,i} - \xi_t, \Sigma)\right\}.$$

The maximum likelihood estimators for the mean vectors ξ_t do not change, whereas the common Σ is estimated by the pooled variance matrix $\widehat{\Sigma}_{M_2} = (1/5)\sum_{t=1}^{5}\widehat{\Sigma}_t$. The log-likelihood evaluated at the maximum likelihood estimators is computed as

$$\ell_{M_2} = \tfrac{1}{2}\{-n\log|\widehat{\Sigma}_{M_2}| - 4n - 4n\log(2\pi)\},$$

with $n = 150$ the total sample size. Since there is now only one covariance matrix to estimate, the total number of parameters for model 2 is equal to $5 \cdot 4 + 10 = 30$. This yields $\text{AIC}(M_2) = -3483.18$ and $\text{BIC}(M_2) = -3573.499$. Both of these values are larger than those corresponding to model 1, indicating a preference for a common covariance structure.

Model 3. As a further simplification we construct a model with a common covariance matrix (as in model 2), and with a common mean vector $\xi_t = \xi$ for all five time periods. This means reluctance to believe that there is a trend over time. The likelihood function equals

$$\mathcal{L}_{M_3} = \prod_{t=1}^{5}\left\{\prod_{i=1}^{30} \phi(Y_{t,i} - \xi, \Sigma)\right\}.$$

The maximum likelihood estimator for the common mean is $\widehat{\xi} = (1/5)\sum_{t=1}^{5}\widehat{\xi}_t = \bar{y}_{..}$, while the covariance matrix is estimated by

$$\widehat{\Sigma}_{M_3} = \widehat{\Sigma}_{M_2} + \sum_{t=1}^{5}\frac{n_t}{n}(\bar{y}_{t,.} - \bar{y}_{..})(\bar{y}_{t,.} - \bar{y}_{..})^t.$$

Inserting these estimators in the log-likelihood function gives

$$\ell_{M_3} = \tfrac{1}{2}\{-n\log|\widehat{\Sigma}_{M_3}| - 4n - 4n\log(2\pi)\}.$$

There are 14 estimated parameters in the model. This yields the values AIC(M_3) = -3512.695 and BIC(M_3) = -3554.844. We compare these values to those obtained for models 1 and 2. For AIC, the value of model 3 is very comparable to that of model 1 (only slightly larger), but much smaller than that of model 2. This still points to a preference for model 2. For BIC, model 3's value is bigger than the previous two values, indicating a preference for the simpler model with common mean and common covariance matrix.

Model 4. Inspired by Figure 9.1, the next models we consider all include a linear trend over time in the mean vector. More specifically, we assume that ξ_t has components $a_j + b_j t$ for $j = 1, 2, 3, 4$, allowing for different intercepts and slopes for each of the four measurements. For ease of computation, we use thousands of years as time scale and write the trend model as $\xi_t = \alpha + \beta(\text{time}_t - \text{time}_1)/1000$, where time_t is the calendar year at time $t = 1, 2, 3, 4, 5$ (these were not equidistant, hence the slightly heavy-handed notation). The likelihood function is

$$\mathcal{L}_{M_4} = \prod_{t=1}^{5}\left\{\prod_{i=1}^{30} \phi(Y_{t,i} - \alpha - \beta(\text{time}_t - \text{time}_1)/1000, \Sigma)\right\}.$$

The covariance matrix is assumed to be the same for the five time periods, but otherwise unspecified. As explicit formulae are not available here, a numerical optimisation algorithm is used, as explained on various earlier occasions; see Exercise 9.1. The number of parameters in this model is equal to four intercepts and four slope parameters, plus ten parameters for the covariance matrix, in total 18 parameters. For the mean structure, we find maximum likelihood estimates $\widehat{\alpha} = (131.59, 133.72, 99.46, 50.22)$ and $\widehat{\beta} = (1.104, -0.544, -1.390, 0.331)$. This further gives AIC(M_4) = -3468.115 and BIC(M_4) = -3522.306. Compared to the previous three values of the information criteria, for both AIC and the BIC there is a large increase. This clearly indicates that including a linear time trend is to be preferred above both an unspecified trend (models 1 and 2) and no trend ($\beta = (0, 0, 0, 0)$ in model 3).

Model 5. We keep the linear time trend as in model 4, but now bring some structure into the covariance matrix. This will reduce the number of parameters. The simplification in model 5 assumes that all four measurements on one skull have equal correlation. Their variances are allowed to differ. In the equicorrelation model for Σ, the total number of parameters is equal to $8 + 5 = 13$. Again we use numerical optimisation to find the value of the log-likelihood function at its maximum. The estimated common correlation between the four measurements equals 0.101. The information criteria values for this model are AIC(M_5) = -3464.649 and BIC(M_5) = -3503.787. Both AIC and the BIC prefer this simpler model above all previously considered models.

Model 6. As a further simplification we take model 5, though now with the restriction that all four variances are the same. This further reduces the number of parameters to $8 + 2 = 10$. The AIC value equals AIC(M_6) = -3493.05 and the BIC value equals

Table 9.1. *Summary of AIC and BIC values for the Egyptian skull data: the model with rank (1) is preferred by the criterion.*

Model	No. of parameters	AIC	Rank	BIC	Rank
M_1	70	−3512.960	(6)	−3723.704	(6)
M_2	30	−3483.180	(3)	−3573.499	(5)
M_3	14	−3512.695	(5)	−3554.844	(4)
M_4	18	−3468.115	(2)	−3522.306	(2)
M_5	13	−3464.694	(1)	−3503.787	(1)
M_6	10	−3493.050	(4)	−3523.156	(3)

$\mathrm{BIC}(M_6) = -3523.156$. Neither AIC nor the BIC prefers this model above the other models.

It is always helpful to provide a summary table of criteria values. For the Egyptian skull data, the results of the AIC and BIC model selection are brought together in Table 9.1. For the BIC, the best three models are those with a linear trend in the mean structure. Amongst those, the model which assumes equal correlation though unequal variances is the best, followed by an unstructured covariance matrix. AIC has the same models in places 1 and 2, but then deviates in choosing the model without time trend and an unstructured covariance matrix on the third place.

9.2 Low birthweight data: FIC plots and FIC selection per stratum

We apply focussed model selection to the low birthweight data set (Hosmer and Lemeshow, 1999), see Section 1.5. In this study of $n = 189$ women with newborn babies, we wish to select the variables that most influence low birthweight of the baby, the latter defined as a weight at birth of less than 2500 grams. As mentioned earlier, we choose to retain in every model the intercept $x_1 = 1$ and the weight x_2 of the mother prior to pregnancy. Hence these form the two components of the covariate vector x. All other covariates listed in Section 1.5, denoted by $z_1, \ldots, z_{11}$, including the two interaction terms $z_{10} = z_4 z_7$ and $z_{11} = z_1 z_9$, form the set of variables from which we wish to select a subset for estimation of the probability of low birthweight. As for Example 6.1, we argue that not all of $2^{11} = 2048$ subsets should be taken into consideration; we may only include the interaction term z_{10} if both main effects z_4 and z_7 are present in the model, and similarly $z_{11} = z_1 z_9$ may not be included unless both of z_1 and z_9 are present. A counting exercise shows that $288 + 480 + 480 = 1248$ of the 2^{11} models are excluded, for reasons explained, leaving us with exactly 800 valid candidate models. We fit logistic regressions to each of these. Table 9.2 shows the five best models

Table 9.2. *Low birthweight data. Variable selection using AIC and the BIC. The five best models are shown, together with the AIC and BIC values.*

Covariates included											AIC	BIC
z_1	z_2	z_3	z_4	z_5	z_6	z_7	z_8	z_9	z_{10}	z_{11}	AIC	BIC
1	1	1	1	1	1	1	0	1	1	1	−213.452	−252.353
1	1	1	1	1	1	1	0	1	0	1	−213.670	−249.329
1	1	1	1	0	1	1	0	1	0	1	−213.861	−246.278
1	1	1	1	0	1	1	0	1	1	1	−214.055	−249.714
1	1	1	1	1	1	1	1	1	1	1	−214.286	−256.428
0	0	0	0	0	1	0	0	0	0	0	−227.142	−236.867
0	0	0	0	1	1	0	0	0	0	0	−223.964	−236.931
0	0	0	0	0	1	1	0	0	0	0	−224.613	−237.580
0	0	0	1	0	1	0	0	0	0	0	−224.858	−237.825
0	1	0	0	1	1	0	0	0	0	0	−222.433	−238.642

ranked by AIC, and similarly the five best models judged by the BIC. In particular, the best BIC model is rather parsimonious, including only z_6 (history of hypertension). The best AIC model includes all of $z_1, \ldots, z_{11}$, including the two interactions, apart from z_8 (ftv1).

Now we wish to make the model search more specific. We consider the following six strata in the data, defined by race (white, black, other) and by smoker/non-smoker. For each of the six groups thus formed, we take a representative subject with average values for the (x_i, z_i) variables. Averages of indicator variables are rounded to the nearest integer. We use focussed model selection for a logistic regression model as described in Section 6.6.1. The results are given as FIC plots in Figure 9.2 and with FIC output in Table 9.3. The best model according to the FIC is the one with the lowest value on the horizontal axis in Figure 9.2. These plots allow us to graphically detect groups of models with similar FIC values. For example, for stratum F there are four models with about equal FIC score and the corresponding estimated probabilities for low birthweight are also about equal. One can then decide to finally select the most parsimonious model of these four. In this example, this corresponds to the narrow model, see Table 9.3, first line of the results for stratum F.

The latter FIC table includes information criterion difference scores $\Delta\text{AIC} = \text{AIC} - \max_{\text{all}} \text{AIC}$ and $\Delta\text{BIC} = \text{BIC} - \max_{\text{all}} \text{BIC}$, for easy inspection of the best FIC models in terms of how well, or not well, they score on the AIC and BIC scales.

The most conspicuous aspect of the results is that the low birthweight probabilities are so markedly different in the six strata. The chance is lowest among white non-smokers and highest among black smokers. The expected association between smoking

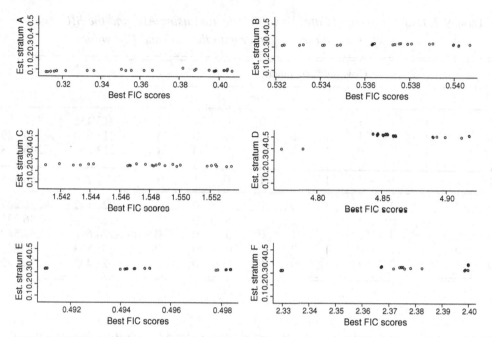

Fig. 9.2. FIC plots for low birthweight probabilities, for six strata of mothers: for each stratum, the 25 best estimates of the low birthweight probability are shown, corresponding to the best 25 FIC scores among the 800 candidate models for selecting among $z_1, \ldots, z_{11}$, given in the text. The lowest FIC score indicates the best FIC model.

and increased chance of low birthweight is found for 'white' and for 'black' mothers, but not, interestingly, for the 'other' group. Secondly, there are striking differences between the subsets of most influential covariates, for the different strata. Strata B (white, smoker), C (black, non-smoker), E (other, non-smoker), F (other, smoker) demand very parsimonious models, which appears to mean that the simplest model that only uses x_1 (intercept) and x_2 (mother's weight) does a satisfactory job, and is difficult to improve on by inclusion of further covariates. This may be caused by inhomogeneities inside the strata in question, or by x_2 being a dominant influence. The situation is different for strata A (white, non-smoker) and D (black, smoker), where there are models that do much better than the simplest one using only the mother's weight. Results for stratum A indicate that nearly all of the z_j variables play a significant role for the low birthweight event, including the interactions. Stratum D is different from the others in that the two best estimates are so markedly different from (and lower than) the others, and inspection of the best models hints that the interaction term $z_1 z_9$, age with `ftv2p` interaction, is of particular importance. Constructing models per subgroup allows us to obtain better predictions, and hence potentially better health recommendations, for new prospective mothers. See also Exercise 9.5.

Table 9.3. *Model selection via the FIC for six strata of mothers, when the focus is estimation of low birthweight probability. The table displays for each stratum the best five among the 800 candidate models described in the text, with the selected covariates among $z_1, \ldots, z_{11}$ indicated to the left; followed by the estimate, the estimated standard deviation, bias and root mean squared error, all given in percent; and the FIC score. For comparison also the ΔAIC and ΔBIC scores are included.*

Covariates included	est	sd	Bias	rmse	FIC	ΔAIC	ΔBIC
A (white, non-smoker)							
111 111 101 11	7.87	3.20	0.24	3.204	0.312	0.000	−15.486
111 111 111 11	7.76	3.21	0.00	3.211	0.312	−0.834	−19.561
111 101 111 11	8.13	3.20	0.50	3.223	0.314	−1.834	−17.366
111 101 101 11	8.19	3.19	0.64	3.238	0.316	−0.603	−12.847
111 111 111 01	8.72	3.14	1.10	3.249	0.317	−1.349	−16.835
B (white, smoker)							
000 000 000 00	31.49	3.88	3.65	3.875	0.532	−19.239	−2.307
000 001 000 00	31.81	3.88	3.51	3.876	0.532	−13.690	0.000
000 000 100 00	31.63	3.88	3.33	3.880	0.533	−17.644	−3.954
000 001 100 00	31.92	3.88	3.12	3.881	0.533	−11.161	−0.713
100 000 000 00	31.58	3.89	3.44	3.887	0.534	−19.671	−5.982
C (black, non-smoker)							
000 000 000 00	24.93	5.16	−1.16	5.156	1.541	−19.239	−2.307
000 000 100 00	26.02	5.16	−0.81	5.161	1.542	−17.644	−3.954
000 000 010 00	24.71	5.17	−1.52	5.166	1.543	−19.232	−5.542
000 000 001 00	24.92	5.17	−1.01	5.168	1.543	−21.202	−7.513
000 000 110 00	25.80	5.17	−1.16	5.171	1.544	−18.009	−7.561
D (black, smoker)							
100 100 011 01	39.59	7.41	−14.06	11.887	4.772	−12.369	−11.646
100 100 001 01	39.65	7.41	−14.09	11.924	4.789	−12.191	−8.227
010 100 000 00	53.06	12.04	−1.35	12.044	4.843	−15.327	−4.879
010 101 000 00	52.60	12.04	−1.44	12.044	4.843	−9.945	−2.739
010 110 000 00	52.09	12.05	−2.05	12.051	4.846	−13.415	−4.153
E (other, non-smoker)							
000 000 100 00	33.07	3.95	4.837	3.948	0.491	−17.644	−3.954
100 000 100 00	33.37	3.95	4.854	3.948	0.491	−18.217	−7.769
000 000 101 00	33.11	3.95	4.862	3.948	0.491	−19.604	−9.156
100 000 101 00	33.37	3.95	4.876	3.948	0.491	−20.216	−13.010
000 010 000 00	32.70	3.97	4.446	3.968	0.494	−15.955	−2.265
F (other, smoker)							
000 000 000 00	32.19	4.45	−10.70	4.450	2.329	−19.239	−2.307
100 000 000 00	32.51	4.45	−10.76	4.451	2.329	−19.671	−5.982
000 000 001 00	32.27	4.45	−10.78	4.454	2.330	−21.202	−7.513
100 000 001 00	32.51	4.46	−10.82	4.455	2.330	−21.671	−11.223
000 010 000 00	35.34	4.67	−8.09	4.669	2.367	−15.955	−2.265

9.3 Survival data on PBC: FIC plots and FIC selection

For the investigation of the survival data on primary biliary cirrhosis, see Section 1.4, we include x_1, age and x_2, the drug-or-not indicator, in every model, and further perform selection amongst the other 13 variables $z_1, \ldots, z_{13}$ using an all-subsets search. This leads to $2^{13} = 8192$ possible models. Table 1.2 already gave the parameter estimates together with their standard errors in the full model containing all 15 variables and assuming the Cox proportional hazards model as in (3.10).

We apply the model selection methods AIC, BIC and FIC to this data set. As in Section 3.4, the criteria are $\text{AIC}_{n,S} = 2\ell^p_{n,S}(\widehat{\beta}_S, \widehat{\gamma}_S) - 2(p + |S|)$ and $\text{BIC}_{n,S} = 2\ell^p_{n,S}(\widehat{\beta}_S, \widehat{\gamma}_S) - (\log n)(p + |S|)$, where $\ell^p_{n,S}$ is the log-partial likelihood based on the subset S model. The model which among all candidates has the largest value of $\text{AIC}_{n,S}$, respectively $\text{BIC}_{n,S}$, is selected as the best model. For the focussed information criterion we use its definition as in Section 6.8, namely, $\text{FIC}(S) = (\widehat{\psi} - \widehat{\psi}_S)^2 + 2(\widehat{\omega} - \widehat{\kappa})^t \widehat{Q}^0_S (\widehat{\omega} - \widehat{\kappa})$, where $\widehat{\psi} = (\widehat{\omega} - \widehat{\kappa})^t \widehat{\delta}$ and $\widehat{\psi}_S = (\widehat{\omega} - \widehat{\kappa})^t \widehat{G}_S \widehat{\delta}$, where we also include the correction to avoid a negative estimate of the bias squared term, as in (6.23). For this illustration we consider three focus parameters.

First, FIC_1 corresponds to selecting the best model for estimating the relative risk $\mu_1 = \exp\{(x - x_0)^t \beta + (z - z_0)^t \gamma\}$, where (x_0, z_0) represents a 50-year-old man with oedema ($z_5 = 1$) and (x_1, z_1) a 50-year-old woman without oedema ($z_5 = 0$), both in the drug-taking group $x_2 = 1$, with values z_2, z_3, z_4 equal to $0, 0, 0$, and with values $z_6, \ldots, z_{13}$ set equal to the average values of these covariates in the men's and women's strata, respectively. Thus the full (x_0, z_0) vector is $(50, 1, 0, 0, 0, 0, 1, 2.99, 366.00, 3.60, 2106.12, 125.42, 238.21, 11.02, 3.12)$ and the full (x, z) vector is $(50, 1, 1, 0, 0, 0, 0, 3.31, 371.26, 3.50, 1968.96, 124.47, 264.68, 10.66, 3.03)$. The second and third criteria FIC_2 and FIC_3 correspond to estimating five-year survival probabilities for the two groups just indicated, i.e. $\mu_2 = S(t_0 \mid x_0, z_0)$ and $\mu_3 = S(t_0 \mid x, z)$ with t_0 equal to five years.

Table 9.4 gives the values of AIC and the BIC, together with the list of variables in the corresponding models. All models are constrained to contain variables x_1, age and x_2, the drug-or-not indicator. The BIC selects variables z_6, z_8, z_{10}, z_{12} and z_{13}; these are the variables serum bilirubin, albumin, sgot, prothrombin time and histologic stage of disease. All of these variables were individually significant at the 5% level. These five variables are all present in each of the 20 best AIC-selected models. The BIC-selected model is ranked 7th amongst the AIC ordered models. The best model according to AIC adds to these five variables the variables z_1 and z_5, which correspond to the patient's gender and to the variable indicating presence of oedema. Variable z_5 was also individually significant (see Table 1.2), but this was not the case for variable z_1.

For the FIC the selections give suggestions rather different from those of AIC and the BIC, and also rather different from case to case. The results for the 20 best models for the three criteria are listed in Tables 9.5 and 9.6, sorted such that the first line corresponds

Table 9.4. *Values of AIC and BIC for the data on primary biliary cirrhosis,*
together with the selected variables among $z_1, \ldots, z_{13}$. The results are shown
for the 20 best models.

Rank	Variables	AIC	Variables	BIC
1	1,5,6,8,10,12,13	−1119.14	6,8,10,12,13	−1146.22
2	5,6,8,10,12,13	−1120.03	6,8,12,13	−1147.43
3	1,5,6,8,10,11,12,13	−1120.22	5,6,8,10,12,13	−1149.11
4	1,5,6,8,9,10,12,13	−1120.62	5,6,10,12,13	−1149.39
5	1,5,6,7,8,10,12,13	−1120.62	6,10,12,13	−1149.43
6	1,4,5,6,8,10,12,13	−1120.70	1,6,8,12,13	−1149.85
7	6,8,10,12,13	−1120.78	1,6,8,10,12,13	−1150.05
8	1,6,8,10,12,13	−1120.97	5,6,8,12,13	−1150.75
9	1,3,5,6,8,10,12,13	−1121.03	6,8,10,12	−1150.99
10	1,2,5,6,8,10,12,13	−1121.06	4,6,8,10,12,13	−1151.20
11	5,6,8,9,10,12,13	−1121.40	6,8,9,10,12,13	−1151.24
12	5,6,7,8,10,12,13	−1121.55	3,6,8,10,12,13	−1151.73
13	5,6,8,10,11,12,13	−1121.64	2,6,8,10,12,13	−1151.76
14	1,4,5,6,8,10,11,12,13	−1121.67	6,7,8,10,12,13	−1151.82
15	4,5,6,8,10,12,13	−1121.67	6,8,10,11,12,13	−1151.83
16	2,5,6,8,10,12,13	−1121.79	1,5,6,8,10,12,13	−1151.85
17	3,5,6,8,10,12,13	−1121.85	1,5,6,8,12,13	−1151.86
18	1,5,6,8,9,10,11,12,13	−1121.98	4,6,8,12,13	−1152.11
19	1,5,6,7,8,10,11,12,13	−1121.99	6,8,9,12,13	−1152.19
20	1,4,5,6,8,9,10,12,13	−1121.99	6,12,13	−1152.19

to the best model. The results are also conveniently summarised via the FIC plots of
Figures 9.3 and 9.4. Again, all models include the protected covariates x_1 and x_2. For the
relative risk parameter μ_1, the best FIC scores form a close race, with the most important covariates being z_3, z_5, z_9, z_{10}, z_{13} (hepatomegaly, oedema, alkaline, sgot, histologic
stage). For the five-year survival probabilities for the two strata, it is first of all noteworthy that the woman stratum without oedema has a clear advantage over the man stratum
with oedema, and that the best estimates of the woman-without-oedema stratum survival
probability are in rather better internal agreement than those for the man-with-oedema
stratum. Secondly, the two probabilities are clearly influenced by very different sets of
covariates. For the woman group, all covariates z_1–z_{13} are deemed important, apart from
z_3, z_4, z_6 (hepatomegaly, spiders, bilirubin); for the man group, rather fewer covariates
are deemed influential, the most important among them being z_9, z_{11}, z_{12}, z_{13} (alkaline,
platelets, prothrombin, histological stage).

For comparison purposes, Tables 9.5 and 9.6 also include the ∆AIC scores, i.e. the
original AIC scores minus their maximum value. We see that FIC_2 for μ_2 suggests submodels that do somewhat poorly from the AIC perspective, but that the best FIC models
for μ_1 and μ_3 do even worse. This is also a reminder that AIC aims at good predictive

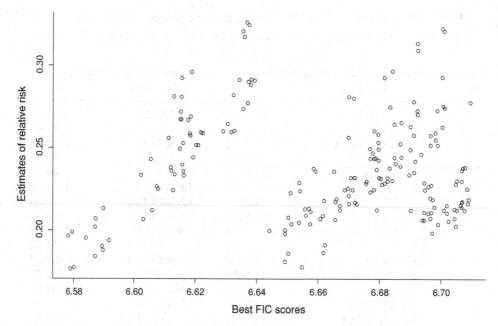

Fig. 9.3. Primary biliary cirrhosis data. FIC plots for the best 200 estimates of relative risk $\exp((x - x_0)^t\beta + (z - z_0)^t\gamma)$, where (x_0, z_0) represents a 50-year-old man with oedema and (x, z) a 50-year-old woman without oedema.

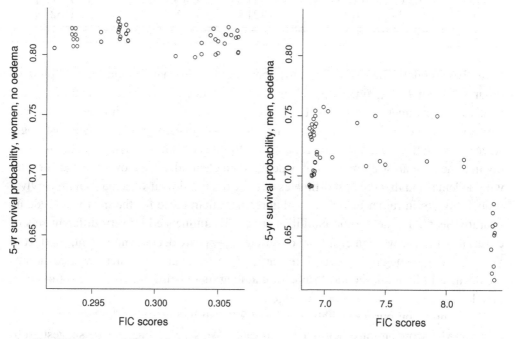

Fig. 9.4. Primary biliary cirrhosis data. FIC plots for the 50 best estimates of five-year survival probabilities, for two strata of patients: women of age 50, without oedema (left panel) and men of age 50, with oedema (right panel).

Table 9.5. *Primary biliary cirrhosis data. The 20 best estimates of the relative risk parameter* $\exp\{(x - x_0)^t + (z - z_0)^t\gamma\}$, *as ranked by the* FIC_1 *scores, together with included covariates, estimated bias, standard deviation, root mean squared error and the* $\Delta AIC = AIC - \max_{all}(AIC)$ *scores.*

Rank	Covariates	Estimate	Bias	sd	rmse	FIC$_1$	ΔAIC
1	3,5	0.196	−0.103	0.076	0.0765	6.578	−78.327
2	3,5,10	0.177	−0.095	0.076	0.0765	6.579	−57.625
3	3,5,9	0.199	−0.099	0.076	0.0765	6.579	−77.763
4	3,5,9,10	0.177	−0.092	0.077	0.0765	6.580	−57.878
5	5,13	0.195	−0.095	0.077	0.0766	6.584	−70.504
6	5,9,13	0.202	−0.087	0.077	0.0767	6.587	−68.436
7	5,10,13	0.184	−0.077	0.077	0.0767	6.587	−44.278
8	3,5,13	0.207	−0.087	0.077	0.0767	6.587	−68.418
9	3,5,10,13	0.190	−0.072	0.077	0.0767	6.589	−44.147
10	5,9,10,13	0.188	−0.070	0.077	0.0767	6.590	−43.251
11	3,5,9,13	0.213	−0.080	0.077	0.0767	6.590	−67.189
12	3,5,9,10,13	0.194	−0.067	0.077	0.0768	6.592	−43.540
13	2,5	0.233	−0.074	0.077	0.0770	6.602	−78.501
14	2,5,10	0.207	−0.066	0.077	0.0770	6.603	−57.144
15	2,5,9	0.243	−0.066	0.077	0.0771	6.605	−75.521
16	2,5,9	0.212	−0.060	0.077	0.0771	6.606	−55.961
17	5,12	0.227	−0.064	0.077	0.0771	6.607	−69.928
18	5,9,12	0.225	−0.062	0.077	0.0771	6.608	−69.148
19	2,5,13	0.256	−0.055	0.077	0.0772	6.611	−61.138
20	5,10,12	0.238	−0.040	0.077	0.0772	6.612	−39.553

performance weighted across all covariates, while the FIC methods are spearheaded criteria aiming at optimal estimation and prediction for specified foci.

We found in Table 1.2 that the variable 'drug' is not significant, when all other covariates are controlled for. If we do not include this covariate as part of the protected covariates in the model, leading to 2^{14} candidate models in the model search, it turns out that the selected models do not change, when compared to the results reported on above.

9.4 Speedskating data: averaging over covariance structure models

Model averaging can be applied without problem to multidimensional data. As a specific example, we use data from the Adelskalenderen of speedskating, which is the list of the best speedskaters ever, as ranked by their personal best times over the four distances 500-m, 1500-m, 5000-m, 10,000-m, via the classical samalogue point-sum $X_1 + X_2 + X_3 + X_4$, where X_1 is the 500-m time, X_2 is the 1500-m time divided by 3, X_3 is the 5000-m time divided by 10, and X_4 the 10,000-m time divided by 20. See also Section 1.7. The correlation structure of the four-vector $Y = (X_1, \ldots, X_4)$ is important for relating, discussing and predicting performances on different distances.

Table 9.6. *Primary biliary cirrhosis data. The 20 best estimates, ranked by* FIC$_2$ *scores,
of five-year survival probabilities* $S(t_0 \mid x, z)$ *for the two strata, women of age 50
without oedema (left) and men of age 50 with oedema (right), together with included
covariates, estimated bias, standard deviation, root mean squared error and the*
ΔAIC = AIC $-$ max$_{\text{all}}$(AIC) *scores.*

Rank	Women: covariates	Estimate	FIC$_2$	ΔAIC	Men: covariates	Estimate	FIC$_3$	ΔAIC
1	5,7,8,9,10,12,13	0.806	0.292	-18.407	none	0.739	6.875	-127.480
2	5,7,8,9,10,11,12,13	0.805	0.292	-20.153	12	0.692	6.875	-83.963
3	2,5,7,10,13	0.822	0.293	-32.438	11	0.734	6.879	-123.790
4	2,5,7,8,19,13	0.817	0.293	-30.912	11,12	0.691	6.879	-84.955
5	2,5,7,10,12,13	0.813	0.293	-18.526	13	0.732	6.886	-93.258
6	2,5,7,8,10,12,13	0.807	0.293	-16.939	12,13	0.692	6.887	-68.348
7	2,5,7,10,11,13	0.822	0.294	-34.398	11,13	0.730	6.893	-93.888
8	2,5,7,8,10,11,13	0.817	0.294	-32.822	9	0.743	6.893	-124.283
9	2,5,7,10,11,12,13	0.813	0.294	-20.022	9,12	0.701	6.894	-83.188
10	2,5,7,8,10,11,12,13	0.807	0.294	-18.383	11,12,13	0.691	6.894	-70.247
11	2,5,7,9,10,12	0.819	0.296	-28.976	9,11	0.738	6.895	-119.094
12	2,5,7,9,10,11,12	0.819	0.296	-30.975	9,11,12	0.700	6.895	-83.206
13	2,5,7,8,9,10,12	0.811	0.296	-25.443	9,13	0.736	6.907	-89.549
14	2,5,7,8,9,10,11,12	0.811	0.296	-27.428	10	0.747	6.907	-103.391
15	2,5,7,9,13	0.828	0.297	-40.628	10,12	0.704	6.907	-52.118
16	2,5,7,9,12,13	0.819	0.297	-31.535	9,12,13	0.701	6.910	-66.418
17	2,5,7,9,11,13	0.830	0.297	-41.439	9,11,13	0.734	6.910	-88.838
18	2,5,7,9,11,12,13	0.821	0.297	-32.884	9,11,12,13	0.701	6.913	-67.750
19	2,5,7,8,9,13	0.823	0.297	-38.806	10,11	0.741	6.917	-102.626
20	2,5,7,8,9,12,13	0.813	0.297	-29.559	10,11,12	0.704	6.919	-54.105

We consider multivariate normal data $Y \sim N_4(\xi, \Sigma)$, where different models for the covariance structure Σ, in the absence of clear a priori preferences, are being averaged over to form estimators of quantities of interest. There are several areas of statistics where covariance modelling is of interest, and sometimes perhaps of primary concern, for example as with factor analysis, and where variations of these methods might prove fruitful.

While there is a long list of parameters $\mu = \mu(\xi, \Sigma)$ that might ignite the fascination of speedskating fans, see e.g. Hjort and Rosa (1999), for this discussion we single out the following four focus parameters: the generalised standard deviation measures $\mu_1 = \{\det(\Sigma)\}^{1/8}$ and $\mu_2 = \{\text{Tr}(\Sigma)\}^{1/2}$, the average correlation $\mu_3 = (1/6)\sum_{i<j} \text{corr}(X_i, X_j)$, and the maximal correlation μ_4 between (X_1, X_2, X_3) and X_4. The latter is the maximal correlation between a linear combination of X_1, X_2, X_3 and X_4, and is for example of interest at championships when one tries to predict the final outcomes, after the completion of the three first distances. It is also equal to $(\Sigma_{10}\Sigma_{00}^{-1}\Sigma_{01}/\Sigma_{11})^{1/2}$, in terms of the blocks

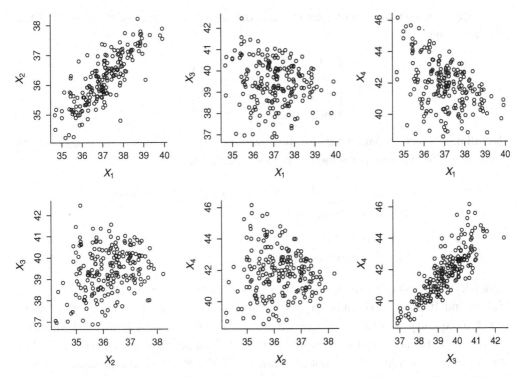

Fig. 9.5. Speedskating data. Pairwise scatterplots of the first 200 entries of the Adelskalenderen, as per the end of the 2005–2006 season. The variables are in seconds per 500-m for the four classical distances: X_1 is 500-m time, X_2 is 1500-m time divided by 3, X_3 is 5000-m time divided by 10, and X_4 is 10,000-m time divided by 20.

of Σ, of size 3×3 for Σ_{00} and so on. We analyse the top of the Adelskalenderen, with the best $n = 200$ skaters ever, as per the end of the 2006 season. The vectors $Y_1, \ldots, Y_n$ are by definition ranked, but as long as one discusses estimators that are permutation-invariant we may view the data vectors as a random sample from the population of the top skaters of the world.

Figure 9.5 gives pairwise scatterplots of the variables X_1, X_2, X_3 and X_4. These give some indication of the correlation values, with a 'non-structured random cloud' indicating correlation close to zero. An upward trend as, for example, for the plots of X_1 versus X_2 and that of X_3 versus X_4 indicates positive correlation. A downward trend would indicate negative correlation. One finds 'sprinters', 'stayers' and 'allrounders' among these 200 skaters.

Write the covariances as $\sigma_{j,k} = \sigma_j \sigma_k \rho_{j,k}$ in terms of standard deviations σ_j and correlations $\rho_{j,k}$ $(j, k = 1, \ldots, 4)$. We shall consider two models for the four standard deviation parameters, namely

$$S_0: \sigma_1 = \sigma_2 = \sigma_3 = \sigma_4 \quad \text{and} \quad S_1: \text{the four } \sigma_j \text{s are free.}$$

These are to be combined with various candidate models for the six parameters of the correlation matrix

$$\begin{pmatrix} 1 & \rho_{1,2} & \rho_{1,3} & \rho_{1,4} \\ \rho_{1,2} & 1 & \rho_{2,3} & \rho_{2,4} \\ \rho_{1,3} & \rho_{2,3} & 1 & \rho_{3,4} \\ \rho_{1,4} & \rho_{2,4} & \rho_{3,4} & 1 \end{pmatrix}.$$

The full model has ten parameters for the covariance matrix, plus four parameters in the mean vector. Some models for the correlation structure are

R_0: $\rho_{2,4} = 0$ and $\rho_{1,2} = \rho_{1,3} = \rho_{1,4} = \rho_{2,3} = \rho_{3,4}$,

R_1: $\rho_{2,4} = 0$ and $\rho_{1,2} = \rho_{3,4}$,

R_2: $\rho_{3,4} = \rho_{1,2}$,

R_3: no restrictions.

Model R_0 assumes equicorrelation, except for the correlation between the times on the 1500-m and 10,000-m, which is set to zero. In model R_1 we keep this zero correlation, as well as equality of correlation between the times on the shortest distances 500-m and 1500-m, with that between the times on the longest distances 5000-m and 10,000-m. The other correlation parameters are not constrained. Model R_2 considers the same equality, but with no restriction on other correlation parameters. Model R_3 does not include any restrictions on the correlation parameters. Several other models could be constructed and included in a model search. We restrict attention to the above-mentioned eight models.

To place this setting into the framework developed earlier, let the full covariance matrix Σ_{S_1,R_3} be equal to

$$\sigma^2 \begin{pmatrix} 1 & \rho\phi_2 & \rho\phi_3(1+v_{1,3}) & \rho\phi_4(1+v_{1,4}) \\ \rho\phi_2 & \phi_2^2 & \rho\phi_2\phi_3(1+v_{2,3}) & \rho\phi_2\phi_4(1+v_{2,4}) \\ \rho\phi_3(1+v_{1,3}) & \rho\phi_2\phi_3(1+v_{2,3}) & \phi_3^2 & \rho\phi_3\phi_4(1+v_{3,4}) \\ \rho\phi_4(1+v_{1,4}) & \rho\phi_2\phi_4(1+v_{2,4}) & \rho\phi_3\phi_4(1+v_{3,4}) & \phi_4^2 \end{pmatrix},$$

where we define $\phi_i = \sigma_i/\sigma_1$; also, let $\sigma = \sigma_1$ and $\rho = \rho_{1,2}$. The parameter $\theta = (\xi, \sigma^2, \rho)$ is present in all of the models, while subsets of

$$\gamma = (\phi_2, \phi_3, \phi_4, v_{1,3}, v_{1,4}, v_{2,3}, v_{2,4}, v_{3,4})^t$$

are present in some of the models. The smallest model corresponds to $\gamma = \gamma_0 = (1, 1, 1, 0, 0, 0, -1, 0)^t$, since with this choice of values we are back at the parameter combination of the smallest model (S_0, R_0). The number of fixed parameters is $p = 6$ and the number of parameters to choose from equals $q = 8$. Listing the models according to their number of parameters in the covariance matrix, we start with the 2-parameter narrow model, then we have (S_1, R_0) and (S_0, R_1), both with 5 parameters, (S_0, R_2) has 6 parameters, (S_0, R_3) contains 7 parameters, (S_1, R_1) has 8 parameters to estimate,

(S_1, R_2) has 9, and the full model (S_1, R_3) contains 10 parameters in the covariance matrix.

We use the criteria AIC, BIC and FIC to select an appropriate covariance structure. For AIC and the BIC we need to fit the eight models, obtain for each the log-likelihood value when the maximum likelihood estimators of the parameters are inserted, and count the number of estimated parameters. The particular focus parameter is ignored in the model search for these two criteria.

The data $Y_1, \ldots, Y_n$ are from $N_4(\xi, \Sigma)$, with different models for the 4×4 covariance matrix. Each of the models has a four-dimensional mean vector, which is considered a nuisance parameter for this occasion. Let us in general write $\Sigma = \Sigma(\phi)$, in terms of a candidate's model parameters ϕ, along with

$$S_n = n^{-1} \sum_{i=1}^{n} (y_i - \widehat{\xi})(y_i - \widehat{\xi})^t \quad \text{and} \quad \widehat{\xi} = \bar{y} = n^{-1} \sum_{i=1}^{n} y_i.$$

The log-likelihood function is

$$\ell_n(\xi, \phi) = \sum_{i=1}^{n} \left\{ -\tfrac{1}{2} \log |\Sigma(\phi)| - \tfrac{1}{2}(y_i - \xi)^t \Sigma(\phi)^{-1}(y_i - \xi) - \tfrac{1}{2} d \log(2\pi) \right\}$$

$$= -\tfrac{1}{2} n \left\{ \log |\Sigma(\phi)| + \text{Tr}[\Sigma(\phi)^{-1}\{S_n + (\bar{y} - \xi)(\bar{y} - \xi)^t\}] + d \log(2\pi) \right\},$$

with $d = 4$ being the dimension. In these models, where there are no restrictions on ξ, the maximum likelihood estimator is indeed $\widehat{\xi} = \bar{y}$, and the profile log-likelihood function is

$$\ell_n(\phi) = -\tfrac{1}{2} n \left[\log |\Sigma(\phi)| + \text{Tr}\{\Sigma(\phi)^{-1} S_n\} + d \log(2\pi) \right].$$

Thus any of these covariance structure models is fitted by letting $\widehat{\phi}$ minimise

$$C_n(\phi) = \log |\Sigma(\phi)| + \text{Tr}\{\Sigma(\phi)^{-1} S_n\} + d \log(2\pi).$$

Also,

$$\text{AIC} = -n[C_n(\widehat{\phi}) + (2/n) \, \text{length}(\phi) + (2/n)d],$$
$$\text{BIC} = -n[C_n(\widehat{\phi}) + \{(\log n)/n\} \, \text{length}(\phi) + (2/n)d].$$

From the AIC perspective, therefore, including one more parameter in a model for Σ is only worth it if the minimum of C_n is pulled down at least $2/n$. For BIC, this quantity should equal at least $(\log n)/n$. For optimisation of the four-dimensional normal profile log-likelihood function, we used the function nlm in R, as for a similar task in Section 9.1.

We now obtain all ingredients of the FIC for this data set and the focus parameters specified above. For more details, see Section 6.5. When using (for example) the R function nlm to fit the models, as output we receive not only the value of the maximised log-likelihood (used to obtain the AIC and BIC values in Table 9.7) but also the parameter estimates, and on request even the Hessian matrix, which is the matrix of second partial

Table 9.7. *Summary of AIC, BIC and FIC values for the speedskating data. The model with rank (1) is preferred by the criterion. See text for the precise definition of the models and focus parameters μ_j. The column 'pars' gives the total number of parameters in that model.*

Models	pars	Information criteria					
		AIC	BIC	FIC(μ_1)	FIC(μ_2)	FIC(μ_3)	FIC(μ_4)
(S_0, R_0)	$4+2$	-2432.93 (8)	-2452.72 (8)	131.80 (8)	31.84 (8)	60.12 (8)	23.54 (8)
(S_1, R_0)	$4+5$	-1950.74 (6)	-1980.43 (6)	126.88 (7)	25.95 (7)	59.57 (7)	23.29 (7)
(S_0, R_1)	$4+5$	-2295.83 (7)	-2325.51 (7)	3.23 (4)	2.11 (4)	1.17 (4)	0.14 (1)
(S_0, R_2)	$4+6$	-1920.97 (5)	-1953.95 (5)	24.80 (5)	3.80 (5)	12.20 (5)	4.89 (6)
(S_0, R_3)	$4+7$	-1890.76 (4)	-1927.04 (4)	32.38 (6)	4.63 (6)	16.07 (6)	3.82 (5)
(S_1, R_1)	$4+8$	-1795.82 (1)	-1835.40 (1)	1.18 (2)	0.12 (2)	0.64 (2)	0.30 (2)
(S_1, R_2)	$4+9$	-1796.39 (2)	-1839.26 (2)	1.13 (1)	0.10 (1)	0.63 (1)	0.43 (3)
(S_1, R_3)	$4+10$	-1798.00 (3)	-1844.17 (3)	1.30 (3)	0.15 (3)	0.71 (3)	0.44 (4)

derivatives of the function to be minimised. This makes it easy to obtain the empirical information matrix $J_{n,\text{wide}}(\widehat{\theta}, \gamma_0)$ as the Hessian divided by n. Note that since we use minus log-likelihood in the minimisation procedure, the Hessian matrix already corresponds to minus the second partial derivatives of the log-likelihood function. This is a matrix of dimension 14×14. Given this matrix, we construct the matrix $\widehat{Q}$ by taking the lower-right submatrix, of dimension 8×8, of $J_{n,\text{wide}}^{-1}$. The next step is to construct the $|S| \times |S|$ matrices $\widehat{Q}_S$ and the $q \times q$ matrices $\widehat{Q}_S^0$.

- For the narrow model (S_0, R_0), or in the previous notation, $S = \emptyset$, we do not include any additional variables and Q_S^0 consists of zeros only.
- The model (S_1, R_0) has compared to the narrow model three extra parameters (ϕ_2, ϕ_3, ϕ_4). The matrix $\widehat{Q}_S$ is formed by taking in $\widehat{Q}$ the upper-left submatrix consisting of elements in the first three rows and first three columns.
- In comparison to (S_0, R_0), the covariance matrix Σ_{S_0, R_1} needs the extra three parameters $(\nu_{1,2}, \nu_{1,3}, \nu_{1,4})$, implying that $\widehat{Q}_S$ consists of the 3×3 submatrix of elements of $\widehat{Q}$ in rows and columns numbered 4, 5, 6.
- For the six-parameter model (S_0, R_2) we use the 4×4 submatrix defined by row and column numbers 4, 5, 6 and 7.
- Model (S_0, R_3) with seven parameters, including $(\nu_{1,3}, \nu_{1,4}, \nu_{2,3}, \nu_{2,4}, \nu_{3,4})$ requires for the 5×5 matrix $\widehat{Q}_S$ the elements in rows and columns numbered 4, 5, 6, 7 and 8.
- The other three models all include S_1. For (S_1, R_1) we have a 6×6 matrix $\widehat{Q}_S$ consisting of the upper-left corner of $\widehat{Q}$ consisting of elements in the first six rows and columns.
- For (S_1, R_2) there is a 7×7 matrix $\widehat{Q}_S$ consisting of the upper-left corner of $\widehat{Q}$ consisting of elements in the first seven rows and columns.
- For the full model (S_1, R_3), $\widehat{Q}_S = \widehat{Q}$.

Using these matrices gives us also the matrices $\widehat{Q}_S^0$ and $\widehat{G}_S$.

Computating estimates of the vector $\omega = J_{n,10} J_{n,00}^{-1} \frac{\partial \mu}{\partial \theta} - \frac{\partial \mu}{\partial \gamma}$ can, in principle, be done explicitly by computing the derivatives of each of the focus parameters μ_j with respect to θ and γ. Since these partial derivatives would be tedious to obtain in this particular situation, we used a numerical derivative procedure, as provided by the function `numericDeriv` in R. First we define a function which computes the focus parameter(s) as a function of the parameters θ, γ. Then, we use the R function `numericDeriv` to evaluate the partial derivatives at $(\widehat{\theta}_{\text{wide}}, \gamma_0)$. For the speedskating models, the situation simplifies somewhat since the mean vector ξ is not used in the computation of the focus parameters, and we basically only need the 10×10 submatrix of $J_{n,\text{wide}}$ corresponding to all parameters but ξ.

Once the J matrix is estimated and the derivatives of the focus parameters are obtained, we construct the ω vector. For this set of data, we have four such ω vectors, one for each focus parameter. Each of the four vectors is of length eight, since we consider eight models. For this set of data we get

$$(\widehat{\omega}_1, \widehat{\omega}_2, \widehat{\omega}_3, \widehat{\omega}_4) = \begin{pmatrix} 0.0489 & 0.1393 & -0.0042 & -0.0031 \\ 0.0060 & 0.0355 & 0.0040 & 0.0030 \\ -0.0464 & -0.1306 & 0.0044 & 0.0032 \\ 0.2351 & -0.3853 & -0.2839 & -0.0208 \\ 0.3251 & 0.3522 & -0.1715 & -0.3159 \\ -0.0256 & 0.34863 & 0.1075 & 0.2686 \\ 0.1723 & -0.1810 & -0.1827 & -0.3242 \\ -0.2363 & -0.2711 & 0.1209 & -0.0997 \end{pmatrix}.$$

Given the vector ω_j, one for each focus parameter, we compute the null variances τ_0^2, again, one for each focus parameter. This results in

$$\widehat{\tau}_0(\mu_1) = 0.242, \quad \widehat{\tau}_0(\mu_2) = 1.011, \quad \widehat{\tau}_0(\mu_3) = 0.129, \quad \text{and } \widehat{\tau}_0(\mu_4) = 0.095.$$

We now have all needed components to construct the FIC values.

Table 9.7 gives the values for the information criteria, together with a rank number for each of the eight models. AIC and the BIC are in agreement here, both preferring model (S_1, R_1) with eight parameters to structure the covariance matrix. This model assumes a zero correlation between the results for the 1500-m and the 10,000-m, as well as equal correlation for the results of the (500-m, 1500-m) and the (5000-m, 10,000-m). The variances of the four components are allowed to be different. Note again that AIC and BIC model selection does not require us to think about what we will do with the selected model; the selected model is the same, no matter what focus parameter we wish to estimate.

The FIC depends on the parameter under focus and hence gives different values for different μ_ks. For this data set, the FIC for parameters μ_1, μ_2 and μ_3 all point to model (S_1, R_2). This model includes one more parameter than the AIC and BIC preferred model. Specifically, it does not set the correlation between the times of the 1500-m and

Table 9.8. *Eleven different estimates of the parameters*
$\mu_1, \mu_2, \mu_3, \mu_4$. *These correspond to AIC-smoothed,*
BIC-smoothed and FIC-smoothed averages and to
each of the eight models considered.

Models	$\widehat{\mu}_1$	$\widehat{\mu}_2$	$\widehat{\mu}_3$	$\widehat{\mu}_4$
sm-AIC	0.732	2.264	0.228	0.877
sm-BIC	0.732	2.257	0.246	0.874
sm-FIC	0.737	2.267	0.226	0.865
(S_0, R_0)	1.099	2.331	0.264	0.426
(S_1, R_0)	1.005	2.798	0.463	0.811
(S_0, R_1)	0.810	2.306	0.256	0.843
(S_0, R_2)	0.794	2.432	0.431	0.865
(S_0, R_3)	0.778	2.436	0.437	0.836
(S_1, R_1)	0.732	2.255	0.252	0.873
(S_1, R_2)	0.732	2.268	0.206	0.879
(S_1, R_3)	0.731	2.279	0.206	0.884

the 10,000-m equal to zero. The FIC for focus parameter μ_4, however, the maximal correlation between (X_1, X_2, X_3) and X_4, points to a much simpler model. The FIC(μ_4) chosen model is model (S_0, R_1) with equal variances, equal values of the correlation of the times for the two smallest distances and for the two largest distances, and zero correlation between the times for the 1500-m and the 10,000-m.

Table 9.8 gives the parameter estimates. Also presented in the table are the model-averaged estimates using smoothed AIC, BIC and FIC weights, where for the latter $\kappa = 1$. The definitions of the weights are as in Examples 7.1, 7.2 and 7.3. Smoothed AIC gives weights 0.478, 0.360, 0.161 to its top three models, the other models get zero weight. Obviously, the highest ranked model receives the largest weight. For smoothed BIC the nonzero weights for the best three models are 0.864, 0.125 and 0.011. For focus parameter μ_1, the smoothed FIC has nonzero weights for the four best models, the corresponding weights are 0.329, 0.317, 0.289 and 0.065. For μ_2, the nonzero weights are for the three best models, with values 0.390, 0.329 and 0.281. FIC for μ_3 assigns nonzero weight to its four highest ranked models, which receive the weights 0.298, 0.295, 0.268 and 0.139. For focus parameter μ_4 the corresponding nonzero weights in the model averaging are 0.366, 0.258, 0.190 and 0.186.

Exercises

9.1 *nlm for skulls:* This exercise gives some practical details for estimating parameters in the multivariate normal models worked with in Section 9.1, using the software package R. Write a function and give it the name `minus.log.likelihood(para)`, requiring the unknown parameter values as input, and returning the negative of the log-likelihood function. There are several possibilities for finding the parameter values maximising the log-likelihood, or equivalently, minimising minus log-likelihood. One of them is the function nlm, which we apply to our own defined function as nlm(minus.log.likelihood, start), where start is a vector of starting values for the search. Other algorithms which can be used are optim and optimise, the latter for one-dimensional optimisation. These functions give the possibility to choose among several optimisation algorithms.

For the skulls data, go through each of the suggested models, and find maximum likelihood parameter estimates and AIC and BIC scores for each. You may also try out other competing models by modelling the mean vectors or variance matrices in perhaps more elaborate ways.

9.2 *State Public Expenditures data:* This data set is available from the Data and Story Library, accessible via the internet address lib.stat.cmu.edu/DASL/Stories/stateexpend.html. The response variable y is 'Ex' in the first column. The variable names are Ex: per capita state and local public expenditures; ecab: economic ability index, in which income, retail sales and the value of output (manufactures, mineral and agricultural) per capita are equally weighted; met: percentage of population living in standard metropolitan areas; grow: percent change in population, 1950–1960; young: percent of population aged 5–19 years; old: percent of population over 65 years of age; and west: Western state (1) or not (0).

(a) For a model without interactions, select the best model according to (i) AIC, (ii) BIC and (iii) Mallows's C_p. Give the final selected model, with estimates for the regression coefficients. Compute also corresponding estimates of $\mu(x_0) = E(Y \mid x_0)$, for a specified position x_0 of interest in the covariate space.

(b) For a model with interactions, do not include all possible subsets in the search but use a stepwise approach, adding one variable at a time in a forward search, or leaving out one variable at a time in a backward search. The R functions step or stepAIC in library(MASS) may, for example, be used for this purpose. Use both stepwise AIC and BIC to select a good model for this data set. Give the final selected model, with estimates for the regression coefficients.

9.3 *Breakfast cereal data:* The data are available at the website lib.stat.cmu.edu/datasets/ 1993.expo. The full data set, after removing cases with missing observations, contains information on nutritional aspects and grocery shelf location for 74 breakfast cereals. The data include the cereal name, number of calories per serving, grams of protein, grams of fat, milligrams of sodium, grams of fibre, grams of carbohydrates, grams of sugars, milligrams of potassium, typical percentage of vitamins, the weight of one serving, the number of cups in one serving and the shelf location (1, 2 or 3 for bottom, middle or top). Also provided is a variable 'rating', constructed by consumer reports in a marketing study.

(a) Construct a parametric regression model that best fits the relation that grams of sugars has on the rating of the cereals.

(b) Fit an additive model (that is, a model without interactions) with 'rating' as the response variable, and with carbohydrates and potassium as explanatory variables. In which way do these two variables influence the ratings? Does the additivity assumption make sense for these variables?

(c) Now include all variables. Select a reasonable model to explain the rating as a function of the other variables. Mention the method you used for the variable search, and give the final model, including parameter estimates and standard errors.

9.4 *The Ladies Adelskalenderen:* Consider the data described in Exercise 2.9. For women's speedskating, the four classic distances are 500-m, 1500-m, 3000-m and 5000-m. Perform an analysis of the correlation structure of the times on these four distances, similar to that obtained for men's speedskating in Section 9.4. Are different models selected for the four focus parameters? Construct the smooth AIC, BIC and FIC weights together with the model-averaged estimates. Also explore joint models for the men and the ladies, aimed at finding potential differences in the correlation structures among distances.

9.5 *Birthweight modelling using the raw data:* In Section 9.2 and earlier we have worked with different logistic regression models for the 0–1 outcome variable of having birthweight less than the cut-off point 2500 g.
  (a) Use the raw data for the 189 mothers and babies, with information on the actual birth-weights, to explore and select among models for the continuous outcome variable. Compare the final analyses as to prediction quality and interpretation.
  (b) There is increasing awareness that also the *too big* babies are under severe risk (for reasons very different from those related to small babies), cf. Henriksen (2007), Voldner *et al.* (2008). Use the raw data to find the most influential covariates for the event that the birthweight exceeds 3750 grams. (This is not a high threshold as such, as health risk typically may set in at level 4500 grams; for the American mothers of the Hosmer and Lemeshow data set there are, however, very few such cases.)

10

Further topics

In this chapter some topics are treated that extend the likelihood methods of the previous chapters. In particular we deal with selection of fixed and random components in mixed models, study some effects of selecting parameters that are at the boundary of their parameter space, discuss finite-sample corrections to earlier limit distribution results, give methods for dealing with missing covariate information, and indicate some problems and solutions for models with a large number of parameters.

10.1 Model selection in mixed models

Mixed models allow us to build a quite complicated structure into the model. A typical example is a model for longitudinal data where repeated observations are collected on the same subject. Often it makes sense to assume a linear regression model for each subject, though with regression coefficients that may vary from subject to subject. Each subject may have its own intercept and slope. In a well-designed study the subjects are a random sample from a population of subjects and hence it is reasonable to assume that also the intercepts and slopes are a random sample. A simple model for such data is the following. For the jth measurement of the ith subject we write the model as

$$Y_{i,j} = (\beta_0 + u_{i,0}) + (\beta_1 + u_{i,1})x_{i,j} + \varepsilon_{i,j},$$

where $j = 1, \ldots, m_i$ and $i = 1, \ldots, n$. The covariate $x_{i,j}$ could, for example, be the time at which the jth measurement of the ith subject is taken. The coefficients β_0 and β_1 are the fixed, common, intercept and slope, while $(u_{i,0}, u_{i,1})$ are the random effects. These random effects are random variables with mean zero (for identifiability reasons) and with a certain variance matrix D. A model that contains both fixed and random effects is called a mixed model.

A general linear mixed model is written in the following way. Let $y_i = (y_{i,1}, \ldots, y_{i,m_i})^t$, $\varepsilon_i = (\varepsilon_{i,1}, \ldots, \varepsilon_{i,m_i})^t$ and denote by X_i the fixed effects design matrix of dimension $m_i \times r$ and by Z_i the random effects design matrix of dimension $m_i \times s$.

Then, for $i = 1, \ldots, n$

$$Y_i = X_i \beta + Z_i u_i + \varepsilon_i, \tag{10.1}$$

with β the vector of fixed effects coefficients and u_i the vector of random effects. This model requires some assumptions on the random effects. In particular, we assume that u_i has mean zero and covariance matrix D and that the error terms ε_i have also mean zero and a covariance matrix R_i. Moreover, the random effects u_i and errors ε_i are assumed to be independent. This implies that Y_i has mean $X_i \beta$ and covariance matrix $Z_i D Z_i^t + R_i$. Observations for different subjects i are assumed to be independent. The above model notation can be further comprised by constructing a long vector y of length $N = \sum_{i=1}^{n} m_i$ that contains $y_1, \ldots, y_n$, a vector ε that contains $\varepsilon_1, \ldots, \varepsilon_n$, u which consists of components $u_1, \ldots, u_n$, a matrix X of dimension $N \times r$ that stacks all X_i design matrices, and a block-diagonal random effects design matrix $Z = \text{diag}(Z_1, \ldots, Z_n)$. The resulting model is written in matrix form as $Y = X\beta + Zu + \varepsilon$.

Model selection questions may be posed regarding both the fixed and random effects. Which subset of fixed effects should be included in the model? Which subset of random effects is relevant?

10.1.1 AIC for linear mixed models

Linear mixed models are mostly worked with under the assumption of normal distributions for both random effects and error terms. For model (10.1) this implies that the conditional distribution $Y_i \mid u_i \sim N_{m_i}(X_i \beta + Z_i u_i, R_i)$ and $u_i \sim N_s(0, D)$, and that the marginal distribution of Y_i is $N_{m_i}(X_i \beta, Z_i^t D Z_i + R_i)$. With this information the likelihood corresponding to the data $y_1, \ldots, y_n$ is obtained as the product over $i = 1, \ldots, n$ of normal density functions, where the parameter vector θ consists of the vector β, together with the variances and covariances contained in the matrices D and R_i. Often the errors are uncorrelated, in which case $R_i = \sigma_\varepsilon^2 I_{m_i}$. When the estimation of θ is performed using the maximum likelihood method, it is straightforward to write AIC as in Section 2.3,

$$\text{AIC} = 2\ell_n(\widehat{\theta}) - 2 \, \text{length}(\theta).$$

Vaida and Blanchard (2005) call this the marginal AIC (mAIC), which is to be used if the goal of model selection is to find a good model for the population or fixed effect parameters. This use of the model treats the random effects mainly to model the correlation structure of the Y_i, since the marginal model for Y_i is nothing but a linear model $Y_i = X_i \beta + \zeta_i$ with correlated errors $\zeta_i = Z_i u_i + \varepsilon_i$.

Sometimes the random effects are themselves of interest. When observations are taken on geographical locations, each such area may have its own random effect. Predictions can be made for a specific area. Another example is in clinical studies where patients are treated in different hospitals and in the model one includes a hospital-specific random effect. The mixed effects model can then be used to predict the effect of a treatment at a certain hospital. For such cases one would want to use the conditional distribution of

$Y_i \mid u_i$ in the information criterion, rather than the marginal distribution. This leads to the conditional Akaike information criterion (cAIC; Vaida and Blanchard, 2005). In the situation where the matrix D and variance σ_ε^2 are known, the definition of cAIC is given as

$$\text{cAIC} = 2 \log f(y \mid \widehat{\theta}, \widehat{u}) - 2\rho,$$

where the conditional distribution of $Y \mid u$ is used to construct the log-likelihood function and where ρ is the effective number of degrees of freedom, defined as $\rho = \text{Tr}(H)$. Here H is the 'hat matrix', that is the matrix such that $\widehat{Y} = X\widehat{\beta} + Z\widehat{u} = HY$. This criterion can be shown to be an unbiased estimator of the quantity

$$2 \, \text{E}_g \int g(y \mid u) \log f(y \mid \widehat{\theta}, \widehat{u}) \, dy,$$

where the expectation is with respect to $(\widehat{\theta}, \widehat{u})$ under the true joint density $g(y, u)$ for (Y, u). In the more realistic situation that the variances are not known, a different penalty term needs to be used. Vaida and Blanchard (2005) obtain that for the situation of a known D but unknown σ_ε^2, instead of ρ, the penalty term for cAIC should be equal to

$$\frac{N(N - r - 1)}{(N - r)(N - r - 2)}(\rho + 1) + \frac{N(r + 1)}{(N - r)(N - r - 2)}.$$

One of the messages given here is that for the conditional approach one cannot simply count the number of fixed effects and variance components to be used in the penalty of AIC.

10.1.2 REML versus ML

The maximum likelihood method for the estimation of variance components in mixed models results in biased estimators. This is not only a problem for mixed models, even in a simple linear model $Y = X\beta + \varepsilon$ with $\varepsilon \sim \text{N}_N(0, \sigma_\varepsilon^2 I_N)$ the maximum likelihood estimator of σ_ε^2 is biased. Indeed, $\widehat{\sigma}_\varepsilon^2 = N^{-1}\text{SSE}(\widehat{\beta})$, while an unbiased estimator is given by $(N - r)^{-1}\text{SSE}(\widehat{\beta})$, with $r = \text{length}(\beta)$. The method of restricted maximum likelihood (REML) produces unbiased estimators of the variance components. It works as follows. Let A be a matrix of dimension $N \times (N - r)$ such that $A^t X = 0$. Consider the transformed random variable $\tilde{Y} = A^t Y$ which has distribution $\text{N}_{N-r}(0, A^t V A)$. Here we denoted $V = \text{Var}(Y)$. The restricted log-likelihood function ℓ_R of Y is defined as the log-likelihood function of $\tilde{Y}$ and the REML estimators are the maximisers of ℓ_R. It can be shown that the REML estimators are independent of the transformation matrix A. Some software packages give values of AIC and BIC constructed with the REML log-likelihood, leading to

$$\text{reml-AIC} = 2\ell_R(\widehat{\theta}) - 2\,\text{length}(\theta),$$
$$\text{reml-BIC} = 2\ell_R(\widehat{\theta}) - \log(N - r)\,\text{length}(\theta).$$

The function lme in R, for example, returns such values. Care has to be taken with the interpretation of these REML-based criteria.

- First, since the restricted log-likelihood is the log-likelihood of a transformation of Y that makes the fixed effects drop out of the model, values of reml-AIC and reml-BIC may only be compared for models that have exactly the same fixed effects structure. Hence these criteria may only be used for the comparison of mixed models that include different random effects but that contain the same X design matrix.
- Second, values of reml-AIC and reml-BIC are *not* comparable to values of AIC and BIC that use the (unrestricted) likelihood function. The reason is that the likelihood of $\tilde{Y}$ is not directly comparable to that of Y.

Different software packages use different formulae for the REML-based criteria. Some discussion and examples are given by Gurka (2006). Another question that is posed here is which sample size should be used for the reml-BIC? There is apparently as yet no decisive answer to this question. Some authors advocate the use of $N - r$ (the subtraction of r from the total number of observations N is motivated by the transformation matrix A that has rank $N - r$), while others would rather suggest using the number of individual subjects, which would be the smaller number n. Also N would be a suggestion, although that is in most software packages only used for the ML version of the BIC.

10.1.3 Consistent model selection criteria

Jiang and Rao (2003) study consistent model selection methods (see Sections 4.1 and 4.3) for the selection of fixed effects and random effects. Different results and proofs are needed for the selection of fixed effects only than for the selection of random effects. The additional problems can be understood in the following way. Consider the mixed model with the random effects part written as separate contributions for each random effect,

$$Y_i = X_i \beta + \sum_{j=1}^{s} Z_{i,j} u_{i,j} + \varepsilon_i.$$

For simplicity assume that $u_j = (u_{1,j}, \ldots, u_{n,j})^t \sim N_n(0, \sigma_{u_j}^2 I_n)$. Leaving out random effect u_j from the model means leaving out all of the components $u_{i,j}$ and is equivalent to having $\sigma_{u_j}^2 = 0$. If the variance component $\sigma_{u_j}^2 > 0$ the random effect is included in the model. Variance components are either positive or zero, the value zero is at the left boundary of its allowable parameter space. The question of whether a random effect should stay in the model or not is more difficult because of the fact that the zero parameter value is not an interior value to the parameter space, cf. the following section.

Jiang and Rao (2003) come to the conclusion that the classical BIC performs well for models that contain only a single random effect. For models with more random effects they propose new conditions for consistency.

Another approach is the fence method of Jiang *et al.* (2008). Similar to other consistent procedures, this method assumes that a true model is within the list of candidate models. The method, roughly, consists in setting up a fence (or upper bound to some criterion values) to remove incorrect models from the search list. Only models that contain the true model as a submodel should remain in the list. Those models might still contain coefficients that are actually zero, thus overfit. A further selection is then performed amongst these models to select a final model. An advantage of this method is that it can also be applied to nonlinear mixed effects models.

10.2 Boundary parameters

The results reached in earlier chapters have required that the parameter value corresponding to the narrow model is an inner point of the parameter space. In this section we briefly investigate what happens to maximum likelihood-based methods and to estimators in submodels when there are one or more 'boundary parameters', where say $\gamma \geq \gamma_0$ is an a priori restriction. There are many such cases of practical interest, for example in situations involving overdispersion, latent variables, variance components, or more simply cases where the statistician decides that a certain model parameter needs to be non-negative, say, on grounds dictated by context as opposed to mathematical necessity.

Traditional likelihood and estimation theory is made rather more complicated when there are restrictions on one or more of the parameters. This makes model selection and averaging more challenging. Here we indicate how the theory may be modified to yield limit distributions of maximum likelihood estimators in such situations, and use this to characterise behaviour of AIC and related model selection methods, including versions of the FIC. Model average methods are also worked with. Before delving into the required theory of inference under constraints, we provide illustrations of common situations where the problems surface.

Example 10.1 Poisson with overdispersion
We shall consider a Poisson regression model that includes the possibility of over-dispersion, where natural questions relating to tolerance level, model selection and model averaging inference arise, and are more complicated than in earlier chapters. The narrow model is a log-linear Poisson regression model, where $Y_i \sim \text{Pois}(\xi_i)$ and $\xi_i = \exp(x_i^t \beta)$, for a p-dimensional vector of β_j coefficients. The overdispersion model takes $Y_i \sim \text{Pois}(\lambda_i)$ and $\lambda_i \sim \text{Gamma}(\xi_i/c, 1/c)$, leading in particular to

$$\text{E}\, Y_i = \exp(x_i^t \beta) \quad \text{and} \quad \text{Var}\, Y_i = \exp(x_i^t \beta)(1 + c).$$

Likelihood analysis can be carried out, for the $p + 1$-parameter model with density function

$$f_i(y, \beta, c) = \frac{1}{y!} \frac{\Gamma(\xi_i/c + y)}{\Gamma(\xi_i/c)} \frac{(1/c)^{\xi_i/c}}{(1 + 1/c)^{\xi_i/c + y}},$$

which when $c \to 0$ is the more familiar density function of the Poisson model, $\exp(-\xi_i)\xi_i^y/y!$. There is a positive probability that the log-likelihood function is max-imised for $c = 0$, as we show below. How much overdispersion may the ordinary Poisson regression methods tolerate? When should one prefer the wider $p + 1$-parameter model to the usual one? ∎

Example 10.2 How much t-ness can the normal model tolerate?

For some applications the normal tails are too light to describe data well. A remedy is to use the t-distribution, with density $f(y, \xi, \sigma, \nu) = g_\nu((y - \xi)/\sigma)/\sigma$. The three pa-rameters may be estimated by numerically maximising the log-likelihood function. But should one use the simpler normal model, or the bigger t-model, when making inference for various interest parameters? Should one use $\bar{Y} + 1.645s$ (with estimates in the narrow model) or the more complicated $\widehat{\xi} + \widehat{\sigma} t_{\widehat{\nu}, 0.95}$ (with estimates from the wider model) to make inference for an upper quantile, for example? How may one decide which model is best, and how can one compromise? ∎

Example 10.3 The ACE model for mono- and dizygotic twins

Dominicus *et al.* (2006) study a genetic model for twins data, where n data of twin pairs $Y_i = (Y_{i,1}, Y_{i,2})^t$ follow a bivariate normal $N_2(\xi, \Sigma)$ distribution, and where the issues of interest are related to the variance matrix

$$
\Sigma = \begin{cases} \begin{pmatrix} \lambda_A^2 + \lambda_C^2 + \lambda_E^2 & \lambda_A^2 + \lambda_C^2 \\ \lambda_A^2 + \lambda_C^2 & \lambda_A^2 + \lambda_C^2 + \lambda_E^2 \end{pmatrix} & \text{for monozygotic pairs,} \\[2em] \begin{pmatrix} \lambda_A^2 + \lambda_C^2 + \lambda_E^2 & \frac{1}{2}\lambda_A^2 + \lambda_C^2 \\ \frac{1}{2}\lambda_A^2 + \lambda_C^2 & \lambda_A^2 + \lambda_C^2 + \lambda_E^2 \end{pmatrix} & \text{for dizygotic pairs.} \end{cases}
$$

Hypotheses about environmental and genetic influences may be formulated in terms of the variance components, or factor loadings, and the three primary models of interest correspond to

$$H_E: \quad \lambda_A = \lambda_C = 0, \lambda_E \text{ free};$$

$$H_{AE}: \quad \lambda_C = 0, \lambda_A \text{ and } \lambda_E \text{ free};$$

$$H_{ACE}: \quad \text{all parameters } \lambda_A, \lambda_C, \lambda_E \text{ free}.$$

We shall illustrate general issues associated with testing H_E inside H_{ACE}. The likelihood ratio test $Z_n = 2 \log LR_n$ is, for example, not asymptotically a χ_2^2, in the present frame-work where λ_A and λ_C are non-negative parameters. The limiting null distribution of Z_n is derived and discussed in Dominicus *et al.* (2006), but we go further in that we also find the local power of this and similar tests, along with properties of AIC. ∎

10.2.1 Maximum likelihood theory with a boundary parameter *

To answer the questions raised by these examples we need to find parallels of results of Sections 5.2–5.4 that cover cases of borderline parameters. The starting assumption is that data $Y_1, \ldots, Y_n$ are i.i.d. from the density

$$f_n(y) = f(y, \theta_0, \gamma_0 + \delta/\sqrt{n}), \quad \text{where } \delta \geq 0 \text{ a priori}. \tag{10.2}$$

We need to study the behaviour of the maximum likelihood estimators, denoted as before $(\widehat{\theta}, \widehat{\gamma})$ in the wide model and $(\widetilde{\theta}, \gamma_0)$ in the narrow model. We work as earlier with the score functions $U(y)$ and $V(y)$ as in (2.29), with the log-derivative for γ now defined as the right derivative. Consider the $(p + 1) \times (p + 1)$ information matrix J for the model, at (θ_0, γ_0), defined as the variance matrix of $(U(Y), V(Y))$. As in Section 5.2 we work with the quantities

$$\tau_0^2 = \left(\tfrac{\partial \mu}{\partial \theta}\right)^{\mathrm{t}} J_{00}^{-1} \tfrac{\partial \mu}{\partial \theta}, \quad \omega = J_{10} J_{00}^{-1} \tfrac{\partial \mu}{\partial \theta} - \tfrac{\partial \mu}{\partial \gamma}, \quad \kappa^2 = J^{11},$$

in terms of the appropriate blocks of J and J^{-1}. We let $D \sim \mathrm{N}(\delta, \kappa^2)$. A central difference from earlier chapters is that $D_n = \sqrt{n}(\widehat{\gamma} - \gamma_0)$ does not tend in distribution to the normal D under the present circumstances; rather, we shall see below that D_n tends to $\max(D, 0)$, with a positive probability of being equal to zero. We first state (without proof) the result that extends Corollary 5.1 to the case of estimation under constraints.

Theorem 10.1 *Suppose that the estimand $\mu = \mu(\theta, \gamma)$ has derivatives at (θ_0, γ_0), where the derivative with respect to γ is taken from the right, and write $\mu_{\mathrm{true}} = \mu(\theta_0, \gamma_0 + \delta/\sqrt{n})$ for the value under model (10.2). With $\Lambda_0 \sim \mathrm{N}(0, \tau_0^2)$ and $D \sim \mathrm{N}(\delta, \kappa^2)$ being independent variables, as before,*

$$\sqrt{n}(\widehat{\mu}_{\mathrm{narr}} - \mu_{\mathrm{true}}) \xrightarrow{d} \Lambda_{\mathrm{narr}} = \Lambda_0 + \omega\delta,$$

$$\sqrt{n}(\widehat{\mu}_{\mathrm{wide}} - \mu_{\mathrm{true}}) \xrightarrow{d} \Lambda_{\mathrm{wide}} = \begin{cases} \Lambda_0 + \omega(\delta - D) & \text{if } D > 0, \\ \Lambda_0 + \omega\delta & \text{if } D \leq 0. \end{cases}$$

We use this theorem to compare limiting risks and hence characterise the tolerance radius of the narrow model versus the wider model. The limiting risk $\mathrm{E}\Lambda_{\mathrm{narr}}^2 = \tau_0^2 + \omega^2\delta^2$ for the narrow method is exactly as before. For the wide method the limiting risk can be calculated by writing $D = \delta + \kappa N$, with N a standard normal variable, and $a = \delta/\kappa$:

$$\begin{aligned} \mathrm{E}\Lambda_{\mathrm{wide}}^2 &= \tau_0^2 + \omega^2 \mathrm{E}\big[(\delta - D)^2 I\{D > 0\} + \delta^2 I\{D \leq 0\}\big] \\ &= \tau_0^2 + \omega^2\kappa^2 \mathrm{E}[N^2 I\{N > -a\} + a^2 I\{N \leq -a\}] \\ &= \tau_0^2 + \omega^2\kappa^2\{a^2\Phi(-a) - a\phi(a) + \Phi(a)\}. \end{aligned}$$

As a corollary to Theorem 10.1, therefore, we learn that narrow-based inference is as good as or better than wide-based inference for all estimands provided $a^2 \leq a^2\{1 - \Phi(a)\} - a\phi(a) + \Phi(a)$, which via numerical inspection is equivalent to $0 \leq a \leq 0.8399$ (since a is non-negative). The large-sample tolerance radius, for a given model in the direction of a new parameter with a borderline, is therefore

$$\delta \le 0.8399\,\kappa, \quad \text{or} \quad \gamma \le \gamma_0 + 0.8399\,\kappa/\sqrt{n}. \tag{10.3}$$

This is the boundary parameter modification of the result in Theorem 5.2 (which is valid for inner parameter points). Note as for the earlier case that the result is valid for all estimands μ that are smooth functions of (θ, γ).

Result (10.3) may be used for the situations of Examples 10.1–10.2. Some work gives $\kappa = \sqrt{2}$ for the Poisson overdispersion case and $\kappa = \sqrt{2/3}$ for the t-model. The latter result translates into the statement that as long as the degrees of freedom ν, is at least $1.458\sqrt{n}$, the normal-based inference methods work as well as or better than those based on the three-parameter t-model, for all smooth estimands; see Hjort (1994a) for further discussion. Further analysis may be given to support the view that these tolerance threshold results do not so much depend on the finer aspects of the local departure from the narrow model. For the Poisson overdispersion situation, for example, the implied tolerance constraint $\mathrm{Var}\,Y_i/\mathrm{E}\,Y_i \le 1 + 0.6858/\sqrt{n}$ provides a good threshold even when the overdispersion in question is not quite that dictated by the Gamma distributions.

The limit distributions found above are different from those of earlier chapters, due to the boundary effect, with further consequences for testing hypotheses and constructing confidence intervals. For testing the hypothesis $\gamma = \gamma_0$ versus the one-sided alternative $\gamma > \gamma_0$, we have

$$D_n = \sqrt{n}(\widehat{\gamma} - \gamma_0) \xrightarrow{d} \max(D, 0) = \begin{cases} D & \text{if } D > 0, \\ 0 & \text{if } D \le 0, \end{cases} \tag{10.4}$$

as indicated above. Thus a test with asymptotic level 0.05, for example, rejects $\gamma = \gamma_0$ when $D_n/\widehat{\kappa} > 1.645 = \Phi^{-1}(0.95)$; the perhaps expected threshold level $1.96 = \Phi^{-1}(0.975)$ most often associated with pointwise 0.05-level tests here corresponds to level 0.025. The local power of these tests may also be found using (10.4). Similarly, for the likelihood ratio statistic

$$Z_n = 2\log \mathrm{LR}_n = 2\ell_n(\widehat{\theta}, \widehat{\gamma}) - 2\ell_n(\widetilde{\theta}, \gamma_0),$$

the traditional chi-squared limit result no longer holds. Instead, one may prove that

$$Z_n \xrightarrow{d} \{\max(0, D/\kappa)\}^2 = \begin{cases} D^2/\kappa^2 & \text{if } D > 0, \\ 0 & \text{if } D \le 0, \end{cases}$$

where it is noted that $D^2/\kappa^2 \sim \chi_1^2(\delta^2/\kappa^2)$. The limiting null distribution, in particular, is 'half a chi-squared', namely $Z_n \to_d \{\max(0, N)\}^2$, writing N for a standard normal.

Next consider compromise or model average estimators of the form

$$\mu^* = \{1 - c(D_n)\}\widehat{\mu}_{\text{narr}} + c(D_n)\widehat{\mu}_{\text{wide}}, \tag{10.5}$$

where D_n is as in (10.4). In particular, $P(\widehat{\gamma} = \gamma_0)$ converges to $\Phi(-\delta/\kappa)$. In the null model case, we can expect the maximum likelihood estimator to be $\widehat{\gamma} = \gamma_0$ in half of the cases. We furthermore find that $\sqrt{n}(\mu^* - \mu_{\text{true}})$ has the limit in distribution

$$\Lambda = \{1 - c(\max(D, 0))\}\Lambda_{\text{narr}} + c(\max(D, 0))\Lambda_{\text{wide}}$$
$$= \begin{cases} \Lambda_0 + \omega\{\delta - c(D)D\} & \text{if } D \geq 0, \\ \Lambda_0 + \omega\delta & \text{if } D \leq 0. \end{cases}$$

The risk function is found from this result to take the form $\tau_0^2 + R(\delta)$, but with a more complicated formula for the second term than on earlier occasions:

$$R(\delta) = \omega^2\delta^2\Phi(-\delta/\kappa) + \omega^2 \int_0^\infty \{c(x)x - \delta\}^2\phi(x - \delta, \kappa^2)\,\mathrm{d}x.$$

Different $c(x)$ functions correspond to different compromise strategies in (10.5), and to different estimators $c(T)T$ of δ in the one-sided limit experiment where $D \sim N(\delta, \kappa^2)$ but only $T = \max(0, D)$ is observed. See Hjort (1994a) for a wider discussion of general model average strategies in models with one boundary parameter.

10.2.2 Maximum likelihood theory with several boundary parameters *

Above we dealt with the case of $q = 1$ boundary parameter. Similar methods apply for the case of $q \geq 2$ boundary parameters, but with more cumbersome details and results, as we now briefly explain. The framework is that of a density $f(y, \theta_0, \gamma_0 + \delta/\sqrt{n})$, with a p-dimensional θ and a q-dimensional γ, as in Section 5.4, but with the crucial difference that there are boundary restrictions on some or all of the γ_js (translating in their turn into boundary restrictions on some or all of the δ_js). For S a subset of $\{1, \ldots, q\}$, we again define maximum likelihood estimators $\widehat{\theta}_S$ and $\widehat{\gamma}_S$ in the submodel indexed by S, but where the likelihoods in question are restricted via the boundary constraints on $\gamma_1, \ldots, \gamma_q$. The task is to derive suitable analogues of results in Chapters 5, 6 and 7.

To prepare for this, we shall rely on definitions and notation as in Section 6.1. Let in particular $D \sim N_q(\delta, Q)$, where Q is the $q \times q$ lower-right submatrix of the inverse information matrix J^{-1}, again computed at the null point (θ_0, γ_0). It will also prove fruitful to work with

$$E = Q^{-1}D \sim N_2(\varepsilon, Q^{-1}), \quad \text{with } \varepsilon = Q^{-1}\delta. \tag{10.6}$$

To state the following crucial result, let us define the set Ω_S, which is the subset of $\mathbb{R}^{|S|}$ corresponding to the parameter space of $\gamma_S - \gamma_{0,S}$. If each of $\gamma_1, \ldots, \gamma_q$ are boundary parameters with $\gamma_j \geq \gamma_{0,j}$, for example, then Ω_S is the quadrant of t for which $t_j \geq 0$

for each $j \in S$; and if there are no restrictions at all, so that γ has γ_0 as an inner point, then $\Omega = \mathbb{R}^{|S|}$.

For S any subset of $\{1, \ldots, q\}$, let $\widehat{\theta}_S$ and $\widehat{\gamma}_S$ be the maximum likelihood estimators in the submodel that includes γ_j for $j \in S$ and has $\gamma_j = \gamma_{0,j}$ for $j \notin S$. Consider also some parameter of interest $\mu = \mu(\theta, \gamma)$, for which using submodel S leads to maximum likelihood estimator $\widehat{\mu}_S = \mu(\widehat{\theta}_S, \widehat{\gamma}_S, \gamma_{0,S^c})$. Let, as in earlier chapters, $\omega = J_{10} J_{00}^{-1} \frac{\partial \mu}{\partial \theta} - \frac{\partial \mu}{\partial \gamma}$, with partial derivatives computed at the null point (θ_0, γ_0), where derivatives are appropriately one-sided for those components that have restrictions. Let finally $\Lambda_0 \sim N(0, \tau_0^2)$ and independent of $D \sim N_q(\delta, Q)$, where $\tau_0^2 = (\frac{\partial \mu}{\partial \theta})^t J_{00}^{-1} \frac{\partial \mu}{\partial \theta}$. The following result (without proof) summarises the major aspects regarding the behaviour of estimators in submodels.

Theorem 10.2 *With E as in (10.6), define $\widehat{t}_S$ as the random maximiser of $E_S^t t - \frac{1}{2} t^t Q_S^{-1} t$ over all $t \in \Omega_S$, with Ω_S determined by any boundary restrictions on the γ_j parameters for $j \in S$. Then*

$$\begin{pmatrix} \sqrt{n}(\widehat{\theta}_S - \theta_0) \\ \sqrt{n}(\widehat{\gamma}_S - \gamma_{0,S}) \end{pmatrix} \xrightarrow{d} \begin{pmatrix} J_{00}^{-1}(A - J_{01,S} \widehat{t}_S) \\ \widehat{t}_S \end{pmatrix}.$$

Also, $\sqrt{n}(\widehat{\mu}_S - \mu_{\text{true}}) \to_d \Lambda_S = \Lambda_0 + \omega^t(\delta - \pi_S^t \widehat{t}_S)$.

It is worthwhile considering the case of no boundary restrictions for γ, which means $\Omega_S = \mathbb{R}^{|S|}$ in the above notation. Then $\widehat{t}_S = Q_S E_S$ and $\pi_S^t \widehat{t}_S = \pi_S^t Q_S \pi_S Q^{-1} D$. This we recognise as $G_S D$, with notation as in Section 6.1, showing how the result above properly generalises the simpler inner-point parameter case of Theorem 6.1.

For an illustration with real boundaries, consider the case of $q = 2$ parameters with restrictions $\gamma_1 \geq \gamma_{0,1}$ and $\gamma_2 \geq \gamma_{0,2}$. Here there are four candidate models: the narrow model '00', the wide model '11', model '10' (γ_1 estimated, but γ_2 set to its null value) and model '01' (γ_2 estimated, but γ_1 set to its null value). Various results of interest now follow as corollaries to the theorem above, using some algebraic work to properly identify the required $\widehat{t}_S$ for each candidate model. We note first that for the wide model of dimension $p + 2$,

$$\begin{pmatrix} \sqrt{n}(\widehat{\theta} - \theta_0) \\ \sqrt{n}(\widehat{\gamma} - \gamma_0) \end{pmatrix} \xrightarrow{d} \begin{pmatrix} J_{00}^{-1}(A - J_{01}\widehat{t}) \\ \widehat{t} \end{pmatrix},$$

where

$$\widehat{t} = \begin{cases} QE & \text{if } E_1 > 0, E_2 > 0, \\ (E_1/Q^{11}, 0)^t & \text{if } E_1 > 0, E_2 \leq 0, \\ (0, E_2/Q^{22})^t & \text{if } E_1 \leq 0, E_2 > 0, \\ (0, 0)^t & \text{if } E_1 \leq 0, E_2 \leq 0, \end{cases}$$

with Q^{11} and Q^{22} being the diagonal elements of Q^{-1}. Furthermore, the four limit distributions associated with the four estimators $\widehat{\mu}_S$ are as follows:

$$\Lambda_{00} = \Lambda_0 + \omega^{\mathrm{t}}\delta,$$

$$\Lambda_{10} = \Lambda_0 + \begin{cases} \omega^{\mathrm{t}}(\delta - (E_1/Q^{11}, 0)^{\mathrm{t}}) & \text{if } E_1 > 0, \\ \omega^{\mathrm{t}}\delta & \text{if } E_1 \leq 0, \end{cases}$$

$$\Lambda_{01} = \Lambda_0 + \begin{cases} \omega^{\mathrm{t}}(\delta - (0, E_2/Q^{22}))^{\mathrm{t}}) & \text{if } E_2 > 0, \\ \omega^{\mathrm{t}}\delta & \text{if } E_2 \leq 0, \end{cases}$$

$$\Lambda_{11} = \Lambda_0 + \begin{cases} \omega^{\mathrm{t}}(\delta - QE) & \text{if } E_1 > 0, E_2 > 0, \\ \omega^{\mathrm{t}}(\delta - (E_1/Q^{11}, 0)^{\mathrm{t}}) & \text{if } E_1 > 0, E_2 \leq 0, \\ \omega^{\mathrm{t}}(\delta - (0, E_2/Q^{22}, 0)^{\mathrm{t}}) & \text{if } E_1 \leq 0, E_2 > 0, \\ \omega^{\mathrm{t}}\delta & \text{if } E_1 \leq 0, E_2 \leq 0. \end{cases}$$

There are several important consequences of these distributional results, pertaining to testing the null model, to the behaviour of the AIC and FIC selection methods, and to that of model average estimators. We briefly indicate some of these.

First, for testing the null hypothesis H_0: $\gamma = \gamma_0$ against the alternative that one or both of $\gamma_1 > \gamma_{0,1}$ and $\gamma_2 > \gamma_{0,2}$ hold, consider the likelihood ratio statistic

$$Z_n = 2 \log \mathrm{LR}_n = 2\ell_n(\widehat{\theta}, \widehat{\gamma}) - 2\ell_n(\widetilde{\theta}, \gamma_0).$$

In regular cases, where γ_0 is an inner point of the parameter region, Z_n has a limiting χ_2^2 distribution under H_0. The story is different in the present boundary situation. The limit distribution is a mixture of a point mass at zero with two χ_1^2 and one χ_2^2 random variables, and these are *not* independent. Indeed,

$$Z_n \xrightarrow{d} Z = 2 \max_{t_1 \geq 0, t_2 \geq 0} (D^{\mathrm{t}}Q^{-1}t - \tfrac{1}{2}t^{\mathrm{t}}Q^{-1}t) \quad \text{where } D \sim \mathrm{N}_2(\delta, Q)$$

$$= 2 \max_{t_1 \geq 0, t_2 \geq 0} (E^{\mathrm{t}}t - \tfrac{1}{2}t^{\mathrm{t}}Q^{-1}t) \quad \text{where } E \sim \mathrm{N}_2(Q^{-1}\delta, Q^{-1}),$$

where $\delta = 0$ is the null hypothesis situation. From results above, we reach

$$Z = \begin{cases} E^{\mathrm{t}}QE & \text{if } E_1 > 0, E_2 > 0, \\ E_1^2/Q^{11} & \text{if } E_1 > 0, E_2 \leq 0, \\ E_2^2/Q^{22} & \text{if } E_1 \leq 0, E_2 > 0, \\ 0 & \text{if } E_1 \leq 0, E_2 \leq 0. \end{cases} \tag{10.7}$$

We note that E_1^2/Q^{11}, E_2^2/Q^{22}, $E^{\mathrm{t}}QE$ are separately noncentral chi-squared variables, with degrees of freedom 1, 1, 2, respectively, and with excentre parameters $\lambda_1 = \varepsilon_1^2/Q^{11}$, $\lambda_2 = \varepsilon_2^2/Q^{22}$, $\lambda = \varepsilon^{\mathrm{t}}Q\varepsilon = \delta^{\mathrm{t}}Q^{-1}\delta$. They are dependent, so the limit Z is not quite a mixture of independent noncentral chi squares. The limiting null distribution of the $2 \log \mathrm{LR}_n$ statistic is as above, but with $\delta = 0$. These results are in agreement with those reached by Dominicus *et al.* (2006) for the case of Example 10.3, who however were concerned only with the limiting null distribution of the likelihood ratio statistic. Carrying out a test in practice at a wished for significance level, say $\alpha = 0.01$, is not difficult, since we easily may simulate the distribution of Z at the estimated $\widehat{Q}$, via simulated copies of the $E \sim N_2(0, Q^{-1})$. Our more general results, for $\delta \neq 0$, provide in addition the limiting local power of the likelihood ratio test.

Second, we briefly examine the behaviour of AIC for boundary situations. The AIC methods are constructed from arguments valid only for regular models at inner parameter points, so we do not expect its direct use in boundary models to be fully sensible. This does not stop statisticians from using AIC also to determine, say, whether the narrow or the wide model is best in the type of situations spanned by the examples indicated in our introduction to this section. For illustration consider the $q = 2$ case, with four models to choose from: '00' (narrow model), '11' (wide model), '10' (γ_1 estimated, γ_2 set to its null value) and '01' (γ_2 estimated, γ_1 set to its null value). There is a somewhat cumbersome parallel to Theorem 5.4, in that limits in distribution of AIC differences take the form

$$
\text{aic}_{11} - \text{aic}_{00} = \begin{cases} E^{\mathrm{t}} Q E - 4 & \text{if } E_1 > 0, E_2 > 0, \\ E_1^2/Q^{11} - 2 & \text{if } E_1 > 0, E_2 \leq 0, \\ E_2^2/Q^{22} - 2 & \text{if } E_1 \leq 0, E_2 > 0, \\ -4 & \text{if } E_1 \leq 0, E_2 \leq 0, \end{cases}
$$

and furthermore

$$
\text{aic}_{10} - \text{aic}_{00} = \begin{cases} E_1^2/Q^{11} - 2 & \text{if } E_1 > 0, \\ -2 & \text{if } E_1 \leq 0, \end{cases}
$$

$$
\text{aic}_{01} - \text{aic}_{00} = \begin{cases} E_2^2/Q^{22} - 2 & \text{if } E_2 > 0, \\ -2 & \text{if } E_2 \leq 0. \end{cases}
$$

It is possible to use these results to 'repair' AIC for boundary effects, by subtracting different penalties than merely two times the number of parameters, but the resulting procedures are cumbersome. It appears more fruitful to use Theorem 10.2 for construction of FIC procedures, via estimates of limiting risk.

Finally we briefly describe model average estimators and their performance. Using again the $q = 2$ case for illustration. It is clear that compromise estimators of the form

$$
\widehat{\mu}^* = c_{00}(T_n)\widehat{\mu}_{00} + c_{01}(T_n)\widehat{\mu}_{01} + c_{10}(T_n)\widehat{\mu}_{10} + c_{11}(T_n)\widehat{\mu}_{11}
$$

can be studied, where T_n is the vector with components $\max(0, D_{n,1})$ and $\max(0, D_{n,2})$. The limit distribution of $\sqrt{n}(\widehat{\mu}^* - \mu_{\text{true}})$ is of the form $c_{00}(T)\Lambda_{00} + \cdots + c_{11}(T)\Lambda_{11}$, where T has components $\max(0, D_1)$ and $\max(0, D_2)$. Its distribution may be simulated at each position of (δ_1, δ_2). Understanding the behaviour of such model-averaging methods is important since every post-selection-estimator is of this form, even for example the estimator that follows from using (incorrect) AIC.

10.3 Finite-sample corrections *

Several of the model selection and average methods we have worked with in earlier chapters stem from results and insights associated with first-order asymptotic theory, more specifically limit distributions of sequences of random variables. Thus AIC and the FIC, and methods related to model average estimators, are at the outset constructed from such first-order results. Here we briefly discuss approaches for fine-tuning these results and methods, aiming for approximations that work better for small to moderate sample sizes.

For AIC we have actually already discussed such fine-tuning methods. In Sections 2.6–2.7 we used a different penalty that depends on the number of parameters in the model as well as on the sample size. These methods are partly general in nature but most often spelled out for certain classes of models, like generalised linear models. We also point out that fully exact results have been reached and discussed inside the multiple linear regression model, relying only on the first and second moments of the underlying error distribution, cf. Sections 2.6 and 6.7. The latter result entails that FIC calculus in linear regression models is not only in agreement with asymptotics but fully exact, i.e. for all sample sizes.

Under mild regularity conditions the basic Theorem 6.1 may be extended to obtain the following expansion of the limiting risk:

$$\text{risk}_n(S, \delta) = n \, \mathrm{E}(\widehat{\mu}_S - \mu_{\text{true}})^2 = a(S, \delta) + b(S, \delta)/\sqrt{n} + c(S, \delta)/n + o(1/n).$$

One may view that theorem as providing the leading constant $a(S, \delta) = \mathrm{E} \, \Lambda_S^2$ in this expansion, and this is what led to the FIC (and later relatives, along with results for model averaging). One approach towards fine-tuning is therefore to derive expressions for, and then estimate, the second most important constant $b(S, \delta)$. In Hjort and Claeskens (2003b) it is essentially shown that if

$$\mathrm{E} \widehat{\mu}_S = \mu_{\text{true}} + B_1(S, \delta)/\sqrt{n} + B_2(S, \delta)/n + o(1/n),$$

then $b(S, \delta) = B_1(S, \delta)B_2(S, \delta)$. This in particular indicates that a more careful approximation to the mean is more important for finite-sample corrections than corresponding second-order corrections to the variance. Here we already know $B_1(S, \delta) = \omega^{\text{t}}(I_q - G_S)\delta$, and a general expression for $B_2(S, \delta)$ is given in Hjort and Claeskens (2003b), using second-order asymptotics results of Barndorff-Nielsen and Cox (1994, Chapters 5 and 6). These expressions depend not only on the specifics of the parametric family but also on the second-order smoothness properties of the focus parameter.

Thus there is actually an implicit method for finite-sample-tuning the FIC scores, of the form

$$\text{FIC}^*(S) = \widehat{\omega}^{\text{t}} \widehat{G}_S \widehat{Q} \widehat{G}_S^{\text{t}} \widehat{\omega} + \widehat{\omega}^{\text{t}}(I - \widehat{G}_S)(D_n D_n^{\text{t}} - \widehat{Q})(I_q - \widehat{G}_S)^{\text{t}} \widehat{\omega} + \widehat{B}_1 \widehat{B}_2/\sqrt{n},$$

for suitable estimates of the two mean expansion terms $B_1(S, \delta)$ and $B_2(S, \delta)$. This method can in particular easily be made operational for the class of generalised linear models. For the special case of the linear model and a linear focus parameter, the $B_2(S, \delta)$ term is equal to zero. More analytic work and simulation experience appear to be required in order to give specific advice as to which versions of $\widehat{B_2}\widehat{B_2}$ might work best here, say for specific classes of regression models. Sometimes, particularly for smaller sample sizes, the added estimation noise caused by taking the B_1B_2 term into account might actually have a larger negative effect on the mean squared error than the positive effect of including a bias correction.

10.4 Model selection with missing data

The information criteria described so far all assume that the set of data is complete. In other words, for each subject i, all of the values of $(y_i, x_{i,1}, \ldots, x_{i,p})$ are observed. In practice that is often not the case. For example, think about questionnaires where people do not answer all questions, or about clinical studies where repeated measurements are taken but after a few measurements the subject does not show up any more (for example due to a move to a different country). These missing observations cause problems in the application of model selection methods.

A simple (but often naive) way of dealing with missing data is to leave out all observations that are incomplete. This is justified only under strict assumptions. Let us denote by R the indicator matrix of the same size as the full data matrix, containing a one if the corresponding data value is observed, and a zero otherwise. The missing data mechanism is called completely at random (MCAR) if R is independent of the values of the observed and missing data. In this case the subset of complete cases is just a random sample of the original full data. No bias in the estimators is introduced when working with the subset of complete cases only. Under the MCAR assumption it is allowed to perform model selection on the subset of complete cases only.

If the missingness process is not MCAR, it is well known (see Little and Rubin, 2002) that an analysis of the subset of complete cases only may lead to biased estimators. In such cases, leaving out the observations that are incomplete and performing the model selection with the subset of complete observations, leads to incorrect model selection results. A less stringent assumption is that of cases being missing at random (MAR). In this situation the missing data mechanism is allowed to depend only on the data that are observed, but not on the missing variables. If the distribution of R also depends on the unobserved variables, the missing data mechanism is called missing not at random (MNAR). Most work on model selection with missing data makes the assumption of MAR.

Several approaches have been developed to deal with incomplete data. Cavanaugh and Shumway (1998) propose an AIC-type criterion for the situation of correctly specified likelihood models. Their AIC criterion aims at selecting a model for the full (theoretical) set of data, consisting of both the observed and the unobserved missing data. Earlier work is by Shimodaira (1994), who proposed the predictive divergence for indirect observation models (PDIO). This method differs from the proposal by Cavanaugh and Shumway (1998) in that it uses the likelihood of the incomplete data as the goodness-of-fit part of the criterion instead of the expected log-likelihood of the full theoretical set of data where the expectation is taken over the missing data. Claeskens and Consentino (2008) construct a version of AIC and TIC by using the expectation-maximisation (EM) algorithm by the methods of weights of Ibrahim *et al.* (1999a,b), that can deal with models with incomplete covariates. Write the design matrix of covariate values as $X = (X_{\text{obs}}, X_{\text{mis}})$, clearly separating the set of fully observed covariates X_{obs} and those that contain at least one missing observation. Assume that the response vector Y is completely observed, that the parameters describing the regression relation between Y and X are distinct from the parameters describing the distribution of X and that the missing data mechanism is MAR. The kth step in the iterative maximisation procedure of the EM algorithm consists in maximising the function

$$Q(\theta \mid \theta_k) = \sum_{i=1}^n Q_i = \sum_{i=1}^n \int w_i \log f(y_i, x_i; \theta) \, dx_{\text{mis},i},$$

where $w_i = f(x_{\text{mis},i} \mid x_{\text{obs},i}, y_i; \theta_k)$ and θ_k is the result of the previous step. Denote the final maximiser by $\widehat{\theta}$. A model robust information criterion, similar to Takeuchi's TIC (see Section 2.5), is for missing covariate data defined as

$$\text{TIC} = 2 \, Q(\widehat{\theta} \mid \widehat{\theta}) - 2 \, \text{Tr}\{\widehat{J}(\widehat{\theta})\widehat{I}^{-1}(\widehat{\theta})\}$$

where

$$\widehat{I}(\widehat{\theta}) = -n^{-1} Q''(\widehat{\theta} \mid \widehat{\theta}) \quad \text{and} \quad \widehat{J}(\widehat{\theta}) = n^{-1} \sum_{i=1}^n Q'_i(\widehat{\theta} \mid \widehat{\theta}) Q'_i(\widehat{\theta} \mid \widehat{\theta})^{\text{t}},$$

and $Q'(\theta_1 \mid \theta_2) = \frac{\partial}{\partial \theta_1} Q(\theta_1 \mid \theta_2)$, with a similar definition for Q'', the second derivative of Q with respect to θ_1. An information criterion similar to Akaike's AIC (Section 2.1) is obtained by replacing the penalty term in TIC by a count of the number of parameters, that is, by the length of θ. Simulations illustrated that leaving out incomplete cases results in worse model selection performance.

A different approach to model selection with missing data is developed by Hens *et al.* (2006), who consider weighting the complete cases by inverse selection probabilities, following the idea of the Horvitz–Thompson estimator. This method requires estimation of the selection probabilities, which can be done either parametrically or nonparametrically, with the latter requiring additional smoothing parameters to be determined.

Similar to the AIC-type information criteria, Bayesian model selection criteria are not directly applicable in case observations are missing. Sebastiani and Ramoni (2001) explain that for model selection purposes different ignorability conditions are needed than for estimation. When only one variable contains missing observations, they show that the missing data can be ignored in case the missingness probability is independent of the observed data. However, when the missingness probability does depend on the observed data, the missing data mechanism is called partially ignorable and adjustments to model selection methods are required.

Bueso *et al.* (1999) propose an extension of the minimum description length method to incomplete data situations. They select the statistical model with the smallest conditional expected stochastic complexity. Because the stochastic complexity is not defined for incomplete data, they approximate the expected stochastic complexity, conditional on the observed data, via an application of the EM algorithm.

Celeux *et al.* (2006) define the deviance information criterion DIC for data sets with missing observations. They propose several versions of this adjusted criterion.

10.5 When p and q grow with n

The methods developed and worked with in this book have for the most part been in the realm of traditional large-sample theory, i.e. the dimension of the model is a small or moderate fraction of the sample size. The basic limit distribution theorems that underlie selection criteria, like AIC, the BIC, the FIC and relatives, have conditions that amount to keeping the set of models fixed while the sample size tends to infinity. The collective experiences of modern statistics support the view that the resulting large-sample approximations tend to be adequate as long as the parameter dimension is a small or moderate fraction of the sample size. In our frameworks this would translate to saying that methods and approximations may be expected to work well as long as n is moderate or large and $p + q$ is a small or moderate fraction of n. Work by Portnoy (1988) and others indicates that the first-order theorems continue to hold with $p + q$ growing with n, as long as $(p + q)/\sqrt{n} \to 0$. Also supportive of methods like AIC and the FIC is the fact that the large-sample approximations are *exactly correct* for linear models, cf. Section 6.7, as long as the empirical covariance matrix has full rank, i.e. $p + q < n$.

With further extensions of already existing fine-tuning tools, as indicated in Section 10.3, we expect that methods like AIC, the BIC and FIC (with modifications) may work broadly and well even for $p + q$ growing faster with n than $\sqrt{n}$. For the FIC, in particular, the essence is approximate normality of the set of one-dimensional estimators, and such projections are known to exhibit near-normal behaviour even in situations where the parameter vector estimators (i.e. $\widehat{\theta}, \widehat{\gamma}_S$) they are based on have not yet reached multivariate normality. There is, of course, a significant and challenging algorithmical aspect of these situations, particularly if one allows a long list of candidate models (that also grows with p and q).

In many modern application areas for statistics the situation is drastically different, however, namely those where $p + q$ is larger than, or in various cases much larger than, n. Examples include modern biometry, micro-arrays, fields of genetics, technological monitoring devices, chemometrics, engineering sensorics, etc. In such cases not even the maximum likelihood or least squares type estimators exist as such, for the most natural models, like that of $Y = X\beta + Z\gamma + \varepsilon$ for n data points in terms of recorded $n \times p$ and $n \times q$ matrices X and Z. The 'p growing with n' and 'p bigger than n' are growth areas of modern statistics. Natural strategies include (i) modifying the estimation strategies themselves, typically via full or partial shrinking; (ii) breaking down the dimension from $p + q > n$ to $p + q < n$ with variable selection or other dimension reduction techniques; (iii) computing start scores for each individual covariate in X and Z and then selecting a subset of these afterwards. There is also overlap between these tentatively indicated categories.

Methods of category (i) include the lasso (Tibshirani, 1997), the least angle regression (Efron *et al.*, 2004), in addition to ridging (Section 7.6) and generalised ridging (Hjort and Claeskens, 2003, Section 8). Some of these have natural empirical Bayes interpretations. Type (ii) methods would include partial least squares and various related algorithms (Helland, 1990; Martens and Næs, 1992), of frequent use in chemometrics and engineering sensorics. A new class of variable selectors with demonstrably strong properties in high-dimensional situations is that exemplified by the so-called Dantzig selector (Candes and Tao, 2007). Bøvelstad *et al.* (2007) give a good overview of methods that may be sorted into category (iii), in a context of micro-array data for prediction of lifetimes, where an initial scoring of individual predictors (via single-covariate Cox regression analyses) is followed by, for example, ridging or the lasso.

The methods pointed to in the previous paragraph are for the most part worked out specifically for the linear model with mean $X\beta$ of $X\beta + Z\gamma$, and the selectors do not go beyond aims of 'good average performance'. Van de Geer (2008) is an instance of a broader type of model for high-dimensional data, in the direction of generalised linear models. It appears fully possible to work out variations and specialisations of several methods towards 'the FIC paradigm', where special parameters, like $\mu = x_0^t\beta + z_0^t\gamma$ with a given (x_0, z_0) position in the covariate space, are the focus of an investigation. It is also likely that classes of estimators of this one-dimensional quantity are approximately normal even with moderate n and high $p + q$, suggesting that suitably worked our variants of the FIC-type methods dealt with in Chapter 6 might apply.

10.6 Notes on the literature

There are several books on the topic of mixed models, though not specifically oriented towards model selection issues. Examples include Searle *et al.* (1992); McCulloch and Searle (2001); Pinheiro and Bates (2000); Vonesh and Chinchilli (1997); Verbeke and Molenberghs (2000); Jiang (2007).

Another use of mixed models is to build flexible regression models through the use of penalised regression splines (see Ruppert *et al.*, 2003). Model selection methods there play a role to determine the smoothing parameter, which can be rewritten as a ratio of variance components. Kauermann (2005) makes a comparison between REML, C_p and an optimal minimisation of mean squared error for the selection of a smoothing parameter.

Two central references for likelihood theory in models with boundary parameters are Self and Liang (1987) and Vu and Zhou (1997). The results summarised in Section 10.2 are partly more general and have been reached using argmax arguments in connection with study of certain random processes that have Gaussian process limits. The literature on one-sided testing in the case of $q = 1$ boundary parameter is large, see e.g. Silvapulle (1994) and Silvapulle and Silvapulle (1995) for score tests. Claeskens (2004) also handles the $q = 2$-dimensional boundary case in a specific context of testing for zero variance components in penalised spline regression models. Problems related to 'how much t-ness can the normal model tolerate?' were treated in Hjort (1994a). There is a growing literature on correct use of, for example, likelihood ratio testing in genetic models where various parameters are non-negative a priori, and where there has been much mis-use earlier; see Dominicus *et al.* (2006). Some of the results reported on in Section 10.2 are further generalisations of those of the latter reference.

There is a quickly expanding literature on methods for high-dimensional ' p bigger than n' situations, including aspects of variable selection. One may expect the development of several new approaches, geared towards specific scientific contexts. Good introductory texts to some of the general issues involved, in contexts of biometry and micro-arrays, include van de Geer and van Houwelingen (2008) and Bøvelstad *et al.* (2007).

Overview of data examples

Several real data sets are used in this book to illustrate aspects of the methods that are developed. Here we provide brief descriptions of each of these real data examples, along with key points to indicate which substantive questions they relate to. Key words are also included to indicate the data sources, the types of models we apply, and pointers to where in our book the data sets are analysed. For completeness and convenience of orientation the list below also includes the six 'bigger examples' already introduced in Chapter 1.

Egyptian skulls

There are four measurements on each of 30 skulls, for five different archaeological eras (see Section 1.2). One wishes to provide adequate statistical models that also make it possible to investigate whether there have been changes over time. Such evolutionary changes in skull parameters might relate to influx of immigrant populations. *Source*: Thomson and Randall-Maciver (1905), Manly (1986).

We use multivariate normal models, with different attempts at structuring for mean vectors and variance matrices, and apply AIC and the BIC for model selection; see Example 9.1.

The (not so) Quiet Don

We use sentence length distributions to decide whether Sholokhov or Kriukov is the most likely author of the Nobel Prize winning novel (see Section 1.3). *Source*: Private files of the authors, collected by combining information from different tables in Kjetsaa *et al.* (1984), also with some additional help of Geir Kjetsaa (private communication); see also Hjort (2007a). The original YMCA-Press Paris 1974 publication that via its serious claims started the whole investigation was called *Stremya 'Tihogo Dona': Zagadki romana* (The Rapids of Quiet Don: The Enigmas of the Novel), and the title of the Solzhenitsyn (1974) preface is *Nevyrvannaya tajna* (The not yet uprooted secret). The full Russian text is available on the internet.

Certain four-parameter models are put forward for sentence lengths, and these are shown to be fully adequate via goodness-of-fit analyses. Then some non-standard versions of the BIC are used to determine the best model. See Example 3.3.

Survival with primary biliary cirrhosis

Data are from a randomised study where patients receive either drug or placebo (see Section 1.4). We used the data set that is, at the time of writing, available at the website `lib.stat.cmu.edu/datasets/pbcseq` (see Murtaugh *et al.*, 1994), after removing cases with missing observations. There are fifteen covariates, one of which is treatment vs. placebo, and the task is to investigate their influence on survival. *Source*: Murtaugh *et al.* (1994), Fleming and Harrington (1991).

We analyse and select among hazard rate regression models of the proportional hazards form; see Example 9.3.

Low birthweight data

Data on low birthweights are used to assess influence of various background factors, like smoking and the mother's weight before pregnancy (see Section 1.5). *Source*: Hosmer and Lemeshow (1999). The data set is available in R under the name `birthwt` in the library MASS. In this often-used data set there are four instances among the $n = 189$ data lines where pairs are fully identical, down to the number of grams, leading to a suspicion of erroneously duplicated data lines. To facilitate comparison with other work we have nevertheless included all 189 data vectors in our analyses, even if some of the babies might be twins, or if there might have been only 185 babies.

Different logistic regression models are used and selected among, using AIC, the BIC (along with more accurate variants) and FIC, the latter concentrating on accurate estimation of the low birthweight probability for given strata of mothers. Also, the difference between smokers and non-smokers is assessed. See Examples 2.1, 2.4, 3.3, 6.1, Section 9.2 and Exercise 2.4.

Football matches

Data on match results and FIFA ranking numbers are used to select good models and make predictions from them (see Section 1.6). Our analyses concern only the full-time results after 90 minutes, i.e. we disregard extra time and penalty shoot-outs. *Source*: Various private files collected and organised by the authors, with information gathered from various official internet sources, e.g. that of FIFA.

We investigate and select among different Poisson rate models, using AIC and the FIC, with different functions of the FIFA ranking numbers serving as covariates. We demonstrate the difference between selecting a good model for a given match and finding a model that works well in an average sense. See Examples 2.8, 3.4 and Section 6.6.4.

Speedskating

We use data from Adelskalenderen, listing the personal bests of top skaters over the four classic distances 500-m, 1500-m, 5000-m, 10,000-m (see Section 1.7). This is an opportunity to investigate both how well results on some distances may be predicted from those on other distances, and the inherent correlation structure. *Source*: Private files collected and organised by the authors; also 'Adelskalender' Wikipedia and various internet sites.

Regression models with quadraticity and variance heterogeneity are investigated, and we select among them using AIC and the FIC, focussing on the probability of setting a new world record on the 10k for a skater with given 5k time. Also, multinormal models are used to investigate different structures for the variance matrices. We use a different speedskating data set, from the European 2004 Championships, to illustrate model-robust inference. See Section 5.6, Examples 5.11 and 5.12, Sections 6.6.5 and 9.4, with robust reparation for an outlier in Example 2.14.

Mortality in ancient Egypt

The age at death was recorded for 141 Egyptian mummies in the Roman period, 82 men and 59 women, dating from around year 100 B.C. The life lengths vary from 1 to 96 years, and Pearson (1902) argued that these can be considered a random sample from one of the better-living classes in that society, at a time when a fairly stable and civil government was in existence. The task is to provide good parametric hazard models for these life lengths. *Source*: Spiegelberg (1901), Pearson (1902).

Different hazard rate models are evaluated and compared, for example via AIC and the BIC; see Examples 2.6 and 3.2.

Exponential decay of beer froth

An article in *The European Journal of Physics* 2002, made famous by earning the author the Ig Nobel Prize for Physics that year, analysed the presumed exponential decay of beer froth for three different German brands of beer, essentially using normal nonlinear regression. *Source*: Leike (2002), Hjort (2007c).

We study alternative models, including some involving Lévy processes, and demonstrate that these work better than Leike's, e.g. via AIC; see Example 2.11.

Blood groups A, B, AB, O

The data are the numbers of individuals in the four blood type categories, among 502 Japanese living in Korea, collected in 1924. This data set is famous for having contributed to determining which of two genetics theories is correct, regarding blood groups in man.

In statistical terms the question is which of two multinomial models is correct, given that one of them is right and the other not. *Source*: Bernstein (1924).

Accurate versions of the BIC (and AIC) are worked out to support the one-locus theory and soundly reject the two-loci theory; see Example 3.6 and Exercise 3.9.

Health Assessment Questionnaires

These HAQ and MHAQ data are from the usual and modified Health Assessment Questionnaires, respectively. *Source*: Ulleval University Hospital of Oslo, Division for Women and Children (from Petter Mowinckel, personal communication).

Both HAQ and MHAQ data live inside strictly defined intervals, leading us to use certain Beta regression methods. This gives better results, in terms of for example AIC, than with more traditional regression models. See Example 3.7.

The Raven

We have recorded the lengths of the 1086 words used in Poe's famous 1845 poem, attempting to learn aspects of poetic rhythm, like succession of and changes between short, middle, long words. *Source*: Files transcribing words to lengths, organised and collected by the authors.

We use model selection criteria to select among Markov chain models of order one, two, three and four; see Example 3.8.

Danish melanoma data

This is the set of data on skin cancer survival that is described in Andersen *et al.* (1993). The study gives information on 205 patients who were followed during the time period 1962–1977.

The challenge is to select the best variables for use in a Cox proportional regression model, for the purpose of estimating relative risk, a cumulative hazard function and a survival function. This is done in Example 3.9.

Survival for oropharynx carcinoma

These survival data for patients with a certain type of carcinoma include several covariates such as patients' gender, their physical condition, and information on the tumour. The task is to understand how the covariates influence different aspects of the survival mechanisms involved. For ease of comparison with previous analyses we have used the data precisely as given in these sources, even though there is an error with data vector #141. *Source*: Kalbfleisch and Prentice (2002, p. 378), also Aalen and Gjessing (2001).

Here we build hazard regression models rather different from those of proportional hazards, and model time to death as the level crossing times of underlying Gamma processes.

We demonstrate via AIC, the BIC and TIC for hazard regressions that these models work better than some that have been applied earlier for these data; see Example 3.10.

Fifty years survival since graduation

Data on 50 years survival since graduation at the Adelaide University in Australia are available, sorted by calender year, gender and field of study. Several different hypotheses can be formulated and investigated for these survival probabilities and odds, reflecting equality or not of different subsets, e.g. for Science vs. Arts students and for men vs. women. *Source*: Dobson (2002, Chapter 7).

Assessment of such hypotheses, and forming overall weighted average estimates of survival rates and odds based on these assessments, is the topic of Examples 3.11 and 7.9.

Onset of menarche

The age of 3918 girls from Warsaw were recorded, and each answered 'yes' or 'no' to the question of whether they had experienced their first menstruation. The task is to infer the statistical onset distribution from these data. *Source*: Morgan (1992).

We fit logistic regression models of different orders, and show that different orders are optimal for estimating the onset age at different quantiles, using the FIC. See Example 6.2.

Australian Institute of Sports data

Data have been collected on the body mass index x and hematocrit level y (and several other quantities) for different Australian athletes, and it is of interest to model the influence of x on y. *Source*: Cook and Weisberg (1994), also available as data set `ais` in R.

Here certain skewed extensions of the normal distribution are used, and we select among different versions of such using the FIC. Different focus parameters give different optimal models; see Section 6.6.3.

CH$_4$ concentrations

The data are atmospheric CH$_4$ concentrations derived from flask samples collected in the Shetland Islands, Scotland. The response variables are monthly values expressed in parts per billion by volume (ppbv), the covariate is time. There are 110 monthly measurements, starting in December 1992 and ending in December 2001, and one wishes to model the concentration distribution over time. *Source*: `cdiac.esd.ornl.gov/ftp/trends/atm_meth/csiro/shetlandch4_mm.dat`.

We investigate nonlinear regressions of different orders, and select among these using weighted FIC. One might be more interested in models that work well for the recent past (and the immediate future) than in a model that is good in the overall sense, and this is reflected in our weighted FIC strategy. See Example 6.14.

Low-iron rat teratology data

A total of 58 female rats were given a different amount of iron supplements, ranging from none to normal levels. The rats were made pregnant and after three weeks the haemoglobin level of the mother (a measure for the iron intake), as well as the total number of foetuses (here ranging between 1 and 17), and the number of foetuses alive, was recorded. One wishes to model the proportion $\pi(x)$ of dead foetuses as a function of haemoglobin level x. This is a nontrivial task since there are dependencies among foetuses from the same mother. *Source*: Shepard *et al.* (1980).

Models are developed that take this intra-dependency into account, and we use order selection lack-of-fit tests to check its validity. See Example 8.2.

POPS data

The data are from the 'Project on preterm and small-for-gestational age infants in the Netherlands', and relate to 1310 infants born in the Netherlands in 1983 (we deleted 28 cases with missing values). The following variables are measured: gestational age, birthweight, $Y = 0$ if the infant survived two years without major handicap, and $Y = 1$ if otherwise. A model is needed for the Y outcome. *Source*: Verloove and Verwey (1983) and le Cessie and van Houwelingen (1993).

It is easy to fit logistic regression models here, but less clear how to check whether such models can be judged to be adequate. We apply order selection-based lack-of-fit tests for this purpose; see Example 8.6.

Birds on islands

The number Y of different bird species found living on paramos of islands outside Ecuador has been recorded (ranging from 4 to 37) for each of 14 islands, along with the covariates island's area, elevation, distance from Ecuador, and distance from nearest island. One wishes a model that predicts Y from the covariate information. *Source*: Hand *et al.* (1994, case #52).

The traditional model in such cases is that of Poisson regression. In Section 6.10 we apply Bayesian focussed information criteria for selecting the most informative covariates. We also check the adequacy of such models via certain goodness-of-fit monitoring processes, in Section 8.8.

References

Aalen, O. O. and Gjessing, H. (2001). Understanding the shape of the hazard rate: a process point of view. *Statistical Science*, **16**:1–22.

Abramovich, F. and Benjamini, Y. (1996). Adaptive thresholding of wavelet coefficients. *Computational Statistics and Data Analysis*, **22**:351–361.

Abramovich, F., Sapatinas, F. and Silverman, B. W. (1998). Wavelet thresholding via a Bayesian approach. *Journal of the Royal Statistical Society, Series B*, **60**:725–749.

Aerts, M., Claeskens, G. and Hart, J. D. (1999). Testing the fit of a parametric function. *Journal of the American Statistical Association*, **94**:869–879.

Aerts, M., Claeskens, G. and Hart, J. D. (2000). Testing lack of fit in multiple regression. *Biometrika*, **87**:405–424.

Aerts, M., Claeskens, G. and Hart, J. D. (2004). Bayesian-motivated tests of function fit and their asymptotic frequentist properties. *The Annals of Statistics*, **32**:2580–2615.

Agostinelli, C. (2002). Robust model selection in regression via weighted likelihood methodology. *Statistics & Probability Letters*, **56**:289–300.

Akaike, H. (1969). Fitting autoregressive models for prediction. *Annals of the Institute of Statistical Mathematics*, **21**:243–247.

Akaike, H. (1970). Statistical predictor identification. *Annals of the Institute of Statistical Mathematics*, **22**:203–217.

Akaike, H. (1973). Information theory and an extension of the maximum likelihood principle. In Petrov, B. and Csáki, F. (editors), *Second International Symposium on Information Theory*, pages 267–281. Akadémiai Kiadó, Budapest.

Akaike, H. (1977). On entropy maximization principle. In *Applications of Statistics (Proceedings of Symposium, Wright State University, Dayton, Ohio, 1976)*, pages 27–41. North-Holland, Amsterdam.

Akaike, H. (1978). A new look at the Bayes procedure. *Biometrika*, **65**:53–59.

Allen, D. M. (1974). The relationship between variable selection and data augmentation and a method for prediction. *Technometrics*, **16**:125–127.

Andersen, P. K., Borgan, Ø., Gill, R. D. and Keiding, N. (1993). *Statistical Models Based on Counting Processes*. Springer-Verlag, New York.

Antoniadis, A., Gijbels, I. and Grégoire, G. (1997). Model selection using wavelet decomposition and applications. *Biometrika*, **84**:751–763.

Augustin, N. H., Sauerbrei, W. and Schumacher, M. (2005). The practical utility of incorporating model selection uncertainty into prognostic models for survival data. *Statistical Modelling*, **5**:95–118.

Bai, Z. D., Rao, C. R. and Wu, Y. (1999). Model selection with data-oriented penalty. *Journal of Statistical Planning and Inference*, **77**:103–117.

Barndorff-Nielsen, O. E. and Cox, D. R. (1994). *Inference and Asymptotics*, Vol. 52. Chapman & Hall, London.

Basu, A., Harris, I. R., Hjort, N. L. and Jones, M. C. (1998). Robust and efficient estimation by minimising a density power divergence. *Biometrika*, **85**:549–559.

Benjamini, Y. and Hochberg, Y. (1995). Controlling the false discovery rate: a practical and powerful approach to multiple testing. *Journal of the Royal Statistical Society, Series B*, **57**:289–300.

Berger, J. (1982). Estimation in continuous exponential families: Bayesian estimation subject to risk restrictions and inadmissibility results. In *Statistical Decision Theory and Related Topics, III, Vol. 1 (West Lafayette, Indiana, 1981)*, pages 109–141. Academic Press, New York.

Berger, J. O. and Pericchi, L. R. (1996). The intrinsic Bayes factor for model selection and prediction. *Journal of the American Statistical Association*, **91**:109–122.

Berger, J. O. and Pericchi, L. R. (2001). Objective Bayesian methods for model selection: introduction and comparions. In Lahiri, P. (editor), *Model Selection*, IMS Lecture Notes – Monograph Series, pages 135–207. IMS, Beechwood, Ohio.

Bernstein, F. (1924). Ergebnisse einer biostatistichen zusammenfassenden Betrachtung über die erblichen Blutstrukturen des Menschen. *Klinische Wochenschriften*, **3**:1467–1495.

Bickel, P. J. (1981). Minimax estimation of the mean of a normal distribution when the parameter space is restricted. *The Annals of Statistics*, **9**:1301–1309.

Bickel, P. J. (1983). Minimax estimation of the mean of a normal distribution subject to doing well at a point. In *Recent Advances in Statistics*, pages 511–528. Academic Press, New York.

Bickel, P. J. (1984). Parametric robustness: small biases can be worthwhile. *The Annals of Statistics*, **12**:864–879.

Bickel, P. J. and Doksum, K. A. (2001). *Mathematical Statistics: Basic Ideas and Selected Topics*, Vol. I (2nd edition). Prentice-Hall, London.

Bøvelstad, H. M., Nygård, S., Størvold, H. L., Aldrin, M., Borgan, Ø., Frigessi, A. and Lingjærde, O. C. (2007). Predicting survival from microarray data – a comparative study. *Bioinformatics*, **23**:2080–2087.

Bozdogan, H. (1987). Model selection and Akaike's information criterion (AIC): the general theory and its analytical extensions. *Psychometrika*, **52**:345–370.

Bozdogan, H., Sclove, S. L. and Gupta, A. K. (1994). AIC-replacements for some multivariate tests of homogeneity with applications in multisample clustering and variable selection. In *Proceedings of the First US/Japan Conference on the Frontiers of Statistical Modeling: An Informational Approach, Vol. 2 (Knoxville, Tennessee, 1992)*, pages 19–20. 199–232. Kluwer Academic Publishers, Dordrecht.

Breiman, L. (1992). The little bootstrap and other methods for dimensionality selection in regression: X-fixed prediction error. *Journal of the American Statistical Association*, **87**:738–754.

Breiman, L. (1995). Better subset regression using the nonnegative garrote. *Technometrics*, **37**:373–384.

Breiman, L. (2001). Statistical modeling: the two cultures. *Statistical Science*, **16**:199–231. With discussion and a rejoinder by the author.

Breiman, L. and Freedman, D. (1983). How many variables should be entered in a regression equation? *Journal of the American Statistical Association*, **78**:131–136.

Brockwell, P. J. and Davis, R. A. (1991). *Time Series: Theory and Methods* (2nd edition). Springer-Verlag, New York.

Brockwell, P. J. and Davis, R. A. (2002). *Introduction to Time Series and Forecasting* (2nd edition). Springer-Verlag, New York.

Buckland, S. T., Burnham, K. P. and Augustin, N. H. (1997). Model selection: an integral part of inference. *Biometrics*, **53**:603–618.

Bueso, M. C., Qian, G. and Angulo, J. M. (1999). Stochastic complexity and model selection from incomplete data. *Journal of Statistical Planning and Inference*, **76**:273–284.

Bunea, F. and McKeague, I. W. (2005). Covariate selection for semiparametric hazard function regression models. *Journal of Multivariate Analysis*, **92**:186–204.

Burnham, K. P. and Anderson, D. R. (2002). *Model Selection and Multimodel Inference: A Practical Information-Theoretic Approach* (2nd edition). Springer-Verlag, New York.

Candes, E. and Tao, T. (2008). The Dantzig selector: statistical estimation when p is much larger than n. *Annals of Statistics*, **35**:2313–2351.

Carroll, R. J. and Ruppert, D. (1988). *Transformation and Weighting in Regression*. Chapman & Hall, London.

Cavanaugh, J. E. and Shumway, R. H. (1998). An Akaike information criterion for model selection in the presence of incomplete data. *Journal of Statistical Planning and Inference*, **67**:45–65.

Celeux, G., Forbes, F., Robert, C. and Titterington, D. (2006). Deviance information criteria for missing data models. *Bayesian Analysis*, **1**(4):651–705.

Chatfield, C. (1995). Model uncertainty, data mining and statistical inference. *Journal of the Royal Statistical Society, Series A*, **158**:419–466.

Chatfield, C. (2004). *The Analysis of Time Series* (6th edition). Chapman & Hall/CRC, Boca Raton, Florida.

Chipman, H., George, E. I. and McCulloch, R. E. (2001). The practical implementation of Bayesian model selection. In *Model Selection*, IMS Lecture Notes – Monograph Series, pages 65–134. Institute of Mathematical Statistics, Beechwood, Ohio.

Chipman, H., Kolaczyk, E. and McCulloch, R. (1997). Adaptive Bayesian wavelet shrinkage. *Journal of the American Statistical Association*, **92**:1413–1421.

Choi, B. S. (1992). *ARMA Model Identification*. Springer-Verlag, New York.

Choi, E., Hall, P. and Presnell, B. (2000). Rendering parametric procedures more robust by empirically tilting the model. *Biometrika*, **87**:453–465.

Claeskens, G. (2004). Restricted likelihood ratio lack of fit tests using mixed spline models. *Journal of the Royal Statistical Society, Series B*, **66**:909–926.

Claeskens, G. and Carroll, R. (2007). An asymptotic theory for model selection inference in general semiparametric problems. *Biometrika*, **2**:249–265.

Claeskens, G. and Consentino, F. (2008). Variable selection with incomplete covariate data. *Biometrics*, **64**:1062–1069.

Claeskens, G., Croux, C. and Van Kerckhoven, J. (2006). Variable selection for logistic regression using a prediction focussed information criterion. *Biometrics*, **62**:972–979.

Claeskens, G., Croux, C. and Van Kerckhoven, J. (2007). Prediction focussed model selection for autoregressive models. *The Australian and New Zealand Journal of Statistics*, **49**(4):359–379.

Claeskens, G. and Hjort, N. L. (2003). The focused information criterion. *Journal of the American Statistical Association*, **98**:900–916. With discussion and a rejoinder by the authors.

Claeskens, G. and Hjort, N. L. (2004). Goodness of fit via nonparametric likelihood ratios. *Scandinavian Journal of Statistics*, **31**:487–513.

Claeskens, G. and Hjort, N. L. (2008). Minimising average risk in regression models. *Econometric Theory*, **24**:493–527.

Clyde, M. A. (1999). Bayesian model averaging and model search strategies. In *Bayesian Statistics VI (Alcoceber, 1998)*, pages 157–185. Oxford University Press, Oxford.

Clyde, M. A. and George, E. (2004). Model uncertainty. *Statistical Science*, **19**:81–94.

Cook, R. D. and Weisberg, S. (1994). *An Introduction to Regression Graphics*. John Wiley & Sons Inc., New York.

Cover, T. M. and Thomas, J. A. (1991). *Elements of Information Theory*. John Wiley & Sons Inc., New York.

Cox, D. R. (1972). Regression models and life-tables. *Journal of the Royal Statistical Society, Series B*, **34**:187–220. With discussion.

de Rooij, S. and Grünwald, P. (2006). An empirical study of minimum description length model selection with infinite parametric complexity. *Journal of Mathematical Psychology*, **50**:180–192.

DeBlasi, P. and Hjort, N. L. (2007). Bayesian survival analysis in proportional hazard models with logistic relative risk. *Scandinavian Journal of Statistics*, **34**:229–257.

Dobson, A. (2002). *An Introduction to Generalized Linear Models* (2nd edition). CRC/Chapman & Hall.

Dominicus, A., Skrondal, A., Gjessing, H., Pedersen, N. and Palmgren, J. (2006). Likelihood ratio tests in behavioral genetics: problems and solutions. *Behavior Genetics*, **36**:331–340.

Donoho, D. L. and Johnstone, I. M. (1994). Ideal spatial adaptation via wavelet shrinkage. *Biometrika*, **81**:425–455.

Donoho, D. L. and Johnstone, I. M. (1995). Adapting to unknown smoothness via wavelet shrinkage. *Journal of the American Statistical Association*, **90**:1200–1224.

Draper, D. (1995). Assessment and propagation of model uncertainty. *Journal of the Royal Statistical Society, Series B*, **57**:45–97. With discussion and a reply by the author.

Efron, B. (1983). Estimating the error rate of a prediction rule: improvement on cross-validation. *Journal of the American Statistical Association*, **78**:316–331.

Efron, B. and Gous, A. (2001). Scales of evidence for model selection: Fisher versus Jeffreys. In *Model Selection*, IMS Lecture Notes – Monograph Series, pages 208–256. Institute of Mathematical Statistics, Beechwood, Ohio.

Efron, B., Hastie, T., Johnstone, I. and Tibshirani, R. (2004). Least angle regression. *The Annals of Statistics*, **32**:407–499. With discussion and a rejoinder by the authors.

Efron, B. and Tibshirani, R. J. (1993). *An Introduction to the Bootstrap*. Chapman & Hall, New York.

Eguchi, S. and Copas, J. (1998). A class of local likelihood methods and near-parametric asymptotics. *Journal of the Royal Statistical Society, Series B*, **60**:709–724.

Einstein, A. (1934). On the method of theoretical physics. *The Philosophy of Science*, **1**:163–169.

Eubank, R. L. and Hart, J. D. (1992). Testing goodness-of-fit in regression via order selection criteria. *The Annals of Statistics*, **20**:1412–1425.

Fan, J. and Gijbels, I. (1996). *Local Polynomial Modelling and its Applications*. Chapman & Hall, London.

Fan, J. and Li, R. (2002). Variable selection for Cox's proportional hazards model and frailty model. *Annals of Statistics*, **30**:74–99.

Fenstad, A. M. (1992). *How Much Dependence Can the Independence Assumption Tolerate?* Cand. Scient. thesis, University of Oslo, Department of Mathematics. In Norwegian.

Firth, D. (1993). Bias reduction of maximum likelihood estimates. *Biometrika*, **80**:27–38.

Fleming, T. R. and Harrington, D. P. (1991). *Counting Processes and Survival Analysis*. John Wiley & Sons Inc., New York.

Foster, D. P. and George, E. I. (1994). The risk inflation criterion for multiple regression. *The Annals of Statistics*, **22**:1947–1975.

Frank, I. E. and Friedman, J. H. (1993). A statistical view of some chemometric regression models. *Technometrics*, **35**:109–135. With discussion contributions and a rejoinder.

Geisser, S. (1975). The predictive sample reuse method with applications. *Journal of The American Statistical Association*, **70**:320–328.

George, E. I. (1986a). Combining minimax shrinkage estimators. *Journal of the American Statistical Association*, **81**:437–445.

George, E. I. (1986b). Minimax multiple shrinkage estimation. *The Annals of Statistics*, **14**:188–205.

Grünwald, P. (2005). A tutorial introduction to the minimum description length principle. In Grünwald, P., Myung, I. J. and Pitt, M. (editors), *Advances in Minimum Description Length: Theory and Applications*. MIT Press, Boston.

Grünwald, P. (2007). *The Minimum Description Length Principle*, MIT Press, Boston.

Gurka, M. J. (2006). Selecting the best linear mixed model under REML. *The American Statistician*, **60**:19–26.

Guyon, X. and Yao, J.-F. (1999). On the underfitting and overfitting sets of models chosen by order selection criteria. *Journal of Multivariate Analysis*, **70**:221–249.

Hampel, F. R., Ronchetti, E. M., Rousseeuw, P. J. and Stahel, W. A. (1986). *Robust Statistics: The Approach Based on Influence Functions*. John Wiley & Sons Inc., New York.

Hand, D. J., Daly, F., Lunn, A., McConway, K. J. and Ostrowski, E. (1994). *A Handbook of Small Data Sets*. Chapman & Hall, London.

Hand, D. J. and Vinciotti, V. (2003). Local versus global models for classification problems: fitting models where it matters. *The American Statistician*, **57**:124–131.

Hannan, E. J. and Quinn, B. G. (1979). The determination of the order of an autoregression. *Journal of the Royal Statistical Society, Series B*, **41**:190–195.

Hansen, B. E. (2005). Challenges for econometric model selection. *Econometric Theory*, **21**:60–68.

Hansen, M. and Yu, B. (2001). Model selection and Minimum Description Length principle. *Journal of the American Statistical Association*, **96**:746–774.

Hart, J. D. (1997). *Nonparametric Smoothing and Lack-of-fit Tests*. Springer-Verlag, New York.

Hart, J. D. (2009). Frequentist-Bayes lack-of-fit tests based on Laplace approximations. *Journal of Statistical Theory and Practice*, **3**:681–704.

Hart, J. D. and Yi, S. (1998). One-sided cross-validation. *Journal of the American Statistical Association*, **93**:620–631.

Harville, D. A. (1997). *Matrix Algebra from a Statistician's Perspective*. Springer-Verlag, New York.

Hastie, T. J. and Tibshirani, R. J. (1990). *Generalized Additive Models*. Chapman & Hall, London.

Hastie, T. J., Tibshirani, R. J. and Friedman, J. (2001). *The Elements of Statistical Learning: Data Mining, Inference, and Prediction*. Springer-Verlag, Heidelberg.

Haughton, D. (1988). On the choice of a model to fit data from an exponential family. *The Annals of Statistics*, **16**:342–355.

Haughton, D. (1989). Size of the error in the choice of a model to fit data from an exponential family. *Sankhyā, Series A*, **51**:45–58.

Helland, I. S. (1990). Partial least squares regression and statistical models. *Scandinavian Journal of Statistics*, **17**:97–114.

Henriksen, T. (2007). Ernæring, vekt og svangerskap (nutrition, weight and pregnacny). *Tidsskrift for den Norske Lægeforening*, **127**:2399–2401.

Hens, N., Aerts, M. and Molenberghs, G. (2006). Model selection for incomplete and design-based samples. *Statistics in Medicine*, **25**:2502–2520.

Hjort, N. L. (1986a). Bayes estimators and asymptotic efficiency in parametric counting process models. *Scandinavian Journal of Statistics*, **13**:63–85.

Hjort, N. L. (1986b). *Statistical Symbol Recognition*. Research Monograph. Norwegian Computing Centre, Oslo.

Hjort, N. L. (1990). Nonparametric Bayes estimators based on Beta processes in models for life history data. *The Annals of Statistics*, **18**:1259–1294.

Hjort, N. L. (1991). Estimation in moderately misspecified models. Statistical research report, Department of Mathematics, University of Oslo.

Hjort, N. L. (1992a). On inference in parametric survival data models. *International Statistical Review*, **40**:355–387.

Hjort, N. L. (1992b). Semiparametric estimation of parametric hazard rates. In *Survival Analysis: State of the Art (Columbus, OH, 1991)*, Vol. 211 of *NATO Advanced Science Institute, Series E, Applied Science*, pages 211–236. Kluwer Academic Publishers, Dordrecht. With a discussion by Mike West and Sue Leurgans.

Hjort, N. L. (1994a). The exact amount of t-ness that the normal model can tolerate. *Journal of the American Statistical Association*, **89**:665–675.

Hjort, N. L. (1994b). Minimum L2 and robust Kullback–Leibler estimation. In *Proceedings of the 12th Prague Conference on Information Theory, Statistical Decision Functions and Random Processes*, pages 102–105, Prague.

Hjort, N. L. (2007a). And Quiet does not Flow the Don: Statistical analysis of a quarrel between two Nobel Prize winners. In Østreng, W. (editor), *Consilience: Interdisciplinary Communications*, pages 134–140. Centre of Advanced Studies at the Norwegian Academy of Science and Letters, Oslo.

Hjort, N. L. (2007b). Focussed information criteria for the linear hazard rate regression model. In Vonta, F., Nikulin, M., Limnios, N. and Huber-Carol, C. (editors), *Statistical Models and Methods for Biomedical and Technical Systems*, Statistics for Industry and Technology, pages 491–506. IMS, Birkhauser.

Hjort, N. L. (2007c). A Lévy process approach to beer froth watching. Statistical research report, Department of Mathematics, University of Oslo.

Hjort, N. L. and Claeskens, G. (2003a). Frequentist model average estimators. *Journal of the American Statistical Association*, **98**:879–899. With discussion and a rejoinder by the authors.

Hjort, N. L. and Claeskens, G. (2003b). Rejoinder to the discussion of 'frequentist model average estimators' and 'the focused information criterion'. *Journal of the American Statistical Association*, **98**:938–945.

Hjort, N. L. and Claeskens, G. (2006). Focussed information criteria and model averaging for Cox's hazard regression model. *Journal of the American Statistical Association*, **101**:1449–1464.

Hjort, N. L., Dahl, F. and Steinbakk, G. (2006). Post-processing posterior predictive p-values. *Journal of the American Statistical Association*, **101**:1157–1174.

Hjort, N. L. and Glad, I. (1995). Nonparametric density estimation with a parametric start. *The Annals of Statistics*, **23**:882–904.

Hjort, N. L. and Jones, M. C. (1996). Locally parametric nonparametric density estimation. *The Annals of Statistics*, **24**:1619–1647.

Hjort, N. L. and Koning, A. J. (2002). Tests of constancy of model parameters over time. *Journal of Nonparametric Statistics*, **14**:113–132.

Hjort, N. L. and Pollard, D. B. (1993). Asymptotics for minimisers of convex processes. Statistical research report, Department of Mathematics, University of Oslo.

Hjort, N. L. and Rosa, D. (1999). Who won? *Speedskating World*, **4**:15–18.

Hjort, N. L. and Varin, C. (2008). ML, PL, QL for Markov chain models. *Scandinavian Journal of Statistics*, **35**:64–82.

Hjorth, U. (1994). *Computer Intensive Statistical Methods: Validation, Model Selection and Bootstrap*. Chapman & Hall, London.

Hoerl, A. E. and Kennard, R. W. (1970). Ridge regression: biased estimation for nonorthogonal problems. *Technometrics*, **12**:55–67.

Hoeting, J. A., Madigan, D., Raftery, A. E. and Volinsky, C. T. (1999). Bayesian model averaging: a tutorial. *Statistical Science*, **14**:382–417. With discussion by M. Clyde, D. Draper and E. I. George, and a rejoinder by the authors. [A version where the number of misprints has been significantly reduced is available at www.research.att.com/~volinsky/bma.html.]

Hosmer, D. W. and Hjort, N. L. (2002). Goodness-of-fit processes for logistic regression: simulation results. *Statistics in Medicine*, **21**:2723–2738.

Hosmer, D. W. and Lemeshow, S. (1999). *Applied Logistic Regression*. John Wiley & Sons Inc., New York.

Huber, P. J. (1981). *Robust Statistics*. John Wiley & Sons Inc., New York.

Hurvich, C. M., Simonoff, J. S. and Tsai, C.-L. (1998). Smoothing parameter selection in nonparametric regression using an improved Akaike information criterion. *Journal of the Royal Statistical Society, Series B*, **60**:271–293.

Hurvich, C. M. and Tsai, C.-L. (1989). Regression and time series model selection in small samples. *Biometrika*, **76**:297–307.

Hurvich, C. M. and Tsai, C.-L. (1990). The impact of model selection on inference in linear regression. *The American Statistician*, **44**:214–217.

Hurvich, C. M. and Tsai, C.-L. (1995a). Model selection for extended quasi-likelihood models in small samples. *Biometrics*, **51**:1077–1084.

Hurvich, C. M. and Tsai, C.-L. (1995b). Relative rates of convergence for efficient model selection criteria in linear regression. *Biometrika*, **82**:418–425.

Ibrahim, J. G., Chen, M.-H. and Lipsitz, S. R. (1999a). Monte Carlo EM for missing covariates in parametric regression models. *Biometrics*, **55**:591–596.

Ibrahim, J. G., Lipsitz, S. R. and Chen, M.-H. (1999b). Missing covariates in generalized linear models when the missing data mechanism is non-ignorable. *Journal of the Royal Statistical Society, Series B*, **61**(1):173–190.

Ing, C.-K. and Wei, C.-Z. (2005). Order selection for same-realization predictions in autoregressive processes. *The Annals of Statistics*, **33**:2423–2474.

Inglot, T., Kallenberg, W. C. M. and Ledwina, T. (1997). Data driven smooth tests for composite hypotheses. *The Annals of Statistics*, **25**:1222–1250.

Inglot, T. and Ledwina, T. (1996). Asymptotic optimality of data-driven Neyman's tests for uniformity. *The Annals of Statistics*, **24**:1982–2019.

Jansen, M. (2001). *Noise Reduction by Wavelet Thresholding*, Vol. 161 of *Lecture Notes in Statistics*. Springer-Verlag, New York.

Jansen, M. and Bultheel, A. (2001). Asymptotic behavior of the minimum mean squared error threshold for noisy wavelet coefficients of piecewise smooth signals. *IEEE Transactions on Signal Processing*, **49**:1113–1118.

Jansen, M., Malfait, M. and Bultheel, A. (1997). Generalized cross validation for wavelet thresholding. *Signal Processing*, **56**:33–44.

Jeffreys, H. (1961). *Theory of Probability* (3rd edition). Clarendon Press, Oxford.

Jiang, J. (2007). *Linear and Generalized Linear Mixed Models and their Applications*. Springer Series in Statistics. Springer-Verlag, New York.

Jiang, J. and Rao, J. S. (2003). Consistent procedures for mixed linear model selection. *Sankhyā*, **65**:23–42.

Jiang, J., Rao, J. S., Gu, Z. and Nguyen, T. (2008). Fence methods for mixed model selection. *The Annals of Statistics*, **36**:1669–1692.

Johnstone, I. M. and Silverman, B. W. (2004). Needles and straw in haystacks: empirical Bayes estimates of possibly sparse sequences. *The Annals of Statistics*, **32**:1594–1649.

Johnstone, I. M. and Silverman, B. W. (2005). Empirical Bayes selection of wavelet thresholds. *The Annals of Statistics*, **33**:1700–1752.

Jones, M. C., Hjort, N. L., Harris, I. R. and Basu, A. (2001). A comparison of related density-based minimum divergence estimators. *Biometrika*, **88**:865–873.

Kabaila, P. (1995). The effect of model selection on confidence regions and prediction regions. *Econometric Theory*, **11**:537–549.

Kabaila, P. (1998). Valid confidence intervals in regression after variable selection. *Econometric Theory*, **14**:463–482.

Kabaila, P. and Leeb, H. (2006). On the large-sample minimal coverage probability of confidence intervals after model selection. *Journal of the American Statistical Association*, **101**:619–629.

Kalbfleisch, J. D. and Prentice, R. L. (2002). *The Statistical Analysis of Failure Time Data* (2nd edition). John Wiley & Sons Inc., New York.

Kåresen, K. (1992). *Parametric Estimation: Choosing Between Narrow and Wide Models*. Cand. Scient. thesis, University of Oslo, Department of Mathematics.

Kass, R. E. and Raftery, A. E. (1995). Bayes factors. *Journal of the American Statistical Association*, **90**:773–795.

Kass, R. E. and Vaidyanathan, S. K. (1992). Approximate Bayes factors and orthogonal parameters, with application to testing equality of two binomial proportions. *Journal of the Royal Statistical Society, Series B*, **54**:129–144.

Kass, R. E. and Wasserman, L. (1995). A reference Bayesian test for nested hypotheses and its relationship to the Schwarz criterion. *Journal of the American Statistical Association*, **90**:928–934.

Kass, R. E. and Wasserman, L. (1996). The selection of prior distributions by formal rules. *Journal of the American Statistical Association*, **91**:1343–1370.

Kauermann, G. (2005). A note on smoothing parameter selection for penalized spline smoothing. *Journal of Statistical Planning and Inference*, **127**:53–69.

Kjetsaa, G., Gustavsson, S., Beckman, B. and Gil, S. (1984). *The Authorship of The Quiet Don*. Solum/Humanities Press, Oslo.

Kolmogorov, A. N. (1965). Three approaches to the definition of the concept 'quantity of information'. *Problemy Peredatsi Informatsii*, **1**(vyp. 1):3–11.

Konishi, S. and Kitagawa, G. (1996). Generalised information criteria in model selection. *Biometrika*, **83**:875–890.

Lahiri, P. (2001). *Model Selection*. Institute of Mathematical Statistics Lecture Notes – Monograph Series, 38. Institute of Mathematical Statistics.

Le Cam, L. (1953). On some asymptotic properties of maximum likelihood estimates and related Bayes' estimates. *University of California Publications in Statistics*, **1**:277–330.

Le Cam, L. and Yang, G. L. (2000). *Asymptotics in Statistics: Some Basic Concepts* (2nd edition). Springer-Verlag, New York.

le Cessie, S. and van Houwelingen, J. C. (1991). A goodness-of-fit test for binary regression models, based on smoothing methods. *Biometrics*, **47**:1267–1282.

le Cessie, S. and van Houwelingen, J. C. (1993). Building logistic models by means of a non parametric goodness of fit test: a case study. *Statistica Neerlandica*, **47**:97–109.

Ledwina, T. (1994). Data-driven version of Neyman's smooth test of fit. *Journal of the American Statistical Association*, **89**:1000–1005.

Lee, M.-L. T. and Whitmore, G. A. (2007). Threshold regression for survival analysis: modeling event times by a stochastic process reaching a boundary. *Statistical Science*, **21**(4):501–513.

Lee, S. and Karagrigoriou, A. (2001). An asymptotically optimal selection of the order of a linear process. *Sankhyā Series A*, **63**:93–106.

Leeb, H. and Pötscher, B. M. (2005). Model selection and inference: facts and fiction. *Econometric Theory*, **21**:21–59.

Leeb, H. and Pötscher, B. M. (2006). Can one estimate the conditional distribution of post-model-selection estimators? *The Annals of Statistics*, **34**:2554–2591.

Leeb, H. and Pötscher, B. M. (2008). Sparse estimators and the oracle property, or the return of Hodges' estimator. *Journal of Econometrics*, **142**:201–211.

Lehmann, E. L. (1983). *Theory of Point Estimation*. John Wiley & Sons Inc., New York.

Leike, A. (2002). Demonstration of the exponential decay law using beer froth. *European Journal of Physics*, **23**:21–26.

Li, K.-C. (1987). Asymptotic optimality for C_p, C_L, cross-validation and generalized cross-validation: discrete index set. *The Annals of Statistics*, **15**:958–975.

Lien, D. and Shrestha, K. (2005). Estimating the optimal hedge ratio with focus information criterion. *Journal of Futures Markets*, **25**:1011–1024.

Linhart, H. and Zucchini, W. (1986). *Model Selection*. John Wiley & Sons Inc., New York.

Little, R. J. A. and Rubin, D. B. (2002). *Statistical Analysis with Missing Data* (2nd edition). Wiley Series in Probability and Statistics. Wiley-Interscience [John Wiley & Sons], Hoboken, New Jersey.

Loader, C. (1996). Local likelihood density estimation. *The Annals of Statistics*, **24**:1602–1618.

Longford, N. T. (2005). Editorial: model selection and efficiency—is 'which model ...?' the right question? *Journal of the Royal Statistical Society, Series A*, **168**:469–472.

Lu, H., Hodges, J. and Carlin, B. (2005). Measuring the complexity of generalized linear hierarchical models. Technical report, Division of Biostatistics, School of Public Health, University of Minnesota.

Machado, J. A. F. (1993). Robust model selection and M-estimation. *Econometric Theory*, **9**:478–493.

Mallows, C. L. (1973). Some comments on C_p. *Technometrics*, **15**:661–675.

Manly, B. F. J. (1986). *Multivariate Statistical Methods*. Chapman & Hall, New York.

Martens, H. and Næs, T. (1992). *Multivariate Calibration*. John Wiley & Sons Inc., New York.

McCullagh, P. (2002). What is a statistical model? *The Annals of Statistics*, **30**:1225–1310. With discussion and a rejoinder by the author.

McCullagh, P. and Nelder, J. (1989). *Generalized Linear Models*. Chapman & Hall, London.

McCulloch, C. E. and Searle, S. R. (2001). *Generalized, Linear, and Mixed Models*. Wiley Series in Probability and Statistics: Texts, References, and Pocketbooks Section. Wiley-Interscience [John Wiley & Sons], New York.

McQuarrie, A. D. R. and Tsai, C.-L. (1998). *Regression and Time Series Model Selection*. World Scientific Publishing Co. Inc., River Edge, New Jersey.

Miller, A. (2002). *Subset Selection in Regression* (2nd edition). Chapman & Hall/CRC, Boca Raton, Florida.

Morgan, B. J. T. (1992). *The Analysis of Quantal Response Data*. Chapman & Hall, London.

Müller, S. and Welsh, A. H. (2005). Outlier robust model selection in linear regression. *Journal of the American Statistical Association*, **100**:1297–1310.

Murata, N., Yoshizawa, S. and Amara, S. (1994). Network information criterion – determining the number of hidden units for artificial natural network models. *IEEE Transactions on Neural Networks*, **5**:865–872.

Murtaugh, P. A., Dickson, E. R., Vandam, G. M., Malinchoc, M., Grambsch, P. M., Langworthy, A. and Gips, C. H. (1994). Primary biliary-cirrhosis – prediction of short-term survival based on repeated patient visits. *Hepatology*, **20**:126–134.

Myung, J. I., Navarro, D. J. and Pitt, M. A. (2006). Model selection by normalized maximum likelihood. *Journal of Mathematical Psychology*, **50**:167–179.

Nason, G. P. (1996). Wavelet shrinkage using cross-validation. *Journal of the Royal Statistical Society, Series B*, **58**:463–479.

Neyman, J. (1937). 'Smooth' test for goodness of fit. *Skandinavisk Aktuarietidskrift*, **20**:149–199.

Nishii, R. (1984). Asymptotic properties of criteria for selection of variables in multiple regression. *The Annals of Statistics*, **12**:758–765.

Paulsen, J. (1984). Order determination of multivariate autoregressive time series with unit roots. *Journal of Time Series Analysis*, **5**:115–127.

Pearson, K. (1902). On the change in expectation of life in man during a period of circa 2000 years. *Biometrika*, **1**:261–264.

Phillips, P. C. B. and Park, J. Y. (1988). On the formulation of Wald tests of nonlinear restrictions. *Econometrica*, **56**:1065–1083.

Picard, R. R. and Cook, R. D. (1984). Cross-validation of regression models. *Journal of the American Statistical Association*, **79**:575–583.

Pinheiro, J. and Bates, D. (2000). *Mixed-Effects Models in S and S-PLUS*. Springer-Verlag, New York.

Portnoy, S. (1988). Asymptotic behavior of likelihood methods for exponential families when the number of parameters tends to infinity. *The Annals of Statistics*, **16**:356–366.

Pötscher, B. M. (1989). Model selection under nonstationarity: autoregressive models and stochastic linear regression models. *The Annals of Statistics*, **17**:1257–1274.

Pötscher, B. M. (1991). Effects of model selection on inference. *Econometric Theory*, **7**:163–185.

Qian, G. (1999). Computations and analysis in robust regression model selection using stochastic complexity. *Computational Statistics*, **14**:293–314.

Qian, G. and Künsch, H. R. (1998). On model selection via stochastic complexity in robust linear regression. *Journal of Statistical Planning and Inference*, **75**:91–116.

Rao, C. R. and Wu, Y. H. (1989). A strongly consistent procedure for model selection in a regression problem. *Biometrika*, **76**:369–374.

Rao, J. S. and Tibsirani, R. (1997). The out-of-bootstrap method for model averaging and selection. Technical report, Department of Statistics, University of Toronto.

Ripley, B. D. (1996). *Pattern Recognition and Neural Networks*. Cambridge University Press, Cambridge.

Rissanen, J. (1983). A universal prior for integers and estimation by minimum description length. *The Annals of Statistics*, **11**:416–431.

Rissanen, J. (1987). Stochastic complexity. *Journal of the Royal Statistical Society, Series B*, **49**:223–239, 253–265. With discussion.

Rissanen, J. J. (1996). Fisher information and stochastic complexity. *IEEE Transactions on Information Theory*, **42**:40–47.

Ronchetti, E. (1985). Robust model selection in regression. *Statistics & Probability Letters*, **3**:21–23.

Ronchetti, E. (1997). Robustness aspects of model choice. *Statistica Sinica*, **7**:327–338.

Ronchetti, E., Field, C. and Blanchard, W. (1997). Robust linear model selection by cross-validation. *Journal of the American Statistical Association*, **92**:1017–1023.

Ronchetti, E. and Staudte, R. G. (1994). A robust version of Mallows' C_P. *Journal of the American Statistical Association*, **89**:550–559.

Ruggeri, F. and Vidakovic, B. (1999). A Bayesian decision theoretic approach to the choice of thresholding parameter. *Statistica Sinica*, **9**:183–197.

Ruppert, D., Wand, M. P. and Carroll, R. J. (2003). *Semiparametric Regression*. Cambridge University Press, Cambridge.

Saunders, C., Gammermann, A. and Vovk, V. (1998). Ridge regression learning algorithm in dual variables. In Shavlik, J. (editor), *Machine Learning: Proceedings of the Fifteenth International Conference*, pages 515–521.

Schwarz, G. (1978). Estimating the dimension of a model. *The Annals of Statistics*, **6**:461–464.

Searle, S. R., Casella, G. and McCulloch, C. E. (1992). *Variance Components*. Wiley Series in Probability and Mathematical Statistics: Applied Probability and Statistics. John Wiley & Sons Inc., New York.

Sebastiani, P. and Ramoni, M. (2001). Bayesian selection of decomposable models with incomplete data. *Journal of the American Statistical Association*, **96**(456):1375–1386.

Self, S. G. and Liang, K. Y. (1987). Asymptotic properties of maximum likelihood and likelihood ratio tests under nonstandard conditions. *Journal of the American Statistical Association*, **82**:605–610.

Sen, P. K. and Saleh, A. K. M. E. (1987). On preliminary test and shrinkage m-estimation in linear models. *The Annals of Statistics*, **15**:1580–1592.

Shao, J. (1997). An asymptotic theory for linear model selection. *Statistica Sinica*, **7**:221–264. With comments and a rejoinder by the author.

Shao, J. (1998). Convergence rates of the generalized information criterion. *Journal of Nonparametric Statistics*, **9**:217–225.

Shao, J. and Tu, D. (1995). *The Jackknife and Bootstrap*. Springer-Verlag, New York.

Shen, X. and Ye, J. (2002). Adaptive model selection. *Journal of the American Statistical Association*, **97**:210–221.

Shepard, T. H., Mackler, B. and Finch, C. A. (1980). Reproductive studies in the iron-deficient rat. *Teratology*, **22**:329–334.

Shibata, R. (1976). Selection of the order of an autoregressive model by Akaike's information criterion. *Biometrika*, **63**:117–126.

Shibata, R. (1980). Asymptotically efficient selection of the order of the model for estimating parameters of a linear process. *The Annals of Statistics*, **8**:147–164.

Shibata, R. (1981). An optimal selection of regression variables. *Biometrika*, **68**:45–54.

Shibata, R. (1982). Correction: 'An optimal selection of regression variables'. *Biometrika*, **69**:492.

Shibata, R. (1984). Approximate efficiency of a selection procedure for the number of regression variables. *Biometrika*, **71**:43–49.

Shimodaira, H. (1994). A new criterion for selecting models from partially observed data. In Cheeseman, P. and Oldford, R. W. (editors), *Selecting Models from Data: Artificial Intelligence and Statistics IV*, pages 21–29. Springer-Verlag, New York.

Silvapulle, M. J. (1994). On tests against one-sided hypotheses in some generalized linear models. *Biometrics*, **50**:853–858.

Silvapulle, M. J. and Silvapulle, P. (1995). A score test against one-sided alternatives. *Journal of the American Statistical Association*, **90**:342–349.

Simonoff, J. S. and Tsai, C.-L. (1999). Semiparametric and additive model selection using an improved Akaike information criterion. *Journal of Computational and Graphical Statistics*, **8**:22–40.

Sin, C.-Y. and White, H. (1996). Information criteria for selecting possibly misspecified parametric models. *Journal of Econometrics*, **71**:207–225.

Smola, A. and Schölkopf, B. (1998). On a kernel-based method for pattern recognition, regression, approximation and operator inversion. *Algorithmica*, **22**:211–231.

Solomonoff, R. J. (1964a). A formal theory of inductive inference. I. *Information and Control*, **7**:1–22.

Solomonoff, R. J. (1964b). A formal theory of inductive inference. II. *Information and Control*, **7**:224–254.

Solzhenitsyn, A. I. (1974). Nevyrvannaya tajna (the not yet uprooted secret). In *Stremya 'Tihogo Dona' (Zagadki romana)*, pages 1–9. YMCA-Press, Paris.

Spiegelberg, W. (1901). *Aegyptische und Griechische Eigennamen aus Mumientiketten der Römischen Kaiserzeit*. Cairo.

Spiegelhalter, D. J., Best, N. G., Carlin, B. P. and van der Linde, A. (2002). Bayesian measures of model complexity and fit. *Journal of the Royal Statistical Society, Series B*, **64**:583–639. With a discussion and a rejoinder by the authors.

Spitzer, F. (1956). A combinatorial lemma and its applications to probability theory. *Transactions of the American Mathematical Society*, **82**:323–339.

Stein, C. (1981). Estimation of the mean of a multivariate normal distribution. *The Annals of Statistics*, **9**:1135–1151.

Stone, M. (1974). Cross-validatory choice and assessment of statistical predictions. *Journal of the Royal Statistical Society, Series B*, **36**:111–147. With a discussion and a rejoinder by the authors.

Stone, M. (1977). An asymptotic equivalence of choice of model by cross-validation and Akaike's criterion. *Journal of the Royal Statistical Society, Series B*, **39**:44–47.

Stone, M. (1978). Cross-validation: a review. *Math. Opeartionsforsch. Ser. Statist.*, **9**:127–139.

Sugiura, N. (1978). Further analysis of the data by Akaike's information criterion and the finite sample corrections. *Communications in Statistics. Theory and Methods*, **7**:13–26.

Takeuchi, K. (1976). Distribution of informational statistics and a criterion of model fitting. *Suri-Kagaku (Mathematical Sciences)*, **153**:12–18. In Japanese.

Thomson, A. and Randall-Maciver, R. (1905). *Ancient Races of the Thebaid*. Oxford University Press, Oxford.

Tibshirani, R. (1996). Regression shrinkage and selection via the lasso. *Journal of the Royal Statistical Society, Series B*, **58**:267–288.

Tibshirani, R. (1997). The lasso method for variable selection in the Cox model. *Statistics in Medicine*, **16**:385–395.

Tsay, R. S. (1984). Order selection in nonstationary autoregressive models. *The Annals of Statistics*, **12**:1425–1433.

Uchida, M. and Yoshida, N. (2001). Information criteria in model selection for mixing processes. *Statistical Inference for Stochastic Processes*, **4**:73–98.

Vaida, F. and Blanchard, S. (2005). Conditional Akaike information for mixed-effects models. *Biometrika*, **92**:351–370.

van de Geer, S. (2008). High-dimensional generalized linear models and the lasso. *Annals of Statistics*, **35**:614–645.

van de Geer, S. and van Houwelingen, J. C. (2008). High-dimensional data: $p \gg n$ in mathematical statistics and biomedical applications. *Bernoulli*, **10**:939–942.

van der Linde, A. (2004). On the association between a random parameter and an observable. *Test*, **13**:85–111.

van der Linde, A. (2005). DIC in variable selection. *Statistica Neerlandica*, **59**:45–56.

van der Vaart, A. W. (1998). *Asymptotic Statistics*. Cambridge University Press, Cambridge.

Verbeke, G. and Molenberghs, G. (2000). *Linear Mixed Models for Longitudinal Data*. Springer Series in Statistics. Springer-Verlag, New York.

Verloove, S. P. and Verwey, R. Y. (1983). *Project on preterm and small-for-gestational age infants in the Netherlands*. PhD thesis, University of Leiden.

Vidakovic, B. (1998). Nonlinear wavelet shrinkage with Bayes rules and Bayes factors. *Journal of the American Statistical Association*, **93**:173–179.

Voldner, N., Frøslie, K. F., Bø, K., Haakstad, L. H., Hoff, C. M., Godang, K., Bollerslev, J. and Henriksen, T. (2008). Modifiable determinants of fetal macrosmia: Role of life style related factors. *Acta Obstetricia et Gynecologica Scandica*, **87**:423–429.

Volinsky, C., Madigan, D., Raftery, A. E. and Kronmal, R. A. (1997). Bayesian model averaging in proportional hazard models: predicting the risk of a stroke. *Applied Statistics*, **46**:443–448.

Vonesh, E. F. and Chinchilli, V. M. (1997). *Linear and Nonlinear Models for the Analysis of Repeated Measurements*, Vol. 154 of *Statistics: Textbooks and Monographs*. Marcel Dekker Inc., New York.

Vu, H. T. V. and Zhou, S. (1997). Generalization of likelihood ratio tests under nonstandard conditions. *The Annals of Statistics*, **25**:897–916.

Vuong, Q. H. (1989). Likelihood ratio tests for model selection and nonnested hypotheses. *Econometrica*, **57**:307–333.

Wand, M. P. and Jones, M. C. (1995). *Kernel Smoothing*, Vol. 60 of *Monographs on Statistics and Applied Probability*. Chapman & Hall, London.

Wasserman, L. (2000). Bayesian model selection and model averaging. *Journal of Mathematical Psychology*, **44**:92–107.

Wei, C.-Z. (1992). On predictive least squares principles. *The Annals of Statistics*, **20**:1–42.

Weisberg, S. (2005). *Applied Linear Regression* (3rd edition). John Wiley & Sons Inc., Hoboken, New Jersey.

White, H. (1982). Maximum likelihood estimation of misspecified models. *Econometrica*, **50**:1–25.

White, H. (1994). *Estimation, Inference and Specification Analysis*. Cambridge University Press, Cambridge.

Windham, M. P. (1995). Robustifying model fitting. *Journal of the Royal Statistical Society, Series B*, **57**:599–609.

Woodroofe, M. (1982). On model selection and the arc sine laws. *The Annals of Statistics*, **10**:1182–1194.

Yang, Y. (2001). Adaptive regression by mixing. *Journal of the American Statistical Association*, **96**:574–588.

Yang, Y. (2005). Can the strengths of AIC and BIC be shared? *Biometrika*, **92**:937–950.

Zhang, P. (1993). On the convergence rate of model selection criteria. *Communications in Statistics. Theory and Methods*, **22**:2765–2775.

Zhao, L. C., Dorea, C. C. Y. and Gonçalves, C. R. (2001). On determination of the order of a Markov chain. *Statistical Inference for Stochastic Processes*, **4**:273–282.

Zou, H. and Hastie, T. (2005). Regularization and variable selection via the elastic net. *Journal of the Royal Statistical Society, Series B*, **67**:301–320.

Author index

Subject index

Printed in the United States
By Bookmasters